S0-EJS-963

THE BACTERIAL CELL WALL

THE BACTERIAL CELL WALL

Milton R. J. Salton

Professor of Microbiology
School of Biological Sciences
University of New South Wales, Australia
formerly
Reader in Chemical Bacteriology
University of Manchester, Gr. Br.

ELSEVIER PUBLISHING COMPANY

AMSTERDAM

LONDON NEW YORK

1964

ELSEVIER PUBLISHING COMPANY
335 JAN VAN GALENSTRAAT, P.O. BOX 211, AMSTERDAM

AMERICAN ELSEVIER PUBLISHING COMPANY, INC.
52 VANDERBILT AVENUE, NEW YORK 17, N.Y.

ELSEVIER PUBLISHING COMPANY LIMITED
12B, RIPPLESIDE COMMERCIAL ESTATE
RIPPLE ROAD, BARKING, ESSEX

LIBRARY OF CONGRESS CATALOG CARD NUMBER 64-11343
WITH 77 ILLUSTRATIONS AND 67 TABLES

To my wife,
for her constant help
and understanding.

PREFACE

The first attempts to obtain some direct information about the chemical composition of bacterial cell walls or envelopes were made in several laboratories during 1950. At that time it seemed inconceivable that the stage would be reached when it would be difficult to condense the new knowledge about cell walls into a monograph of the size of the present volume. Indeed, so much information with direct and indirect bearings on the problems of the nature and structure of microbial surfaces has now accumulated that the writing of a completely exhaustive record of the field would be a lengthy task. Some degree of selectivity has therefore been inevitable and this book represents an attempt to summarize the results of certain facets of the investigations into bacterial cell walls from the development of suitable methods of isolation to the current interest in the biosynthesis of these complex and fascinating structures. The treatment of each specialized part of this volume has been largely 'historical' and accordingly I have not hesitated in including some of the earlier electron micrographs and illustrative material which will convey the sequence of events in the development of this field and which has formed the basis for more recent investigations. In the opening chapter, an effort has been made to put the 'bacterial cell wall' into anatomical perspective by discussing its relationship to the cell surface structure as a whole.

Although the book is primarily concerned with walls of bacteria, some mention has been made of the closely related blue-green-algae. The walls of yeasts, fungi and other microbial groups have not been specifically dealt with in this volume for, apart from chemical studies on fungal walls carried out in the laboratories of Dr. W. J. Nickerson, little additional material on fine structure and biosynthesis has appeared since the summary presented in the published account of the 1960 CIBA Lectures on 'Microbial Cell Walls'. It is hoped that this will stimulate rather than deter new researches into the nature of the surface structures of the more neglected groups of microorganisms.

With the rapid growth of scientific literature even in this small specialized field, it is inevitable that this book will be 'out of date' in some sections by the time it is published. The two major problems in the cell wall field, the structure and biosynthesis of wall 'mucopeptides' are attracting increasing attention and rapid progress is bound to be made in solving some of the obvious gaps in our knowledge. The inclusion of new important material presents a constant headache to publisher and author alike. Indeed, since the completion of the original manuscript much new material has appeared. Fortunately it has been possible to insert brief references to several important aspects of wall structure and biosynthesis and I would especially like to thank Drs. J. T. Park, R. W. Jeanloz, N. Sharon, J. M. Ghuysen, J. L. Strominger and P. MEADOW for kindly giving me permission to refer to their results prior to their publication.

The preparation of a monograph is rarely a 'one man show' and I would like to express my thanks to many friends and colleagues who have helped in a direct way, for their valuable discussions, interest and contributions to the development of this field over the years. I should also like to thank especially, Miss Gay LYNCH for the excellence of the typing of the manuscript and tables and for her valuable assistance throughout the preparation, checking and proof reading of the book.

ACKNOWLEDGEMENTS

I am grateful to many authors, editors of journals and publishers for giving permission to reproduce the material presented in this book. I am especially indebted to my colleagues and former collaborators, Dr. J. A. Chapman of the University of Manchester, England, and Robert W. Horne of the Agricultural Research Council's Laboratory, Babraham, England, and to Professor R. G. E. Murray of the University of Western Ontario, Canada, for their generous help in providing many electron micrographs. I should also like to thank the following authors, editors and publishers for their kind co-operation: Dr. J. P. Duguid for fig. 1c; Dr. W. van Iterson for figs. 1d and 25b and the Editor of Koninklijke Nederlandse Akademie van Wetenschappen, Amsterdam, for fig. 1d; the Editors of Nature for figs. 4 and 70; Dr. A. D. Brown for fig. 5a; the Editor of Journal of Ultrastructure Research for figs. 5c and 8; Prof. E. Kellenberger and Dr. W. Hayes for fig. 5d; Prof. R. G. E. Murray and the Editor of the Canadian Journal of Microbiology for figs. 6c, 26b–f and fig. 69; the Editors of the Journal of General Microbiology for figs. 9, 10, 11, 13, 14, 43 and 44; Dr. P. D. Cooper for figs. 13 and 14; Dr. G. Shockman for figs. 15 and 16 and Pergamon Press for fig. 16; Dr. E. Ribi for figs. 17 and 33 and the Editors of Proceedings of the Society for Experimental Biology and Medicine for fig. 33; Dr. A. L. Houwink for figs. 24 and 25 and Dr. V. Mohr for fig. 25a; the Editors of the Proceedings of the Royal Society (London) for figs. 6d, 30 and 31; Dr. K. Takeya for figs. 32, 34 and the Editors of the Journal of Cell Biology for fig. 34; Drs. H. H. Martin, H. Frank, Prof. W. Weidel and the Editors of Zentr. Bakt. Parasitenk. for fig. 35. Dr. H. H. Martin for figs. 36, 37, 41, 42, the Editor of the Journal of Theoretical Biology for fig. 36 and the Editor of Z. für Naturforschung for figs. 41 and 42; Dr. A. Schocher for figs. 39, 40, 60 and the Editor of the Canadian Journal of Microbiology for fig. 60; Prof. J. L. Strominger for figs. 61, 71, 72 and the Editors of Biochemical and Biophysical Research Communications for fig. 61; Prof. J. Baddiley for figs. 63, 64, 65 and 66 and the Editors of the Biochemical Journal; the Editors of the Society for General Microbiology Symposium No. 6 for fig. 67; Dr. M. H. Richmond for fig. 70; Dr. R. M. Cole and the Editor of Science for fig. 74; Dr. J. W. May for figs. 75, 76, 77 and the Editor of Experimental Cell Research for fig. 75; Academic Press for permission to reproduce figs. 2, 5c, 8, 32, 36, 61, 75 from their publications; John Wiley for figs. 38, 52, 58 and 73, illustrations reproduced with their permission as publishers of 'Microbial Cell Walls', 1960 (by the present author).

CONTENTS

CHAPTER 1

The anatomy of the bacterial surface

INTRODUCTION

It is now just a little over twenty years ago that the application of electron microscopy to the problems of cell structure heralded two very fruitful decades of research culminating in our present knowledge of the fine structure and anatomy of cells derived from a wide variety of organisms. Because of the small dimensions of bacteria, electron microscopic studies have assumed special importance in resolving their detailed anatomy. Some of the early electron micrographs of bacterial cells were indeed little better than the photomicrographs a good cytologist could produce with staining techniques and the light microscope, but now even the most ardent bacterial cytologist would have to concede that many of the fine structures of microorganisms could not have been detected without the combined use of shadow casting, thin sectioning and negative staining with electron microscopy. Thus in two decades we have emerged from the rather vague world of the bacterial cytologist and entered the exciting world of the bacterial anatomist where cell structures and functions can be more accurately resolved and described at the macromolecular level.

The growth of our detailed knowledge of the anatomy of microorganisms has of course been but one facet of the general advance in cell biology. With the perfection of the thin sectioning technique it has become possible to compare the principal anatomical features of cells derived from a wide variety of animal, plant and microbial species. The information gained from such comparative investigations of cellular structure together with our biochemical and chemical knowledge of cells presents us with a clear picture of some of the essential similarities and differences between cells from the major classes of organisms.

All types of cells capable of undergoing division and growth possess certain common structures and organelles. These include ribonucleic acid (RNA) – protein particles (ribosomes) of about the same dimensions, plasma membranes of the so-called 'unit membrane' type with an overall thickness of approximately 75 Å (ROBERTSON, 1959; SJÖSTRAND, 1960) and a nucleus or chromatinic body. Bacteria and blue-green algae differ from animal and plant cells in several respects, viz.: the structure of the nuclear body, the absence of organized mitochondria possessing a limiting membrane and enclosed cristae and the absence of an endoplasmic reticulum. The chromatinic bodies or nuclear structures of bacteria and blue-green algae are not surrounded by nuclear membranes (KELLENBERGER, 1960; HOPWOOD AND GLAUERT, 1960; RIS and SINGH, 1961; MURRAY, 1962) a feature which distinguishes them from those of higher plant and animals cells, yeasts and fungi. A well-defined mitochondrial structure similar to that found in animal and plant cells as well as in fungi and yeasts has not been detected in bacteria

The tables are printed together at the end of the book.

and blue-green algae although organelles with equivalent biochemical functions are undoubtedly present. The 'mesosome' structures of bacteria (FITZ-JAMES, 1960; SALTON AND CHAPMAN, 1962) also referred to as the 'intracytoplasmic membranous elements' by GLAUERT (1962) and a 'remarkable organelle' (VAN ITERSON, 1961) appear to possess the enzyme systems normally found in mitochondria isolated from other types of cells. It was formerly believed that a single plasma or protoplast membrane system in bacteria was the mitochondrial equivalent but recent investigations have shown that so-called plasma membrane preparations contain the mesosome membranes as well (SALTON AND CHAPMAN, 1962). As FITZ-JAMES (1960) has pointed out the mesosome is produced by the invaginated growth of the plasma membrane and it may well be that both membrane elements form a continuous and homogeneous system. The closest resemblance to the membrane-mesosome system of bacteria so far reported in other microbial groups is the multimembrane system recently observed by LINNANE, VITOLS AND NOWLAND (1962) in anaerobically grown cells of the yeast *Torulopsis utilis*. In aerobically grown *Torulopsis utilis* the mitochondria were structurally and enzymically normal. It was suggested that the membrane system of the anaerobic cells was concerned with the morphogenesis of the mitochondria. At the chemical level diphosphatidyglycerol has been found in the membrane-mesosome fractions of *Micrococcus lysodeikticus* as well as in mammalian mitochondria (MACFARLANE, 1961; MARINETTI, ERBLAND AND STOTZ, 1958) and it will be of great interest to see if this type of lipid is a characteristic component of membranes and membranous organelles.

A comparison of the anatomy of the photosynthetic apparatus (chloroplast) in plants with the equivalent organelle (chromatophores) in photosynthetic bacteria and blue-green algae bears a similar structural relationship to that seen for mitochondria. The complex chloroplast structure of plants is replaced in bacteria and blue-green algae by a membrane system of lamellae or a chromatophore network, both of which are probably derived from the invagination of the plasma membrane (GIESBRECHT AND DREWS, 1962). Thus in bacteria and blue-green algae the biochemical functions of the highly organized mitochondria and chloroplasts are found in the simple 'unit membranes' with a lower degree of structural complexity.

At the chemical level, bacteria differ from animal, plant and other types of microbial cells in that they have not so far been shown to contain sterols. One possible exception to this general rule seemed to be found in the *Mycoplasma* spp. which require an exogenous supply of cholesterol for growth. However, recent investigations (RODWELL, personal communication) indicate that cholesterol is taken up by the cells but is not chemically modified in any way. The presence of a sterol in *Chlorobium limicola* has been suggested by AARONSON AND BAKER (1961) but this report awaits further confirmatory evidence. The other major chemical difference between most bacteria and blue-green algae, and cells derived from other groups of organisms is the possession of the cell-wall mucopeptides (glycosaminopeptides, glycopeptides) the characteristic and most conspicuous components of which are the substances muramic acid, α, ε-diaminopimelic acid (DAP) and D-isomers of certain amino acids.

Thus structurally and chemically bacterial cells possess much in common with other types of cells. There are, however, several outstanding differences which place the majority of bacterial species apart from other microorganisms and cells of plant and animal origin. One of the very interesting chemical and biochemical differences is the inability of bacteria to produce sterols. So far there have been no reports of sterols in blue-green algae (FOGG, 1953), a fact which again emphasizes their general similarities to the bacteria. Many of the other structural and chemical differences arise from the characteristics of the surface components of the bacterial cell, in particular the mucopeptide nature of the walls. The remaining portion of this chapter will therefore be devoted to a detailed discussion of various aspects of the anatomy of the bacterial surface.

Some of the principal structural, chemical and biochemical properties of bacteria and blue-green algae are contrasted with those of cells of animals, plant and other microbial groups in Table 1. It is of considerable interest that bacteria, once regarded as the most primitive forms of life, are almost as structurally complex and are generally as biochemically sophisticated as 'higher' cells.

CELL SURFACE STRUCTURE

Electron microscopy has not only added a tremendous amount of detail to our knowledge of the fine structure of the bacterial cell as a whole, but it has also contributed to a more precise definition of the anatomy of the cell surface structures. Although the early cytologists had established the presence of flagella and the principal surface layers of capsules, walls and membranes (KNAYSI, 1951), without the electron microscope the isolation and chemical characterization of many of these cellular components would not have been possible. The detection of fine structure in bacterial walls, capsules and flagella and the discovery of fimbriae (DUGUID, SMITH, DEMPSTER AND EDMUNDS, 1955) was of course entirely dependent on the resolution of the electron microscope.

The anatomy of the bacterial surface has been the subject of detailed discussions in 'The Bacteria' (GUNSALUS AND STANIER, 1960) and by WILKINSON (1958); WILKINSON AND DUGUID (1960); DUGUID AND WILKINSON (1961); SALTON (1961). The surface components of the bacterial cell can be separated anatomically into two groups: (1) the surface appendages, (2) the surface layers.

SURFACE APPENDAGES

In general the surface appendages can be readily distinguished from the surface layers. Thus flagella and fimbriae of various bacteria and the filamentous appendages of *Gallionella ferruginea* (VAN ITERSON, 1958) are seen as quite separate entities as illustrated in Fig. 1. Of these three types of surface appendage more is known about the nature and structure of bacterial flagella following the classical work of WEIBULL (1948). He isolated homogeneous preparations of these structures and laid the foundations of our present knowledge of the chemistry

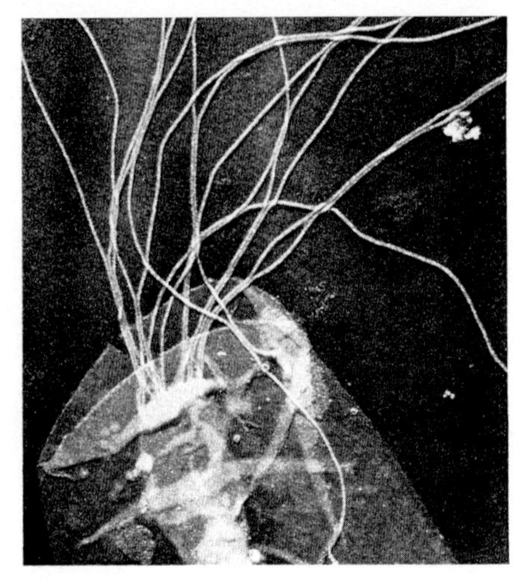

a

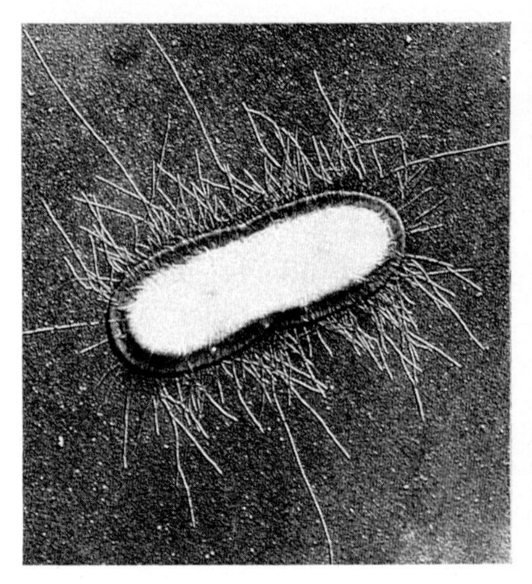

c

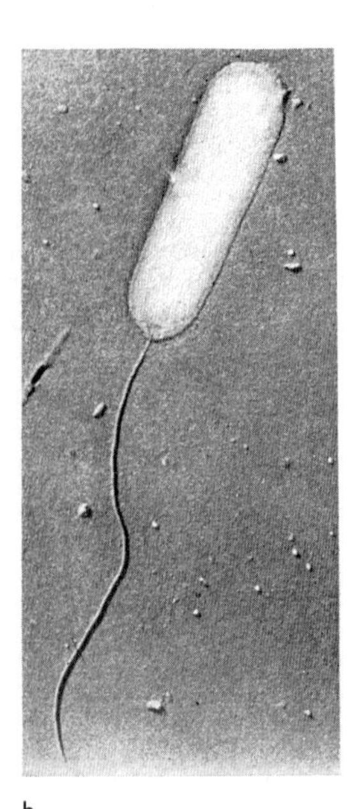

b

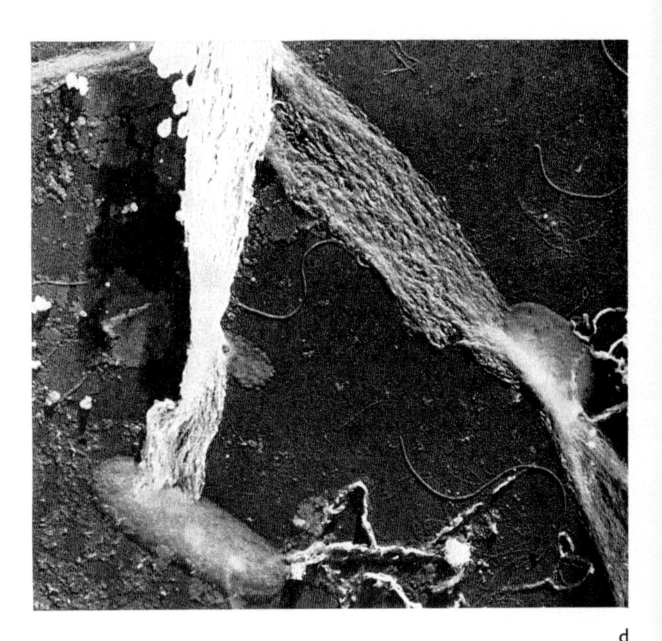

d

FIGURE 1. *Surface appendages of bacterial cells.*
(a) Flagella tuft attached to envelope of autolysed *Spirillum serpens*. × 30,000.
(b) Flagellum of *Pseudomonas aeruginosa*. × 12,500.
(c) Fimbriae surrounding the cell surface of *Shigella flexneri*.
(d) Ferruginous strands attached to cells of *Gallionella ferruginea*. × 8,000.

of flagella (STOCKER, 1956; WEIBULL, 1960; KERRIDGE, 1961). Bacterial fimbriae have also been isolated but at present so far as the author is aware they have not been clearly defined in chemical terms (BRINTON, 1959). The filamentous appendages of *Gallionella ferruginea* are known to contain some organic material as well as iron oxide (VAN ITERSON, 1958).

The decision as to whether the stalks of the *Caulobacter* spp. should be regarded as 'surface appendages' or more properly as specialized extensions of the surface layers and cell contents must await further investigations.

SURFACE LAYERS

The surface layers of the bacterial cell can be visualized as a series of simple or complex concentric shells differentiated into the following principal regions:

1. ionic layer
2. capsules, microcapsules
3. adsorbed slimes and gums
4. cementing layer in cell aggregates
5. cell wall or outer envelope component
6. cell membranes, plasma membranes.

The cytologically demonstrable surface components will be discussed in some detail but passing mention must be made of the ionic layer of the bacterial cell.

The outermost 'layer' of the bacterial surface is an ionic one with the various charged substances of the surface components contributing to the net charge. Although studies of the electrophoretic mobilities of bacteria before and after treatment with various drugs, antibiotics and antibacterial agents and enzymes have yielded interesting information (MCQUILLEN, 1951a, b; DOUGLAS, 1957; GEBICKI AND JAMES, 1962) they have only been of limited value in elucidating the precise nature of surface components and the fate of these components during the various treatments. Investigations of the surface charge of bacteria have thus been largely of diagnostic value and of use in substantiating more direct chemical studies on the outer layers. One interesting example of the latter use of electrophoresis was the demonstration of the differences in mobilities of whole cells of *Bacillus megaterium* and the isolated protoplasts derived by treatment with lysozyme (DOUGLAS AND PARKER, 1958). Such a difference in surface charge found by DOUGLAS AND PARKER (1958) was compatible with the known differences in chemical composition of walls and membranes of *Bacillus megaterium* (WEIBULL AND BERGSTRÖM, 1958).

Capsules and microcapsules

Capsules form the outermost layer of certain bacterial species. It has long been recognized that a capsule is not an essential structural element of the bacterial cell and its production is subject to both phenotypic and genetic variation. Thus environmental conditions may markedly affect the ability of a given bacterial species to produce a capsule detectable by the usual cytological methods (DUGUID AND WILKINSON, 1961). Under suitable conditions of nitrogen and phosphorus deficiency the capsule of *Aerobacter aerogenes* (*Klebsiella aerogenes*) possessed a diameter of up to 4.3 μ (DUGUID AND WILKINSON, 1953, 1954). The retention of a capsule may also be dependent on other factors such as the absence of capsule-degrading enzymes. In certain streptococcal groups (especially Groups A and C) capsules of hyaluronic acid are detectable in the early exponential phase, but as the organisms continue to grow logarithmically, hyaluronidase is produced and the capsules are no longer detectable (BAZELEY, 1940; KASS AND SEASTONE, 1944).

In addition to the role of environmental factors on capsule production, DUGUID AND WILKINSON (1953, 1954) have shown that mutant strains derived from the fully encapsulated strain of *Klebsiella aerogenes* (A3) may possess slime layers of identical polysaccharide, readily removable from the cell surface by washing with water; other mutants may be completely devoid of capsule or slime. The latter strains would be equivalent to the 'rough' variants of the Gram-positive pneumococci.

AVERY AND DUBOS (1931) were the first to demonstrate the selective removal of a capsular layer from bacterial cells. Removal of the capsular polysaccharide from the pneumococcal cells was achieved without impairing the viability of the cells and thus established the anatomical and functional differentiation of the capsule and wall of these organisms (AVERY AND DUBOS, 1931). Similar studies have been extended to other bacterial species and enzymic 'decapsulation' without loss of cell viability has been achieved with *Klebsiella pneumoniae* (ADAMS AND PARK, 1956), *Bacillus anthracis* and *Bacillus megaterium* (TORII, 1955). Although the surface M and T proteins of group A streptococci can also be removed by digestion with trypsin without loss of viability (LANCEFIELD, 1943) these components are not present as recognizable capsular or slime layers.

The selective removal with enzymes of the capsular, slime and other layers external to the rigid cell wall therefore offers an extremely valuable method for investigating the anatomical relationships of the surface layers of the bacterial cell. The prior enzymic removal of capsular structures when chemical investigations are to be performed on the cell wall proper has obvious advantages in determining the nature of the bacterial wall. It can be concluded then that bacterial capsules and slime layers are morphological entities physically distinguishable from the underlying cell envelope structures of many bacteria. By growing bacteria under conditions which prevent capsule formation and by enzymic removal of the fully formed capsular and slime layers it has been shown that the cells retain their morphological integrity. These observations have established the dispensability of capsules, slimes and sheaths and indicate that the walls or envelopes are responsible for cell shape and are more intimately involved in the viability of the cell.

Based largely on stained preparations, it has been widely believed that capsules are homogeneous accumulations of amorphous, viscous gel-like materials around the bacterial cell-wall surface. The possibility that they may be physically and chemically heterogeneous was first suggested when TOMCSIK (1951) and TOMCSIK AND GUEX-HOLZER (1951) applied immunological reactions to *Bacillus anthracis* and other members of the genus Bacillus and examined the antibody-treated cells under the phase-contrast microscope. Cells exposed to antibody against isolated capsular γ-glutamyl polypeptide and antibody to capsular polysaccharide showed a complex disposition of the latter within the glutamyl capsular polypeptide (TOMCSIK, 1951, 1956). LABAW AND MOSLEY (1954) examined encapsulated cells of the Lisbonne strain of *Escherichia coli* in the electron microscope and detected striated fibrillar structures embedded in an amorphous capsular matrix. Discontinuities in the capsular surface of *Bacillus megaterium* were reported by IVANOVICS AND HORVATH (1953). These variations in physical structure of bacterial capsules are illustrated diagrammatically in Fig. 2 (taken from SALTON, 1960).

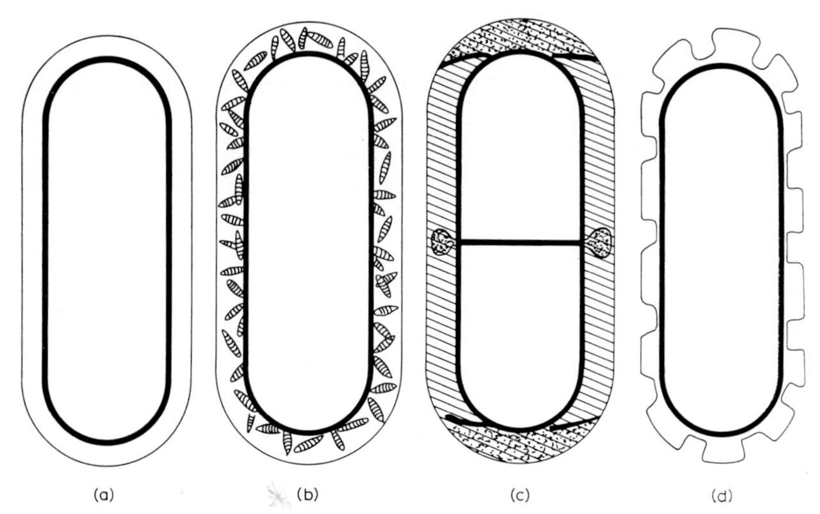

FIGURE 2. A diagrammatic representation of the types of capsular structures found in bacteria.
(a) Capsule forming a continuous layer around the cell.
(b) Capsular layer with banded fibrils as in *Escherichia coli* Lisbonne.
(c) Complex capsule with localized patches of polysaccharide and polypeptide as in *Bacillus megaterium* M.
(d) Discontinuities in the capsular surface as in *Bacillus megaterium*. (SALTON, 1960a).

So far the discussion has referred largely to capsules and slime layers as defined and demonstrated cytologically by DUGUID (1951). Bacteria producing loose slime and extracellular gums and polysaccharides may have such materials adsorbed on the cell wall or cell envelope surfaces. Many of these substances have been easily removed from the cells by repeated washing, Some components may be strongly adsorbed and it was of interest to note that certain halophilic organisms possessed a layer of deoxyribonucleic acid (DNA) which could not be washed from the surface (SMITHIES AND GIBBONS, 1955). This strongly adsorbed DNA was later shown to be of intracellular origin, its presence upon the surface being a consequence of the instability of the cell walls when the organism was grown on media containing less than 0.7 *M* sodium chloride (TAKAHASHI AND GIBBONS, 1957). Both the loosely adherent surface polysaccharides and the more strongly bound DNA of intracellular origin have a different anatomical status to the components described as 'microcapsular' materials (WILKINSON, 1958).

To overcome some of the difficulties of differentiating and defining certain surface components of the bacterial cell WILKINSON (1958) has collectively defined the smooth antigenic substances of Gram-negative bacteria and other related materials as 'microcapsules'. The term microcapsule has thus been taken to indicate the presence of surface components (usually detectable by immunological reactions) which are often difficult to differentiate from other layers of the bacterial envelope. In addition to the smooth O somatic antigens of Gram-negative bacteria, the M-proteins of haemolytic Streptococci have also been placed within this general grouping. It should be kept in mind however, that the latter never assume the dimensions of capsules and it is even difficult to detect any difference in the appearance of isolated streptococcal walls after removal of the M-protein substances (SALTON, 1953; SLADE, 1957).

In discussing the surface layers of the bacterial cell SALTON (1960a) also used the term 'microcapsule' to describe the O antigens of Gram-negative bacteria. Although there is little doubt that the O antigens are anatomically on the surface of the bacterial cell it is now extremely doubtful that the term 'microcapsule' has any valid meaning for these components. The immunologically specific polysaccharides are part of lipid-polysaccharide-protein complexes localized in the cell envelope fraction of Gram-negative bacteria. It is only after treatment with warm 45% phenol solution that the lipopolysaccharides are released from the complexes in the bacterial surface (WESTPHAL, LUDERITZ AND BISTER, 1952). Moreover the investigations of WEIDEL and his colleagues have clearly established the differentiation of the isolated wall or envelope of *Escherichia coli* into a phenol-insoluble rigid layer (mucopeptide, glycosaminopeptide) and the phenol-soluble layers of the lipopolysaccharide-protein complexes (WEIDEL, FRANK AND MARTIN, 1960). These features, together with the present knowledge that there appears to be no enzyme systems available for the selective release of the O antigens from the surface of Gram-negative bacteria (contrast with the capsules of both Gram-positive and Gram-negative bacteria discussed above) lead to the conclusion that the smooth O antigens are part of the multilayered wall or envelope structure. The author therefore feels that the term 'microcapsule' confuses rather than clarifies the anatomical status of the lipo-polysaccharide-protein complexes of Gram-negative bacteria and that these components should not be described as 'microcapsular'. However, in describing the Vi antigens and components such as the M-proteins of Streptococci as 'microcapsules', it may serve to distinguish them from other well-defined capsules or slime.

Although the M proteins and Vi antigens have not been detected in electron micrographs, evidence for a microcapsular layer in *Nocardia calcarea* has been presented by GLAUERT (1962). In her excellent thin sections of *Nocardia calcarea*, GLAUERT (1962) has resolved an extremely uniform layer of 50 Å thickness surrounding the cell wall. The microcapsular layer is much more electron transparent than the underlying wall.

At the chemical level there is usually little confusion between the capsular substances and the constituents of the cell wall. The chemical composition of bacterial capsules, especially the polysaccharides has been discussed in monographs and review articles from time to time and the reader is referred to HEIDELBERGER, 1956; WILKINSON, 1958; SALTON, 1960; STACEY AND

BARKER, 1960, for detailed information. Many of the capsular polysaccharides such as those containing uronic acids are readily distinguishable from the cell-wall structures in which uronic acids have been found fairly infrequently.

There are however, several instances of chemical overlap between capsules and walls as shown in Table 2. One of the first interesting examples of this similarity in chemical constitution and immunological specificity of capsular polysaccharide and cell-wall polysaccharide was revealed in the studies of GUEX-HOLZER AND TOMCSIK (1956). The capsular polysaccharide material appeared to be immunologically identical to the cell-wall substance of *Bacillus megaterium* and on isolation the capsular substance was found to contain glucosamine, galactosamine and an unknown amino sugar presumed to be muramic acid. Unfortunately so far as the author is aware the full identity of the unknown (muramic acid?) in the polysaccharide has never been reported by these workers. The material was isolated from encapsulated cells by treating with lysozyme. WELSHIMER (1953), in earlier experiments, had also observed an apparent action of lysozyme on the capsule of another strain of *Bacillus megaterium*. Whether the material detectable at the polar caps and cross-walls of cells of *Bacillus megaterium* by TOMCSIK'S (1956) treatment with antibody to cell-wall and 'capsular polysaccharide' represents a true capsular substance or an over-production and local accumulation of wall material would have to be decided upon after further investigation.

With the discovery of the teichoic acids (ARMSTRONG, BADDILEY, BUCHANAN, CARSS AND GREENBERG, 1958) of bacterial walls, a new constituent, ribitol phosphate, appeared to be confined to the cell-wall structure. However, REBERS AND HEIDELBERGER (1959) detected ribitol phosphate as a component of the specific polysaccharide of type VI pneumococcus. Unlike the teichoic acid of the cell wall, the pneumococcal capsular polysaccharide contained galactosyl-glucosyl-rhamnosyl-ribitol units joined together through phosphodiester linkages involving the ribitol of one end and the galactose of the neighbouring unit (REBERS AND HEIDELBERGER, 1961).

Another instance of the detection of typical wall compounds in a capsular fraction from *Staphylococcus aureus* was reported recently by WILEY AND WONNACOTT (1962). Apart from the common occurrence of the D-isomer of glutamic acid in the capsular polypeptides and cell-wall glycosaminopeptides the presence of glucosamine and the 'wall' amino acids alanine, glutamic acid, lysine and glycine in this fraction of *Staphylococcus aureus* is the first example of anything like a complete 'peptide' being found in a capsular component. The isolated capsular fraction was also rich in phosphorus accountable for as glycerophosphate.

In addition to possessing capsules and slime, some organisms have a sheath enclosing the cells. *Sphaerotilus natans* has a capsule composed of glucose, glucuronic acid, galactose and fucose (GAUDY AND WOLFE, 1962) surrounding a more rigid sheath-like structure (GAUDY AND WOLFE, 1961). Presumably the cells enclosed in the sheath also have a cell-wall structure. Until thin sections are available and attempts are made to isolate the surface structures it will be difficult to say what structural and chemical relationships exist between walls and sheaths of such organisms.

Cementing and cell adherence layers

In addition to capsules, slimes and gums there are other surface components involved in the adherence of bacterial cells to one another. Cell adherence may differ in type from chain-forming organisms obviously devoid of any capsular, slime or gum layer, to aggregates of cells such as those of *Sarcina ventriculi* (CANALE-PAROLA, BORASKY AND WOLFE, 1961) and *Lampropedia hyalina* (MURRAY, 1963a). The materials responsible for cementing *Lampropedia hyalina* cells into a regular aggregate and the cells of *Sarcina ventriculi* into a less orderly array, are anatomically and functionally different from 'capsules'. This type of surface component therefore appears to be concerned with cell to cell adhesion but in common with capsules they are dispensable components for mutant strains survive and grow quite satisfactorily without such layers (CANALE-PAROLA, BORASKY AND WOLFE, 1961; MURRAY, 1963a).

The adherence of cells after division is of course well known and is responsible for the typical habit of chain formation. Such post-division adherence seems to occur without there being any specialized anatomical structure responsible for chain formation in Streptococci and *Bacillus* spp. (LOMINSKI, CAMERON AND WYLLIE, 1958). The available evidence suggests that the formation of chains is due to 'normal' cell-wall material in Gram-positive bacteria and not an additional surface component. LOMINSKI, CAMERON AND WYLLIE (1958) have made the exceedingly interesting observation that Suramin[Hexasodium sym.-bis (*m*-amino-benzoyl-*m*-amino-*p*-methyl benzoyl-1-naphthylamino-4,6,8-trisulphonate) carbamide] which inhibits certain carbohydrases, including egg-white lysozyme (LOMINSKI AND GRAY, 1961; REITER AND ORAM, 1962) will induce the formation of chains. It will be recalled that WEBB (1948) observed the breaking up of chains of heat-killed streptococci after treatment with egg-white lysozyme. It seems probable therefore that a lysozyme type of enzyme could be responsible for cell separation after division by rupturing glycosidic bonds which may join adjacent portions of the newly synthesized wall material. Such enzymes may have a somewhat different substrate specificity to egg-white lysozyme otherwise it would be difficult to visualize the enzyme restricting its activity to a limited portion of the cell-wall surface. It could also be that the action of such enzymes is limited by potentially sensitive groups in the walls of some bacterial species being covered with protective groups e.g. O-acetylation effectively inhibits the action of egg-white lysozyme (BRUMFITT, WARDLAW AND PARK, 1958).

Although the exact mechanism of cell adherence or the cementing of cells together is not known it is evident from the work with *Sarcina ventriculi* that cellulose is the intercalating material (CANALE-PAROLA, BORASKY AND WOLFE, 1961) and that in some Gram-positive bacteria chain formation involves some polysaccharide or glycosaminopeptide component of the surface of the wall.

CELL ENVELOPES AND WALLS

It is apparent from the foregoing discussion and has moreover been obvious for quite a long time that capsules, slime and microcapsules contribute very little, if anything at all, to the morphological integrity and shape of the bacterial cell. The principal structure of the bacterial cell responsible for its mechanical rigidity and shape has generally been called the 'cell wall'. A number of investigators have preferred to use terms such as 'envelopes', 'hulls', 'shells', 'coats' and 'membranes' for the outermost structural component of bacteria. All of these words (wall, envelope, hull, shell, coat, membrane) have essentially the same meaning in the context in which they have been used in various publications. They all indicate the presence of a *covering, framework, continuous sheet,* or *wrapper* around the bacterial cell. The terms do, however, convey variations in the degrees of rigidity of the outer covering and for that reason some authors may prefer a word which implies less structural rigidity than the name 'wall'. Other authors may feel that 'wall' suggests an inert, rigid structure so for various reasons the term 'membrane' is used to infer the possession of functions other than purely mechanical ones. It is frequently rather difficult to distinguish, at both the physical and chemical level, between what have been described as classical 'walls' and 'membranes' (of the type currently defined by ROBERTSON, 1959, as 'unit membranes'). Thus judged on the appearance and dimensions of the profile seen in thin sections of many fixed Gram-negative bacteria the outer components of the cell envelopes are indistinguishable from 'unit membranes' seen in other cells. However, some of the outer 'membranes' of such bacteria have been shown to possess a rigid layer of mucopeptide, a feature which has not been reported for the classical 'unit membranes'.

With these difficulties in mind, the term *wall* is used in a broad sense throughout this book to mean the outer structure which confers shape upon the cell and which can be isolated as a single morphological entity from the bacterial cell. Used in this sense it would exclude plasma membranes which are the 'outer shape-conferring structures' of bacterial protoplasts of Gram-positive bacteria (WEIBULL, 1953a). The difficulty in defining the wall or a membrane too rigidly is again emphasized by the suggestion from BROWN, DRUMMOND AND NORTH (1962) that in the marine pseudomonad it is the inner membrane which is structurally more rigid. In cases of ambiguity or where two components (wall-membrane or double membrane system) are being discussed the term *envelope* will be used.

It is evident from the studies of the surface anatomy of a large variety of bacteria drawn from different taxonomic groups that there is considerable variation in the degrees of differentiation and complexity of the cell envelope. Thus the cell envelope structure of bacteria may range from the single membrane type forming the only detectable outer component of *Mycoplasma* spp. and *Halobacterium* spp., to the well differentiated walls and plasma membranes of Gram-positive organisms, to the double 'membrane' systems of Gram-negative bacteria, eventually culminating in the high degree of complexity found in the envelope of certain cocci and *Lampropedia hyalina*. The anatomy of these various types of bacterial envelopes will be described in detail in the following sections of this Chapter.

1. *Single surface 'membrane'*

The 'simplest' level of organization of the cell surface so far detected in bacteria is shown in pleuropneumonia-like organisms (PPLO) *Mycoplasma* spp. (VAN ITERSON AND RUYS, 1960;

RODWELL, personal communication) and in the halophilic organisms, *Halobacterium halobium* and *Halobacterium salinarium* (BROWN AND SHOREY, 1962). In both of these groups the organisms are surrounded by single 'membranes' of about 75 Å in thickness for *Mycoplasma* spp. (VAN ITERSON AND RUYS, 1960) and a slightly larger membrane probably of the type defined as a 'compound membrane' by SJÖSTRAND (1960) in the halophilic bacteria (BROWN AND SHOREY, 1962, 1963). Some of the cultures of the PPLO's examined by VAN ITERSON AND RUYS (1960) contained small cocci which did have a cell wall of ~ 150 Å thickness surrounding an underlying limiting membrane (~75Å) the latter being the surface structure for the majority of the cells in these strains and all of the cells in the coccus-free strains of *Mycoplasma* spp.

It would thus appear likely that 'wall' and 'membrane' functions are condensed into a single structure in these bacteria. Correlated with this 'simplicity' in surface organization is the absence of the typical components of the cell-wall glycosaminopeptides or mucopeptides. PLACKETT (1959) was unable to detect any of the mucocomplex substances in *Mycoplasma mycoides* and BROWN AND SHOREY (1962) could detect neither muramic acid nor DAP in whole cells and isolated 'membranes' of the two halophilic species. So far as the author is aware these are the only two examples of organisms normally regarded as belonging to the bacterial world that are apparently devoid of the mucopeptide components. Both organisms have rather specialized natural habitats, and are found in environments which provide the organisms with materials needed for cell stability (e.g. cholesterol, glycerol for *Mycoplasma* spp., RODWELL AND ABBOT, 1961; high salt concentrations for the obligate halophiles). It would seem likely that as these organisms evolved and became adapted to such environments there was no longer a need for the rigid components.

The similarity between the surface envelope of *Mycoplasma* spp. and the bacterial protoplast membrane is further emphasized by the response of both structures to the lytic action of alcohols and sodium dodecyl sulphate (RAZIN AND ARGAMAN, 1962).

2. *Wall and membrane*

The next level of increasing complexity and higher degree of surface differentiation is illustrated by the vast majority of Gram-positive bacteria. These organisms possess a relatively thick wall varying in width from about 150 Å to as much as 800 Å in *Lactobacillus acidophilus* (GLAUERT, 1962). The walls are generally devoid of fine structure and are rather amorphous in appearance. Thin sections of Gram-positive organisms show an underlying plasma membrane with an overall thickness of ~75 Å. The conversion of a number of Gram-positive bacteria to protoplasts by enzymic digestion of the wall with lysozyme or other specific wall-degrading enzymes clearly established the fact that in these organisms the wall and plasma membrane are separate and not integrated structures (WEIBULL, 1953a, 1958a; MCQUILLEN, 1956, 1960).

Complete separation of the plasma membrane from the wall has also been elegantly demonstrated in thin sections of plasmolysed cells of *Bacillus subtilis* by VAN ITERSON (1961). She has shown that under plasmolysis conditions the intact plasma membrane recedes with the cell cytoplasm. These experiments are in accord with what was expected from the reversible plasmolysis of isolated protoplasts of *Bacillus megaterium* reported by WEIBULL (1955). Both of these findings are contrasted with the report by HUGHES (1962), that he was unable to release the membrane component of cell wall-membrane preparations of Gram-positive bacteria when they were subjected to further mechanical disintegration procedures.

The electron density of the walls in thin sections of Gram-positive bacteria is generally rather uniform. In some electron micrographs there is a suggestion of a multilayered appearance but the separation of the wall of Gram-positive bacteria into electron dense and transparent layers is rarely as clear cut as the layering (electron dense layers on either side of an electron transparent layer) seen in the profiles of the plasma membranes of these organisms. The presence of cell-wall layers of different chemical constitution is likely (e.g. glycosaminopeptide, teichoic acid, polysaccharide) and could account for differences in electron density seen in thin sections of fixed cells stained with uranyl salts.

3. Double 'membrane' or 'wall' membrane

Apart from the exceptions already mentioned, the majority of the thin sections of Gram-negative bacteria show two multilayered parallel 'membranes' in the envelope structure. The outermost structure has usually been referred to as the wall and the inner one as the plasma or cytoplasmic membrane. The thin sections prepared by KELLENBERGER AND RYTER (1958) were the first to establish the multilayered outer component in *Escherichia coli*. Independent confirmation of the layered nature of the outer wall of *Escherichia coli* came from the studies of WEIDEL, FRANK AND MARTIN (1960) who achieved a separation of the rigid layer from the more 'plastic' lipopolysaccharide-protein components of the 'wall'.

Similar profiles to those observed in *Escherichia coli* envelopes have been seen in thin sections of many Gram-negative bacteria including *Spirillum serpens* (MURRAY, 1962), *Rhodospirillum molischianum* (GIESBRECHT AND DREWS, 1962) and the marine pseudomonad studied by BROWN, DRUMMOND AND NORTH (1962). All of these organisms have shown a multilayered outer component, referred to as a 'wall' by some authors and as a membrane by others and an underlying multilayered membrane. MARTIN AND FRANK (1962) have conclusively demonstrated that the isolated 'wall' of the *Spirillum* sp. first studied by HOUWINK (1953) can be resolved into the spherical macromolecular components and the rigid continuous layer of glycosaminopeptide or mucopeptide (see illustrations Chapter 3). Thus for both *Escherichia coli* and *Spirillum* sp. the rigid layer probably occurs on the inside of the outer multilayered structure (wall or compound membrane) and is apparently not part of the underlying membrane (plasma membrane). BROWN, DRUMMOND AND NORTH (1962) concluded from their studies of the autolysis of isolated envelopes (inner and outer multilayered structures) of a marine pseudomonad that the inner component was the more rigid of the two. These authors also suggest that in this organism the glycosaminopeptide is associated with both internal and external membranes, there being a greater amount of DAP and amino sugar in the inner membrane structure. This finding is in sharp contrast with the occurrence of the rigid mucopeptide in the outer 'wall' of the other species so far examined (WEIDEL, FRANK AND MARTIN, 1960; MARTIN AND FRANK, 1962). The results for the marine pseudomonad studied by BROWN, DRUMMOND AND NORTH (1962) also differ from the earlier work of FEW (1954) who showed that further treatment of isolated envelopes (wall-membrane) of *Pseudomonas denitrificans* in the Mickle apparatus, resulted in the loss of the inner component leaving the more rigid, rod-shaped outer 'wall'. Unfortunately it was not stated whether the mucopeptide components of the *Pseudomonas denitrificans* envelope were released on removal of the inner component or whether they were exclusively in the outer structure.

Because of the similarity in the profiles of the two multilayered structures forming the envelope of certain Gram-negative bacteria it has been suggested that both may be designated as 'unit membranes' (BROWN, DRUMMOND AND NORTH, 1962; CLARKE AND LILLY, 1962). However, from the data available, it is the present author's view that if the word 'membrane' is to be used instead of 'wall' to describe the outer component then it would perhaps be better to call it a 'compound membrane' (cf. SJÖSTRAND, 1960). Where both multilayered outer and inner components exist together in the bacteria examined up to the present time there seems to be several properties which distinguish the outer bacterial 'wall' or 'compound membrane' structure from the 'unit membrane' as defined by ROBERTSON (1959). So far as the author is aware neither hexagonally packed spherical macromolecules (80–120 Å diameters) nor a continuous sheet of the structural heteropolymer, glycosaminopeptide (mucopeptide), has been detected in 'unit membranes' from bacteria or any other type of cell. The evidence from electron microscopy suggests that the spherical macromolecular layer of the envelopes of organisms such as *Spirillum* sp. (HOUWINK, 1953; MARTIN AND FRANK, 1962) and *Rhodospirillum rubrum* (SALTON AND WILLIAMS, 1954) is in the outer multilayered component and not in the inner plasma membrane. Recent studies by Murray (1963b) have shown that although the 'profile' of the outer wall of *Spirillum serpens* has the appearance of a 'unit membrane' under some fixation conditions, it can now be resolved into a more complex structure revealing the spherical subunits in thin sections and a very complicated arrangement of layers in negatively-stained wall preparations. Moreover, the plasma or cytoplasmic membrane of the photosynthetic organism, *Rhodospirillum rubrum*, is the structure from which the chromatophores arise and so far the spherical macromolecular components (c. 100 Å periodicity) of the outer structure have not been detected in the isolated chromatophores.

From observations at present available, there is reason to believe that in certain Gram-negative bacteria there are differences in the structure and the chemical components of the outer multilayered and inner multilayered 'membranes'. The preferential autolysis of the outer membrane reported by BROWN, DRUMMOND AND NORTH (1962) also appears to be in accord with these conclusions. However, it is of interest to note that CLARKE AND LILLY (1962) have suggested that the surface layers of Gram-negative bacteria consist of two 'unit' membranes, each with the structure of protein-lipid-lipid-polysaccharide, with a rigid layer (mucopeptide) in between. They have suggested that this type of structure would allow for a plasma membrane surrounded by a rigid wall and an outer membrane with its layers reversed to give polysaccharide and lipid-lipid-protein. The suggestion that outer and inner membrane components of envelopes of Gram-negative bacteria are chemically similar is not new and had already been contemplated by SALTON (1961). The conclusion by CLARKE AND LILLY (1962) 'that a double membrane surface structure may be characteristic of all Gram-negative bacteria' is obviously too generalized, for instances where this is not so have already been presented above.

It is apparent that our knowledge of the relationship between the two multilayered structures seen in thin sections of Gram-negative bacteria is inadequate and further investigations are needed before we can be certain about the distribution of the various chemical constituents in each major component. Such studies should enable us to decide whether the only difference between the outer and inner components is due to the presence of glycosaminopeptide in one and not the other or whether the rigid layer is 'sandwiched' between two identical membranes.

As pointed out by SALTON (1961) one of the major outstanding problems in the structural and biochemical analysis of Gram-negative bacteria is the lack of suitable enzyme systems capable of selective removal of the outer wall structure (both rigid glycosaminopeptide and 'plastic' lipopolysaccharide-protein layers) of these organisms. Such an enzyme system would soon lead to a clarification of the similarity or differences between the components of the 'double membrane' envelope of Gram-negative bacteria. BROWN, DRUMMOND AND NORTH (1962) have recently reported a preferential removal of the 'outer membrane' by autolysis thus enabling them to isolate the inner 'cytoplasmic membrane'. If the complete removal of the external 'membrane' component could be achieved and its absence established by immunochemical methods then this system should provide much new information about the relationships of outer and inner components of the envelope and enable some more definitive work to be performed on the chemical properties and distribution of the two structures.

'Protoplasts' (spheroplasts) of Gram-negative bacteria

The relationship between the two surface structures of Gram-negative bacteria has often been clearly shown during the formation of 'protoplasts' or spheroplasts (MCQUILLEN, 1960). For a long time the bacterial cytologist regarded the round and bizarre shaped cells as life-cycle stages but following the investigations of the action of penicillin (DUGUID, 1946; LEDERBERG, 1956) it soon became apparent that such cell forms arise as a result of structural deformation of the 'walls' or envelopes.

Following WEIBULL's (1953a) isolation of the protoplasts of *Bacillus megaterium*, the formation of spherical cells under the action of penicillin or lysozyme on *Escherichia coli* (LEDERBERG, 1956; ZINDER AND ARNDT, 1956) initially led to the view that these were analogous to the protoplasts of Gram-positive bacteria. It soon became apparent however, that the formation of spherical protoplasts from rod-shaped Gram-negative bacteria resulted from a weakening of the outer wall structure rather than its complete removal. The weakening of the 'wall' of the Gram-negative bacterium could thus be brought about by direct action of lysozyme on the glycosaminopeptide components or indirectly by the inhibition of biosynthesis of wall mucopeptide by penicillin.

By direct chemical examination of the isolated walls or envelopes from penicillin spheroplasts or 'protoplasts' of *Vibrio metchnikovi* and *Salmonella gallinarum* it was established that they contained approximately the same amounts of polysaccharide, lipid and protein as present in the original walls but the quantities of amino sugar and diaminopimelic acid were substantially reduced (SALTON AND SHAFA, 1958) as shown in Table 3. These studies together with investigations of the immunological reactions and detection of bacteriophage receptors indicated the presence of the major components of the original cell walls (BRENNER *et al.*, 1958; SHAFA, 1958; HOLME, MALMBORG AND COTA-ROBLES, 1960).

Formation of spheroplasts and the production of stable L-forms of Gram-negative bacteria can be brought about in a variety of ways; penicillin has been more widely used for this purpose. Many of the factors involved in these processes have been discussed by MCQUILLEN (1960), WEIBULL (1958a) and WELSCH (1958). Active in 'protoplast' or 'spheroplast' formation

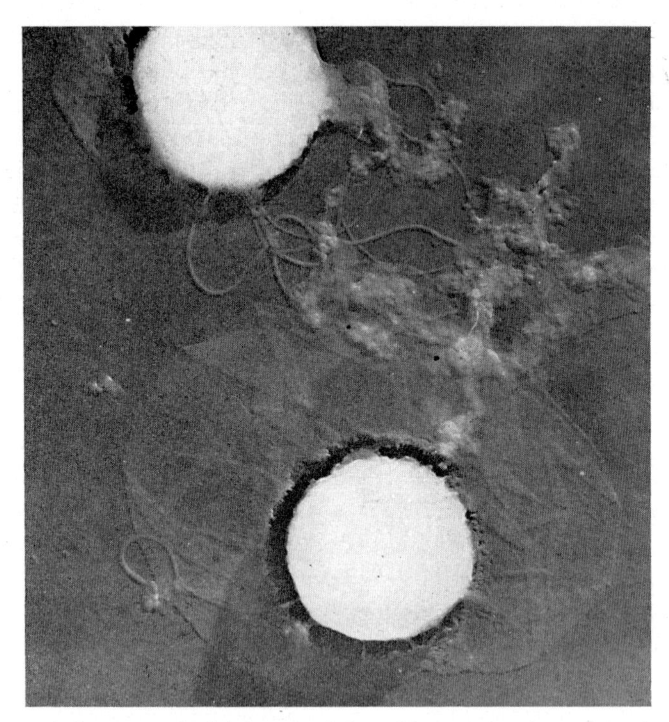

FIGURE 3. *Vibrio metchnikovi* 'protoplasts' (spheroplasts) formed by growing the organism in the presence of penicillin. The true protoplast is surrounded by an envelope of much greater diameter giving the spheroplast the typical 'poached egg' appearance. × 10,000.

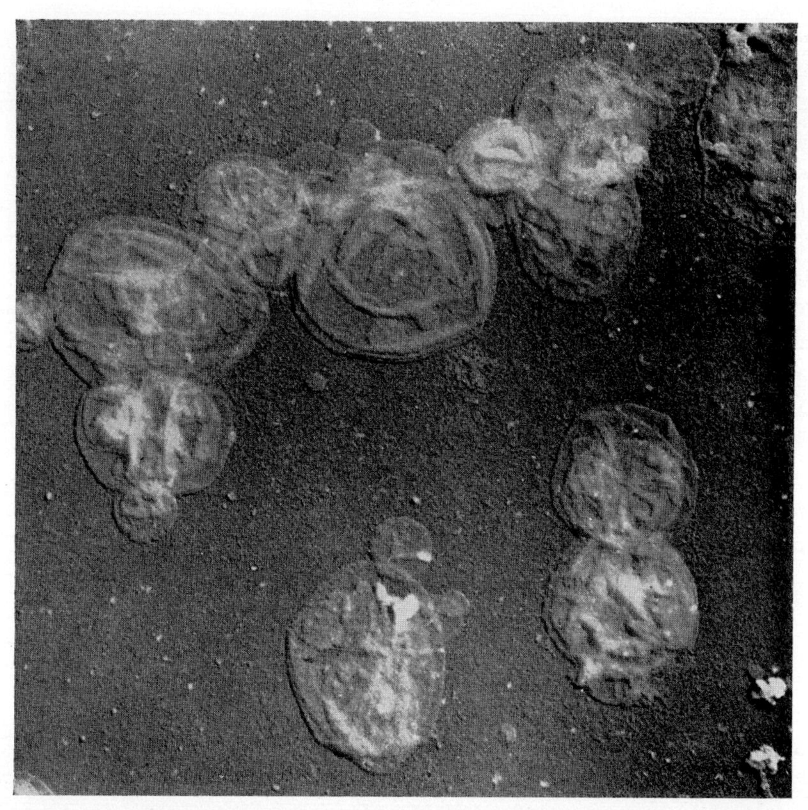

FIGURE 4. 'Double membranes' found on lysis or disintegration of spheroplasts of *Vibrio metchnikovi* formed by growth in the presence of penicillin. The outer thin structure is interpreted as originating from the wall and the inner one corresponding to the plasma membrane. $\times$26,000 (SALTON AND SHAFA, 1958).

are agents such as penicillin and related antibiotics, lysozyme acting in the presence of ethylenediaminetetraacetic acid (EDTA), glycine and deprivation of α,ε-diaminopimelic acid from mutant strains of *Escherichia coli* exacting towards this 'wall' amino acid (MCQUILLEN, 1960). Spheroplasts of galactose-negative mutants of *Salmonella typhimurium* are formed when this organism is grown in the presence of galactose thus providing another fascinating example of biochemical induction of 'protoplast' formation (FUKASAWA AND NIKAIDO, 1959).

Spheroplasts of Gram-negative bacteria, especially the penicillin and glycine induced forms, generally show a separation of the outer envelope component from the inner plasma membrane resulting in a 'poached egg' appearance as shown in Fig. 3 for the 'protoplast' of *Vibrio metchnikovi*. Lysed spheroplasts of this organism gave preparations containing 'double membranes' as shown in Fig. 4 the external one being regarded as the weakened cell wall and the inner structure interpreted as the plasma membrane (SALTON AND SHAFA, 1958). JEYNES (1961) on the other hand believed the formation of glycine-induced 'protoplasts' of *Vibrio cholerae* involved complete loss of the outer 'wall' component giving naked protoplasts analogous to those obtained enzymically from Gram-positive bacteria. Sections of glycine spheroplasts of the marine pseudomonad studied by BROWN, DRUMMOND AND NORTH (1962) showed two membrane systems.

One feature commonly encountered in spheroplasts induced by DAP deprivation, or by growth in the presence of penicillin or glycine is the formation of a large vacuolar region where the outer 'wall' and the inner plasma membrane have separated. The vacuolar region seems to be devoid of cytoplasm and thin sections of spheroplasts (THORSSEN AND WEIBULL, 1958; HOFSCHNEIDER, 1960) have shown the separation of the outer 'wall' from the plasma membrane which in turn surrounds the protoplasm. The bacterial cytoplasm of the spheroplast may be held within the plasma membrane as a crescent shaped body, surrounded in its entirety by an external wall. No completely satisfactory explanation for this type of structural abnormality has been advanced. LARK (1958) suggested that a specific component of the growth medium is responsible for the phenomenon in *Alcaligenes faecalis* spheroplasts. Differences in the rates of formation of the inner and outer components could also explain the structural abnormalities in the glycine induced forms (BROWN, DRUMMOND AND NORTH, 1962), an explanation which could apply equally well to other types of spheroplasts.

The greater external diameter of the outer component of the spheroplast envelope suggests that its growth and extension occurs at a greater rate than the underlying plasma membrane or that it is produced at the same rate but is of greater elasticity. If the latter were true it would imply that an internal pressure is exerted upon the outer 'wall' or 'membrane' of the spheroplast. In deed the fact that the external 'membrane' of the spheroplast is generally so spherical led BROWN, DRUMMOND AND NORTH (1962) to conclude that both outer and internal membranes of spheroplasts possessed permeability properties. However, at present there is no experimental evidence enabling us to say whether or not both the outer and inner envelope components of intact Gram-negative bacteria and their spheroplasts possess permeability properties.

Although spheroplasts or 'protoplasts' of Gram-negative bacteria have revealed an outer multilayered structure (weakened 'wall' or 'compound membrane') often well separated by an extensive vacuolar area from the plasma membrane enclosing the cell cytoplasm no further conclusion can be made about the structural relationships and functions of the two separable components of the envelope.

Thin sections of penicillin-induced and lysozyme – EDTA spheroplasts of *Escherichia coli* were prepared by HOFSCHNEIDER (1960) and although both types of 'protoplast' were surrounded by two multilayered membranes it was difficult to be certain that any change had occurred in the overall appearance of inner or outer membranes. Thus the loss of mucopeptide on formation of EDTA – lysozyme protoplasts could not be correlated with any major alteration in the appearance of thin sections of the surface structures. Finer resolution of the multilayered membranes, together with the specific labelling of components will be required before it will become possible to localize the mucopeptide in the envelope and determine its fate during spheroplast formation.

Both true protoplasts of Gram-positive bacteria (BRENNER *et al.*, 1958) and spheroplasts or 'protoplasts' of Gram-negative bacteria form the starting point for the L-phase cultures of bacteria (LEDERBERG AND ST. CLAIR, 1958; FITZ-JAMES, 1958; FREIMER, KRAUSE AND MCCARTY, 1959). The origin and biological properties of L-forms and the PPLO have been dealt with extensively by KLIENEBERGER-NOBEL (1960, 1962) and no further discussion of this aspect of the topic will be given here. Structurally, the stable L-forms of Gram-positive bacteria are devoid of the typical components of the glycosaminopeptides of the cell walls of the parent organisms as first shown by SHARP, HIJMANS AND DIENES (1957).

The presence of typical wall constituents in the L-forms of Gram-negative bacteria has been investigated independently by KANDLER AND ZEHENDER (1957) and by WEIBULL (1958b), THORSSON AND WEIBULL (1958) and MORRISON AND WEIBULL (1962). KANDLER AND ZEHENDER (1957) were unable to detect DAP in the stable L-form of Proteus, but reported the presence of DAP in the unstable L cells. WEIBULL (1958a) on the other hand found considerable quantities of DAP in the L 9 strain of Proteus and further investigation showed that other stable L strains contained smaller but detectable amounts of DAP (MORRISON AND WEIBULL, 1962). Thus it would appear from these studies that L-forms of Proteus do indeed contain constituents of 'cell-wall' glycosaminopeptides but at a lower level to that found in the organisms from which the L-forms have been derived. The sensitivity of Proteus L-forms to bacteriophages (TAUBENECK, BÖHME AND SCHUMANN, 1958) would suggest that other components of the outer envelope component are also present, thus adding further weight to the conclusions deduced from the appearance of thin sections of such forms. Whether or not the stable L-forms of Proteus possess a double component envelope cannot be said with certainty. The preservation of membranes in some of the earlier thin sections is not quite sufficient to be certain. However, under conditions which showed the double envelope of penicillin spheroplasts of *Escherichia coli*, THORSSON AND WEIBULL (1958) showed L-form cells of Proteus L 9 surrounded by a single multilayered structure.

Studies with L-form cultures of bacteria have therefore shown that Gram-positive bacteria can tolerate a complete loss of the rigid wall and continue their existence as L-forms, while Gram-negative bacteria survive in the L phase after partial loss of at least the glycosaminopeptide component of the wall.

4. Complex surface envelopes

The vast majority of bacteria sub-divided by the Gram reaction fall into two broad groups (b and c), (Fig. 7) with respect to surface anatomy. On the one hand most of the Gram-positive bacteria characterized by a surface envelope with a relatively thick wall (150–800 Å) and a separate plasma membrane structure ($\sim$75 Å), and on the other hand, the majority of Gram-negative bacteria possessing a surface envelope composed of two multilayered components. Contrasted with the two major types of surface envelope is the single 'membrane' surrounding *Mycoplasma* spp. and the halophilic bacteria at the one extreme of structural 'simplicity' and at the other extreme the very complex anatomy of the surface of *Lampropedia hyalina*, *Micrococcus radiodurans* and certain tetrad-forming cocci (MURRAY, 1962).

One of the distinguishing features of the organism *Lampropedia hyalina* isolated as a pure culture by PRINGSHEIM (1955) is its ability to form regular sheets, tablets or rafts of cells. It seemed conceivable that this organism may have had some special surface properties and independent investigations were performed by MURRAY in Canada and CHAPMAN AND SALTON in England. Indeed the surface structure of *Lampropedia hyalina* proved to be quite unlike that of any of the other organisms so far studied and some preliminary reports on the anatomy of this organism have appeared (CHAPMAN AND SALTON, 1962; MURRAY, 1962) and full details of these have been presented by CHAPMAN, MURRAY AND SALTON (1963) and MURRAY (1963).

Surrounding the aggregate of cells of *Lampropedia hyalina* is a complex surface structure possessing a mesh-like layer and another punctate layer. These two layers form a 'corset' around the tablet of cells, with individual cells cemented together by an intercalating substance. So far as can be judged at present each cell of the sheet also possesses a 'wall' and underlying plasma membrane encasing the bacterial protoplasm. This organism, *Lampropedia hyalina* appears to be the only bacterial species in which a 'colony' of cells is surrounded by a specialized envelope (MURRAY, 1963). Fragments isolated from *Micrococcus radiodurans* possess structured layers rather similar to those seen in *Lampropedia hyalina*, but the former organism does not form sheets (GLAUERT, 1962). Although the cementing or intercalating layer in sheets of *Lampropedia hyalina* seems to perform a similar function to the cellulose substance cementing cells of *Sarcina ventriculi* into aggregates (CANALE-PAROLA, BORASKY AND WOLFE, 1961) it cannot be concluded at present that is is of identical chemical composition. Cell aggregates of *Lampropedia hyalina* are broken up by Schweizer's reagent (CHAPMAN, MURRAY AND SALTON, 1963) but further evidence will be needed before it can be concluded that the material responsible for holding the cells in the tablet form is cellulose.

Surface envelopes of bacterial endospores obviously fall within this group of complex structures. As seen in thin sections the spore integument is a multilayered structure possessing a great number of layers external to the spore cortex and inner components of the spore (MAYALL AND ROBINOW, 1957; WARTH, OHYE AND MURRELL, 1963). In some spores the outermost layer will be the sporangium or cell wall of the vegetative cell.

From this discussion of the anatomical status of the bacterial cell envelope and the relationships between cell walls and membranes we can conclude that the term 'wall' is only meaningful in a broad general sense, as a variety of structures can perform the functions of providing the cell with an encasing wall or membrane of varying degrees of rigidity. In some bacteria a single structure of marked similarity in profile and overall dimensions to a 'unit' or plasma membrane will provide the cell with its 'wall' or envelope component. Although such cells may have the 'simplest' type of surface structure it is quite possible that they have achieved a higher degree of structural and biochemical sophistication, rather than being the primitive progenitors of the cells with more complex walls and envelopes. A rigid cell wall analogous to the thick, robust structures of yeasts, fungi and plants is largely confined to the Gram-positive group of bacteria. It is likely from chemical studies (SALTON, 1960b) that there will be grada-

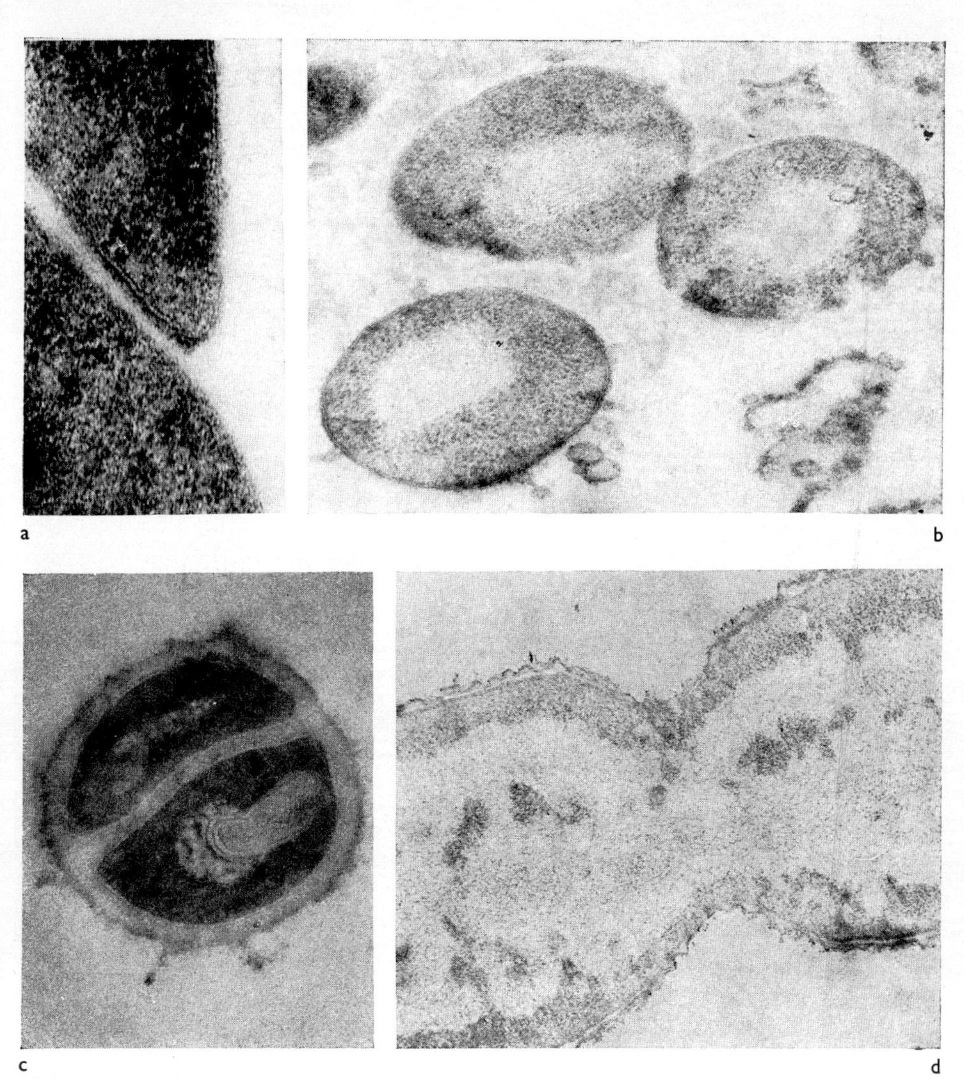

FIGURE 5. Thin sections of bacteria illustrating the types of surface structure.
(a) *Halobacterium halobium* with a single, multilayered surface membrane (BROWN AND SHOREY, 1962). ×100,000.
(b) Protoplasts of *Micrococcus lysodeikticus* bounded by a single membrane (the plasma membrane). ×60,000.
(c) *Micrococcus lysodeikticus* cells typical of Gram-positive bacteria with a thick outer wall and an inner plasma membrane (SALTON AND CHAPMAN, 1962). ×60,000.
(d) *Escherichia coli* cells infected with phage and treated with chloramphenicol, showing a multilayered surface 'wall' (or compound membrane) and an underlying plasma membrane (KELLENBERGER, 1960). ×40,000.

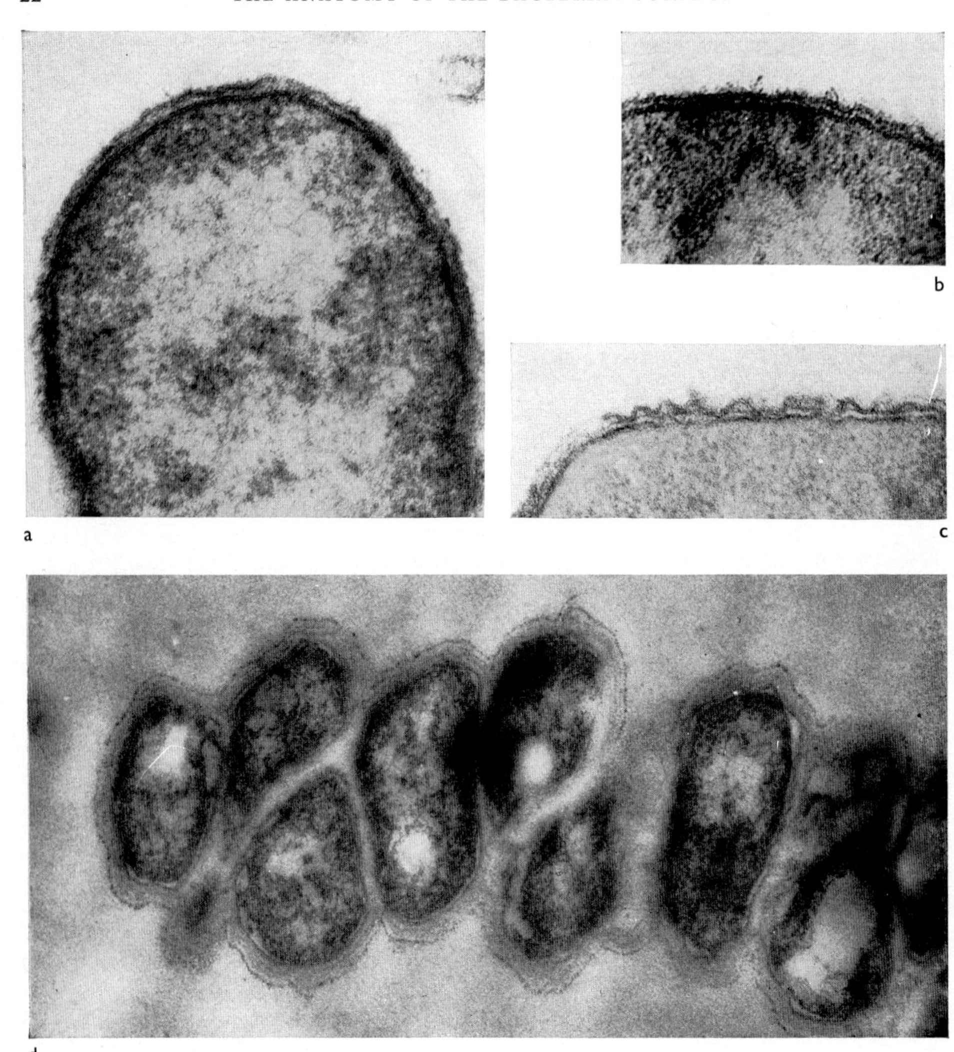

FIGURE 6. Thin sections of bacteria illustrating the types of surface structure.
(a) Portion of a cell of the marine pseudomonad, NCMB 845, showing a double multilayered membrane envelope (BROWN, DRUMMOND AND NORTH, 1962). ×80,000.
(b) The surface membranes of a spheroplast of marine pseudomonad, NCMB 845, showing no detectable difference from those seen in the normal cell (a). ×80,000. (BROWN, DRUMMOND AND NORTH, 1962).
(c) Section of *Spirillum serpens* showing a complex surface envelope with a plasma membrane and wall which appears to consist of a thin and dense layer and a wavy outer 'unit membrane' (MURRAY, 1963). ×85,000.
(d) Part of a tablet of cells of *Lampropedia hyalina* showing a complex surface envelope, the outer structured layer appearing as a dense margin of dots. (CHAPMAN AND SALTON, 1962). ×30,000.

tions in the thickness of the rigid component in Gram-negative bacteria. Studies of the comparative anatomy of the envelope structure of bacteria indicate that four broad groups can be identified and some of the features of these groups and variations in wall and membrane structure are summarized in Table 4. Electron micrographs illustrating the types of envelope, wall and membrane structure are shown in Figs. 5 and 6. The structures of the principal types of bacterial envelopes are represented diagrammatically in Fig. 7.

Localization of enzymes in bacterial envelopes

Studies of the localization of enzymes in fractions of the bacterial cell have been and will be of considerable value in defining the functions and anatomy of the principal surface structures. There is fairly general agreement from work with isolated protoplasts and from cell fractionation studies that the walls of Gram-positive bacteria are probably devoid of enzymic activities. By enzymic removal of the wall for protoplast formation it has been deduced that the bulk of the biochemical apparatus of the cell is left relatively unimpaired in the organized protoplast (McQuillen, 1956, 1960; Weibull, 1958a). The composition of the walls of Gram-positive bacteria also leads us to the conclusion that enzymically active proteins do not make a significant contribution to the cell-wall substance and are probably not integral components of the wall. On the other hand the early studies on protoplast membranes or 'ghosts' isolated from

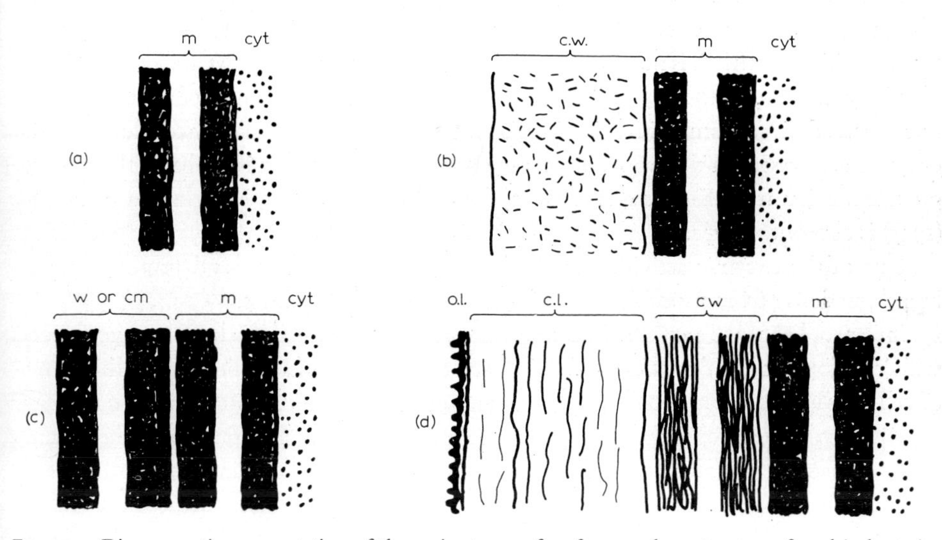

Figure 7. Diagrammatic representation of the major types of surface envelope structures found in bacteria.
(a) Single 'unit membrane' (m) type found at surface of certain bacteria and naked protoplasts (Group I, Table 4).
(b) Thick amorphous cell wall (c.w.) and underlying plasma membrane (m) as in many Gram-positive bacteria (Group 2).
(c) Multilayered 'wall' or compound membrane (c.m.) and underlying membrane as found in Gram-negative bacteria.
(d) Complex surface envelope as found in tablets of *Lampropedia hyalina* cells with outer structured layer (o.l.), cementing layer (c.l.), 'wall' (c.w.) and membrane (m).

lysed protoplasts clearly indicated that these structures were biochemically active (WEIBULL, 1953b; MITCHELL AND MOYLE, 1956; STORCK AND WACHSMAN, 1957). Cytochromes and enzymes involved in the electron transport system were localized in the 'plasma or protoplast membranes' of Gram-positive bacteria. Unpublished experiments of the author (SALTON) add confirmation to these results, for the clean preparations of isolated walls of *Bacillus megaterium* and *Micrococcus lysodeikticus* on the other hand were devoid of the enzymes found in the 'membrane' fractions.

Investigations of enzyme localization in Gram-positive bacteria have been largely confined to determining the distribution of enzymes between the protoplast membrane fraction prepared by lysis of bacterial protoplasts and the 'soluble' and small particles of the cell supernatant fractions. Since most of these studies were performed before the discovery of the intracellular membrane or mesosome systems it is now difficult to say whether the activities are in the plasma membrane or the mesosome structure formed by the invagination of the plasma membrane, or in both.

HUGHES (1962) has prepared cell wall – membrane fractions from several Gram-positive bacteria disintegrated in the HUGHES press and such preparations contained the enzymic activities reported in the isolated 'protoplast membrane' fractions studied by earlier investigators.

From the recent studies by YAMAGUCHI (1960); LUKOIANOVA, GELMAN AND BIRIUSOVA (1961); SALTON AND CHAPMAN (1962) on the morphology of the isolated membranes from Gram-positive bacteria it is no longer possible to say that certain enzymes are localized in the plasma membrane. Until it is possible to differentiate between the plasma membrane and those membranes derived from its invagination (i.e. the 'mesosome') we can only conclude that the enzymes are present in the whole membrane fraction of the cell. It is probable that the red granule fraction from *Bacillus stearothermophilus* studied by GEORGI, MILITZER AND DECKER (1955) is also of this type.

Some of the enzymic activities localized in 'cell wall – membrane' and 'protoplast membrane' fractions of Gram-positive bacteria are summarized in Table 5.

Similar studies of the enzyme distribution on Gram-negative bacteria have been performed with 'envelope' or 'hull' fractions and the supernatant components obtained from disintegrated cells (MARR AND COTA-ROBLES, 1957; MARR, 1960a 1960b; HUNT, RODGERS AND HUGHES, 1959). HUGHES (1962) obtained cell wall – membrane (envelope) fractions from other Gram-negative bacteria. All of these preparations appear to contain both inner and outer multilayered structures so at present there is not enough evidence for us to conclude whether the envelope enzymes are confined to one or more of the membrane components. The fractions from *Azotobacter agilis* investigated by PANGBORN, MARR AND ROBRISH (1962) contained the intracellular membranes as well. Enzymic activities so far known to be localized in envelope fractions of Gram-negative bacteria are presented in Table 6. With these 'envelope', 'hull' or cell-wall membrane fractions of Gram-negative bacteria prepared by sonic disintegration or rupture in the HUGHES press is has been difficult to achieve any selective removal of 'membrane' material by further shaking with glass heads (HUNT, RODGERS AND HUGHES, 1959;

HUGHES, 1962). However, there are several pieces of evidence suggesting that some of the enzymes of the envelope fraction may be located in the outer multilayered component. HUNT, ROGERS AND HUGHES (1959) found that lysozyme and EDTA treatment of the isolated 'wall-membrane' fractions of *Pseudomonas fluorescens* released the total nicotinic acid hydroxylase activity into the supernatant fraction. Alkaline phosphatase activity of *Escherichia coli* K 12 was also liberated quantitatively when the cells were converted into 'protoplasts' or spheroplasts by lysozyme – EDTA action (MALAMY AND HORECKER, 1961). It was not known whether the alkaline phosphatase was localized between the wall and membrane or associated with the wall itself. Further investigations with these types of systems would not only give valuable information about the site of enzymes in the envelopes of Gram-negative bacteria but might also contribute to our knowledge of the chemical anatomy of these structures.

PROTOPLAST MEMBRANES

The existence of cell walls and plasma membranes as separate structures in bacteria has long been assumed but the experimental evidence for a plasma membrane rested solely on staining reactions until WEIBULL (1953a, b) isolated protoplasts of *Bacillus megaterium* and conclusively demonstrated the presence of a limiting membrane which possessed the properties of an osmotic barrier (WEIBULL, 1955). This clear distinction between cell wall and plasma membrane has only been demonstrated in Gram-positive organisms possessing walls susceptible to complete digestion with enzymes (SALTON, 1957). Formation of protoplasts by complete dissolution of the cell wall has thus provided the starting point for the isolation of the plasma membranes. However, complete dissolution of the wall is not an absolute condition for protoplast formation as MITCHELL AND MOYLE (1957) were able to isolate protoplasts of *Staphylococcus aureus* after an autolytic enzyme had broken the walls into hemispherical shells thereby releasing the intact protoplasts.

WEIBULL (1953a, b) referred to the structures left after osmotic lysis of bacterial protoplasts as 'ghosts'. The insoluble fractions obtained from lysed protoplasts have generally been called protoplast membranes by most investigators (ABRAMS, MCNAMARA AND JOHNSON, 1960; GILBY, FEW AND MCQUILLEN, 1958; MCQUILLEN, 1960; MITCHELL AND MOYLE, 1956; STORCK AND WACHSMAN, 1957; WEIBULL, 1958a).

It should be pointed out that despite the investigations on the isolated protoplasts and membrane structures and the unequivocal resolution of a membrane (similar in thickness and profile to the 'unit' membrane) at the surface of the bacterial cytoplasm (RYTER AND KELLENBERGER, 1958; MURRAY, 1962; GLAUERT, 1962), COLOBERT (1957) and COLOBERT AND DIRHEIMER (1961) suggest that protoplast formation and the detection of a stable membrane may be the consequence of insoluble complexes between lysozyme, cytoplasmic components and breakdown products of the cell wall. WEIBULL, ZACHARIAS AND BECKMAN (1959) determined the amount of lysozyme retained by the 'ghosts' of *Bacillus megaterium* after using the normal procedure for protoplast formation (WEIBULL, 1953a). In contrast to COLOBERT'S (1957) suggestions that the membrane is a cytological artefact, WEIBULL, ZACHARIAS AND BECKMAN (1959) concluded that their results 'imply that the lysozyme present in the 'ghost' fraction (2.3% of their dry weight) can scarcely represent a morphologically or structurally important component of the 'ghosts'.'

That the bacterial protoplast presented a different surface to that found on the original intact cell was readily demonstrated in a number of ways. Bacteriophage would not infect bacterial protoplasts (WEIBULL, 1953a) because the receptor is in the cell wall (SALTON, 1956). The wall and protoplast surface can be readily differentiated by immunological reactions (TOMCSIK AND GUEX-HOLZER, 1954; VENNES AND GERHARDT, 1956). Differences in electrophoretic behaviour have been shown by DOUGLAS AND PARKER (1958) and indicate the presence of different ionic groupings in the surface after the wall is removed and the cells are converted to protoplasts. An additional property which helps to distinguish between walls and membranes is the enzymic constitution of the membrane fraction (see Table 5).

The characteristic components of the cell-wall glycosaminopeptides and polysaccharides have been shown to be virtually absent from the isolated protoplasts and their membranes (MCQUILLEN, 1956, 1960; FREIMER, KRAUSE AND MCCARTY, 1959). Chemical analyses have been performed on isolated protoplast membrane fractions from *Bacillus megaterium* by WEIBULL AND BERGSTROM (1958), from *Micrococcus lysodeikticus* by GILBY, FEW AND MC-QUILLEN (1958) and from *Sarcina lutea* by BROWN (1961). More recent investigations of the chemical composition of the membranes isolated from *Streptococcus faecalis* have been made by SHOCKMAN *et al.* (1963) and they have found some differences in overall composition during the transition of log-phase cells into threonine- or valine-depleted cells. The lipid contents varied from 28.3 to 40.2% and the total peptide substance from 55.4 to 42.3%. FREIMER (1963) has studied the chemical and immunological properties of group A streptococcal membranes in some detail. These membranes were composed of 72% protein, 26% lipid and 2% carbohydrate and antigenic material common to group A membranes was found. The antigenic components did not cross-react with membrane antigens of other Gram-positive cocci. As shown in Table 7 there are marked differences in the overall chemical properties of the isolated cell wall and the separated 'membrane' fractions.

At the time analyses of the isolated protoplast membranes were performed it was believed or inferred that these preparations contained only the limiting plasma membrane structure. However, WEIBULL (1953b) and WEIBULL AND THORSSON (1957) did report the presence of additional cytoplasmic material in the form of granules in the 'ghost' fractions of *Bacillus megaterium*. With the detection of intracytoplasmic membranes in bacteria (RYTER AND KELLENBERGER, 1958; SHINOHARA, FUKUSHI AND SUZUKI, 1957; GLAUERT AND HOPWOOD, 1960; FITZ-JAMES, 1960) and the proposed mode of origin of the 'mesosomes' (FITZ-JAMES, 1960) has complicated this simple interpretation and led to a re-examination of the nature of the 'protoplast membrane' preparations of *Micrococcus lysodeikticus*. SALTON AND CHAPMAN (1962) have shown that the 'mesosome' structure is co-fractionated with the plasma membrane. It is now evident that the 'plasma membrane' preparations previously used undoubtedly

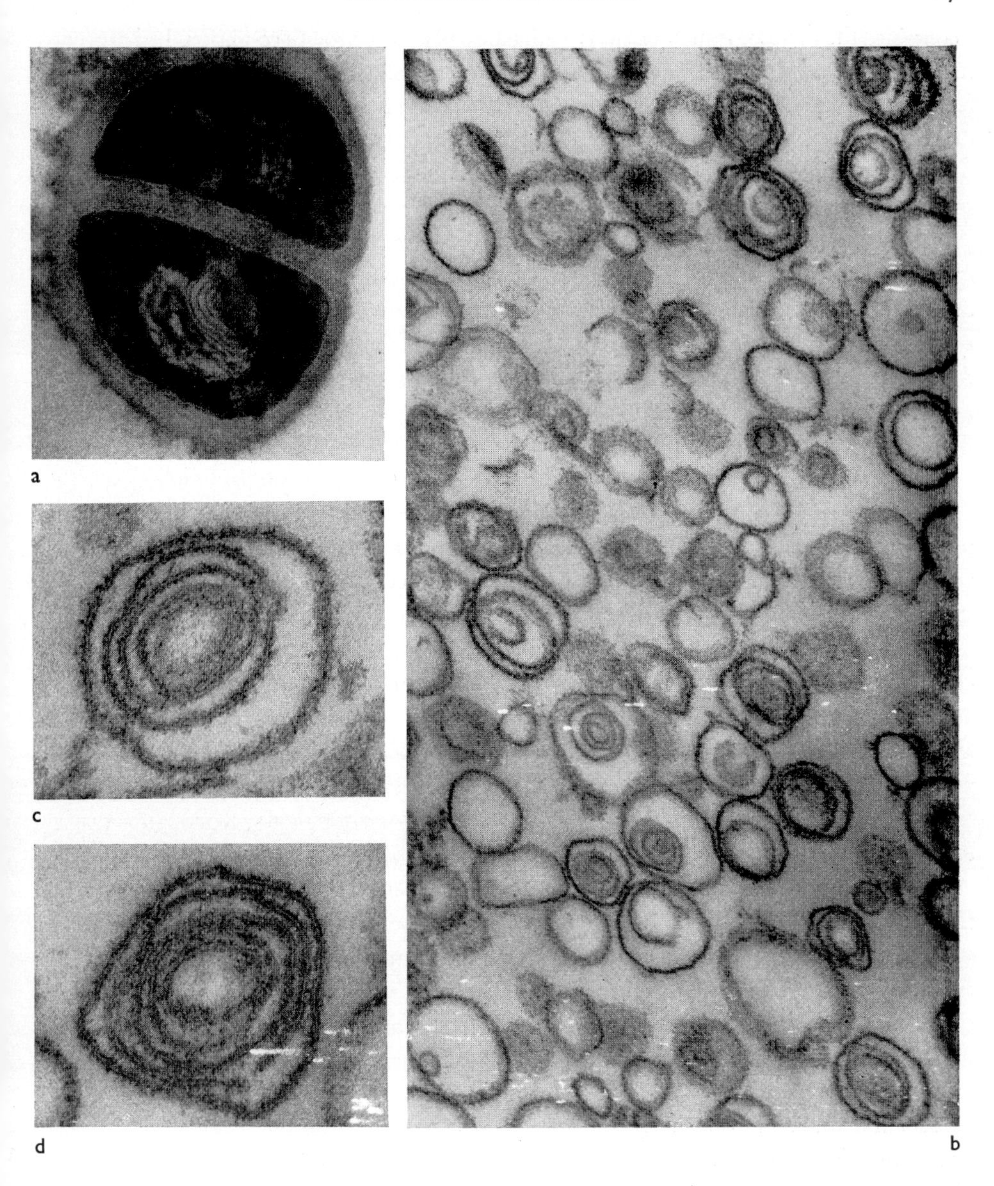

FIGURE 8. Thin sections of *Micrococcus lysodeikticus* cell.
(a) Showing the presence of the intracellular membrane structure ('mesosome'). × 60,000.
(b) The isolated membrane fraction containing the plasma membranes and intracellular ('mesosome') membrane system. × 37,000.
(c) and (d) Membrane structures at higher magnifications demonstrating the maximum number of concentric shells. × 105,000. (SALTON AND CHAPMAN, 1962).

contained the mesosome membrane as well, unless of course growth conditions had prevented invagination of the plasma membrane and the formation of the mesosome structure.

Thin sections of the membrane fractions isolated from *Micrococcus lysodeikticus* were prepared from lysozyme lysates of whole cells and from lysed protoplasts. The mesosome organelle is shown together with the appearance of the isolated membrane fractions in Fig. 8. As many as five concentric membranes may be seen in some cross sections and it is believed that the membrane with the greatest external diameter represents the plasma membrane and that the majority of the smaller diameter membranes are derived from the mesosome. Unlike mitochondria the mesosome does not appear to be enclosed by a complete membrane envelope except perhaps in the vesicular type of intracytoplasmic membrane structure (VAN ITERSON, 1960). The membranes of the mesosome are coiled up and compressed in the intact bacterial cell and on isolation assume the 'expanded' state apparent in Fig. 8. No differences in the appearance of the membrane fractions were detectable when the plasma membrane-mesosome fractions were isolated in the presence of agents inhibiting mitochondrial swelling. There is no evidence, at least at present, to suggest that the morphology of the isolated membrane fractions is the result of swelling analogous to that encountered in mitochondria. Although most of the intracytoplasmic membrane systems of bacteria show a connection with the plasma membrane the precise three-dimensional relationships between the mesosome membranes and the limiting membrane have not been established.

The discovery that what was assumed to be a homogeneous preparation of the limiting plasma membrane, contains in addition the mesosome membranes complicates the interpretation of the chemical analyses of the plasma membranes and the enzymic properties of this structure. However, since the mesosome appears to develop as an invaginated growth of the plasma membrane (FITZ-JAMES, 1960; GLAUERT AND HOPWOOD, 1960; KOIKE AND TAKEYA, 1961; GLAUERT, 1962; MURRAY, 1962) it is possible that all the membrane material isolated in such preparations is chemically and biochemically identical and that the sole distinction between the plasma and mesosome membrane is its location in the cell. Further evidence will have to be sought to clarify the true chemical and enzymic constitution of plasma and mesosome membranes. For the present discussion however, there is no doubt that the cell walls of certain Gram-positive bacteria can be differentiated chemically from the membrane fraction no matter what the distribution of chemical constituents between plasma and mesosome membrane may be.

It will be evident from the details presented in the section on the relationships between 'walls' and 'membranes' in Gram-negative, that the isolation of the 'plasma membrane' (the inner component of the envelope) is a difficult problem. So far as the author is aware the partially purified preparations of the inner membrane component of the pseudomonad studied by BROWN, DRUMMOND AND NORTH (1962) are the closest approach to what has been achieved with certain Gram-positive bacteria. The successful isolation of membranes from Gram-negative bacteria will depend on the development of suitable enzyme systems for selectively degrading the whole of the external envelope structure, which includes a variety of chemical components e.g. lipopolysaccharides, protein, lipid and glycosaminopeptides. Thus as pointed out by SALTON (1955) a multiplicity of enzymes may be needed to achieve this. The enzyme lysozyme may also be a necessary component of any test or screening system. A further complication in the development of such methods for enzymic fractionation of Gram-negative bacteria could arise from the similarity in chemical composition of 'wall' and 'membrane' (SALTON, 1961). BLASKETT (personal communication) has isolated a microorganism producing a lytic system active against intact cells and isolated walls of *Salmonella* spp. Such lytic enzymes should be extremely valuable in future studies on the structural analysis of Gram-negative bacteria.

The extent to which membrane analysis will be complicated by the intrusion of the membrane into the cytoplasm to form the intracytoplasmic organelles or mesosomes in Gram-negative bacteria remains to be determined. Thin sections of *Escherichia coli* from a number of laboratories (KELLENBERGER, RYTER AND SECHAUD, 1958) and others mentioned in a recent review by GLAUERT (1962) have not shown any evidence of mesosome structures. However, small cytoplasmic intrusions (mesosomes) have been observed in sections of *Spirillum serpens* (MURRAY, 1962) and also in *Escherichia coli* (VANDERWINKEL AND MURRAY, 1962). The origin of the chromatophores and photosynthetic lamellae in the photosynthetic bacteria from the plasma membrane structures indicates the extent to which the plasma membrane of Gram-negative bacteria may also contribute to internal membranous organelles (DREWS, 1960; GIESBRECHT AND DREWS, 1962).

Our discussion of the nature and anatomical status of the protoplast or plasma membrane has been confined to those organisms belonging to Groups 1–3 in Table 4. However, the similarities between plasma membranes and the single membrane envelope of the PPLO's, *Mycoplasma* spp., can be emphasized further. The behaviour of both protoplasts of *Micrococcus lysodeikticus* and intact cells of Mycoplasma to lysis with synthetic detergents is very similar. In addition a polyglycerophosphate compound has been found in *Mycoplasma mycoides* (PLACKETT, 1961) and has also been reported in membrane fractions of *Micrococcus lysodeikticus* (GILBY, MCQUILLEN AND FEW, 1958). It will be of considerable interest to see whether there are other chemical constituents common to membranes of other widely separated groups of bacteria.

SURFACE LAYERS OF THE BACTERIAL CELL AND THE GRAM REACTION

The one remaining major topic relevant to the anatomy of the bacterial surface is the problem of the site and mechanism of the Gram stain reaction. It has long been assumed that the Gram reaction is due to some difference in the surface structure of the two groups of bacteria and there have been many attempts to define the difference and thus provide an explanation of the mechanism of the Gram stain. The whole subject of the Gram stain has been reviewed extensively by BARTHOLOMEW AND MITTWER (1952). This excellent contribution covers the historical, technical and basic aspects of the stain procedure and the mechanism of the Gram reaction from the time of its inception by GRAM in 1884, so no attempt will therefore be made to cover the bulk of this material. The discussion here is not intended to provide a review of the Gram reaction but it will be confined to a reassessment of the mechanism in terms of what is now known about the chemistry and anatomy of the bacterial cell.

As pointed out by SALTON (1961) most of the possible mechanisms of the Gram reaction have been proposed at some time or other during the long history of this staining procedure. Moreover, to add to the confusion, every major class of chemical substance found in microbial cells has been implicated in the mechanism of the reaction. The principal theories proposed as explanations of the Gram reaction can be classified into the following two groups:
(i) a chemical mechanism based on the possession of a single substance or group of chemical substances unique to the Gram-positive bacteria. The theories in this group usually involve the formation of a complex between the crystal violet and iodine of the Gram reagents and the chemical substance of the Gram-positive cell responsible for its positivity, such a complex being insoluble and undissociable in the differentiating agent (95% ethanol, acetone, etc.).

The various chemical constituents of the bacterial cell believed to be responsible for the Gram-positive reaction have included lipids (EISENBERG, 1910), special lipids and a glycerophosphate complex (SCHUMACHER, 1928), lipo-proteins (STEARN AND STEARN, 1930), polyglycerophosphate ('positic acid' of MITCHELL AND MOYLE, 1958), carbohydrate and nucleic acid (WEBB, 1948), nucleic acids and nucleoproteins (DEUSSEN, 1921; DUBOS AND MACLEOD, 1938; HENRY, STACEY AND TEECE, 1945; HENRY AND STACEY, 1946). Of the various classes of substances believed to be responsible for the Gram reaction greatest popularity was achieved with the ribonucleic acids in the theory proposed by HENRY, STACEY AND TEECE (1945). The last attempt to explain the Gram reaction on the basis of possession of a Gram-positive substance was made by MITCHELL AND MOYLE (1950) who later suggested the term 'positic acid' for the polyglycerophosphate component found in *Staphylococcus aureus* (MITCHELL AND MOYLE, 1958).

Although convincing at first sight, the last two attempts to base the Gram stain on Mg-ribonucleate (HENRY AND STACEY, 1946) or polyglycerophosphate (MITCHELL AND MOYLE, 1950, 1958) have broken down. We now know that the ribonucleic acid (RNA) of both Gram-positive and Gram-negative bacteria is largely localized in the RNA – protein particles, ribosomes (MCQUILLEN, 1961) and that it is not in the surface envelope as suggested by HENRY AND STACEY (1946. Moreover, MITCHEL AND MOYLE (1954) were unable to find any correlation between RNA and DNA contents of bacteria and the Gram reaction. MITCHELL AND MOYLE (1954) did however find a correlation between the presence of 'XP' (later described as 'positic acid' in *Staphylococcus aureus*) and the Gram reaction. This correlation was later shown to be questionable when JONES, RIZVI AND STACEY (1958) found that certain Gram-negative bacteria also contained 'XP'. SHUGAR AND BARANOWSKA (1957) found that many substances likely to occur in bacterial walls can stain as Gram-positive material. They suggested that the possession of such materials may be a reflection of the general physiological differences between the two groups, rather than being directly responsible for the Gram reaction.

The exact manner in which the possession of these Gram-positive components is alleged to render the cell 'positive' has not always been stated. The theories of STEARN AND STEARN (1924) involved differences in the iso-electric points of the cell proteins or lipoprotein complexes. MITCHELL AND MOYLE (1954) proposed an ion exchange mechanism based on the possession of the polyglycerophosphate substances in the Gram-positive bacteria. Many of the chemical substances thought to be responsible for the 'positive' reaction have been believed to be located in the envelope or periphery of the cell.

(ii) a permeability difference between Gram-positive and Gram-negative bacteria as the basis of the Gram reaction. Although a great deal of effort had been expended in early attempts to isolate a Gram-positive substance and correlate the Gram reaction with a definite chemical entity of the bacterial cell, there were concurrent theories suggesting that permeability differences between walls or cell envelopes of the two groups were responsible for the differential behaviour towards the Gram stain procedure. It hardly needs emphasizing that the term 'permeability' is not used here in a physiological sense. It has been used to indicate the differential extractability of the Gram stain reagents in certain organic solvents, due to the possession of a barrier or a substance or structure retaining the crystal violet-iodine complex within the Gram-positive cell.

BURKE AND BARNES (1929) concluded that the Gram reaction was due to differences in cell-wall permeability. As so little was then known about the nature of walls or envelopes of the organisms subdivided by the Gram reaction it was impossible for BURKE AND BARNES (1929) to offer any chemical evidence supporting the proposed theory. On the other hand the proponents of the theory of a definite Gram-positive substance were able to show the loss of Gram-positivity when certain materials were extracted from the bacterial cell. Just what the extraction procedure did to the various structural components of the cell was never specified and as indicated below this is a matter of some relevance to the problem of the mechanism.

Further support for the permeability theory has come from the studies of KAPLAN AND KAPLAN (1933) and more recently from BARTHOLOMEW, CROMWELL AND FINKELSTEIN (1959). The latter authors have shown a marked difference in 'iodine permeability'; the iodine is retained against elution with 95% ethanol in the Gram-positive bacteria and not in the Gram-negative organism.

The excellent studies of WENSINCK AND BOEVÉ (1957) provided further evidence that the difference in response to the Gram reaction was due to physico-chemical factors. In their careful analysis of the Gram reaction they found that the quantitative uptake of crystal violet and iodine was approximately the same for both Gram-positive and Gram-negative bacteria. They found the main difference was in the pattern of extractability with ethanol of the crystal violet-iodine complex from the bacteria. Thus above about 90% ethanol in water the Gram-positive organism retained the crystal violet-iodine complex in the cell whereas the material from the Gram-negative organisms became more readily extractable in the range 90–100% ethanol. The crystal violet-iodine complex formed by mixing solutions of the Gram reagents is quite soluble in ethanol in the range 90–100% ethanol in water. If it is assumed that there is no Gram-positive substance to bind the crystal violet-iodine complex and retain it inside the cell then the only alternative explanation of the Gram reaction would seem to involve a 'permeability' barrier.

The loss of Gram-positivity following mechanical crushing has been known for a long time (BENIANS, 1920). Although this evidence would appear to provide strong support for the 'permeability' theories of the mechanism of the Gram stain, it may be argued that it supports the theory of a Gram-positive substance equally well. It is thus conceivable that mechanical crushing could result in the enzymic destruction of the Gram-positive substance, a masking of it with other cell constituents, or an altered state of combination in the cell 'mush'. Other treatments, such as exposure of heated cells to lysozyme action (WEBB, 1948) or conversion of cells to protoplasts (GERHARDT, VENNES AND BRITT, 1956) involve a specific enzymic attack on the cell wall. In both instances there is a loss of Gram-positivity and this could be explained by either loss of the Gram-positive substance or loss of the barrier retaining the crystal violet-iodine complex within the cell.

CHELTON AND JONES (1959) have reported that disintegrated yeast cells can under certain conditions give a 'positive' Gram reaction, in contrast with the general agreement in the past that there is a concomitant loss of Gram reaction with the loss of cellular integrity (BARTHOLOMEW AND MITTWER, 1952). Despite the fact that isolated walls of a number of Gram-positive organisms do indeed have a purple colour when subjected to the Gram reaction, in the opinion of the author such a colouration could hardly be called a 'positive' reaction (SALTON, unpublished observations). More quantitative information about the amount of crystal violet-iodine complex retained by the walls will be needed before it is possible to assess the results of CHELTON AND JONES (1959) in terms of the Gram reaction of whole cells.

The nature of the cell wall and the gram stain

Since many of the theories of the mechanism of the Gram reaction were proposed before the last ten year's work on the chemistry of bacterial cell walls, it seemed worthwhile to re-examine the problem in terms of the present knowledge of the surface structure of bacteria. The principal classes of chemical constituent found in cell walls of Gram-positive and Gram-negative organisms are summarized in Table 8 (SALTON, 1963).

One of the most conspicuous differences between the walls of Gram-positive and Gram-negative bacteria observed in early studies (SALTON, 1953) was the high lipid content of the walls of the latter group. SALTON (1959) suggested that a wall with high lipid may be a positive factor contributing to the negative Gram reaction. The amino sugar contents of the walls of Gram-positive bacteria were generally higher than those of Gram-negative organisms. Such a difference was probably a reflection of the total amount of mucopeptide (glycosaminopeptide) component of the wall. In contrast to the Gram-positive bacteria, yeast cell walls have low contents of amino sugars.

In considering the chemical nature of the cell wall in relation to the Gram reaction (see Table 8) it became apparent that Gram-positivity could not be correlated with the presence of a particular type of substance in the cell wall. Thus the teichoic acids (ARMSTRONG, BADDILEY, BUCHANAN, CARSS AND GREENBERG, 1958) are present in only certain Gram-positive bacteria, and similarly, polysaccharide components are found in some and not in others (SALTON, 1960b). All of the Gram-positive bacteria contain mucopeptides and all of the yeast walls are rich in polysaccharides and polysaccharide complexes (NORTHCOTE AND HORNE, 1952; KESSLER AND NICKERSON, 1959). Although the walls of Gram-negative bacteria were rich in lipids, the lipid content of yeast walls may be as high as 10% and as low as 1% (KESSLER AND NICKERSON, 1959). These considerations of the chemistry of cell walls led to the conclusion that the Gram reaction is not due to the presence of any specific substance in the wall and that several types of polymeric substances may serve equally well as wall structures for Gram-positive organisms (SALTON, 1963).

All attempts to relate the Gram reaction to the presence af a unique component either in the cell or in the outer envelope have so far failed. As for more recently studied specific cellular components some further information on the distribution of the intracellular teichoic acid of the glycerol type (BADDILEY, 1961; BADDILEY AND DAVISON, 1961) and the significance of the polyamines found in greatest amounts in Gram-negative bacteria (HERBST, WEAVER AND KEISTER, 1958) would be needed before we could conclude that they are not involved in the mechanism of the Gram reaction.

In view of the impasse reached in attempting to relate the Gram stain to unique cellular components and considering the evidence of the experiments of WENSINCK AND BOEVÉ (1957), SALTON (1963) made an attempt to obtain further information on possible 'permeability' differences between Gram-positive and Gram-negative bacteria under conditons giving the Gram differentiation (i.e. extraction of crystal violet-iodine complex with 95% ethanol). It appeared likely that the differentiation brought about by decolorizing with 95% ethanol may result from a dehydration of the wall structure and consequent decrease in pore-size, thus impeding the passage of small molecules across the wall or envelope of the Gram-positive cell. If this was involved in the Gram stain it would have the effect of 'trapping' the crystal violet-iodine complex within the cell.

To test the possibility that the passage of small molecules across the outer envelope is impeded when Gram-positive organisms are exposed to ethanol concentrations which bring about the Gram differentiation, the release of ^{32}P compounds from ^{32}P labelled organisms suspended in aqueous ethanol solutions (25–100%, v/v) was investigated by SALTON (1963). The influence of ethanol concentration on the leakage of ^{32}P compounds from a variety of Gram-positive and Gram-negative bacteria was studied; the results for two Gram-positive organisms (*Saccharomyces cerevisiae* and *Streptococcus faecalis*) and two Gram-negative bacteria (*Escherichia coli* and *Proteus vulgaris*) are illustrated in Figs. 9 and 10 respectively (SALTON, 1963). These differences in the release of ^{32}P compounds normally contained inside the cells follow, to a remarkable degree, the patterns of extractability of the crystal violet-iodine complex in the investigations of WENSINCK AND BOEVÉ (1957).

That the differences in release of ^{32}P in ethanol solutions were not due to a slower release from Gram-positive bacteria was shown by following the time course of leakage. The results for *Escherichia coli* and *Staphylococcus aureus* presented in Fig. 11 show that the fraction released in 100% ethanol escapes with the same rapidity from both types of cells.

The release of ^{32}P labelled compounds from a collection of Gram-positive and Gram-negative bacteria was studied by SALTON (1963) and the results are presented in Table 9. The results give a 'spectrum' of values from Gram-positive to Gram-negative bacteria. As suggested by SALTON (1963) quantitative differences of this kind may be correlated with the 'spectrum' of mucopeptide components in the walls (SALTON, 1958) for it is well known that the Gram reaction is not an 'all or none' phenomenon and that gradations between extremes exist (NEIDE, 1904; CHURCHMAN AND SIEGEL, 1928; SHUGAR AND BARANOWSKA, 1958).

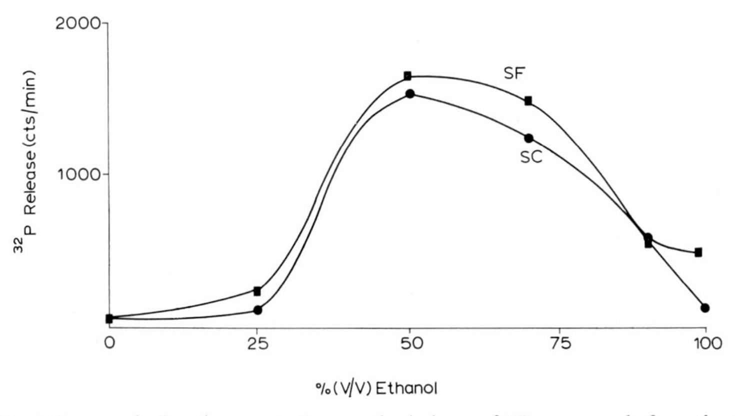

FIGURE 9. The influence of ethanol concentration on the leakage of ^{32}P compounds from the two Gram-positive organisms, *Streptococcus faecalis* and *Saccharomyces cerevisiae*. Note the marked decrease in the release in ethanol solutions above 50% v/v (SALTON, 1963).

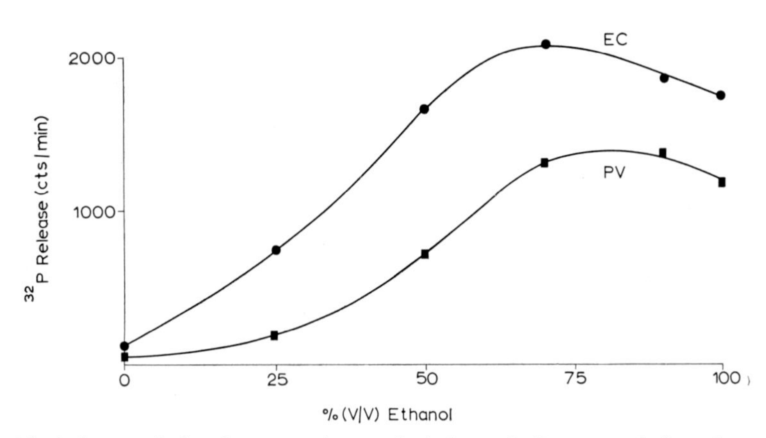

FIGURE 10. The influence of ethanol concentration on the leakage of ^{32}P compounds from the two Gram-negative organisms, *Escherichia coli* and *Proteus vulgaris*. (SALTON, 1963).

The differences between the Gram-positive and Gram-negative organisms shown in Table 9 only establish that the passage of certain small molecules (inorganic phosphate, nucleotides, etc.) across the wall or wall-membrane of Gram-positive bacteria is impeded when they are suspended in high concentrations of ethanol, whereas many of the Gram-negative bacteria were affected to a lesser degree. This finding in itself gives no more than a clue about a likely mechanism of the Gram reaction and suggests that the crystal violet-iodine complex is 'trapped' inside the organisms when the 'permeability' of the outer wall is decreased on treatment with concentrations of ethanol exceeding 90%.

If, as the results of SALTON (1963) and the iodine permeability effects reported by KAPLAN AND KAPLAN (1933) and BARTHOLOMEW, CROMWELL AND FINKELSTEIN (1959) suggest, the pore size of the wall of Gram-positive bacteria is decreased by mordanting with iodine and dehydration with 95% ethanol, thereby trapping the crystal violet-iodine complex within its boundary, then mechanical rupture or enzymic removal of the wall after Gram staining should have rendered the complex accessible to extraction. This indeed proved to be the case as reported by SALTON (1963). TCHAN (1963, personal communication) has elegantly confirmed these results by showing that it is possible to decolorize a Gram-positive organism simply by cutting the outer wall with a glass knife operated by a micromanipulator.

Strong support for the 'permeability' theory for the mechanism of the Gram stain has thus come from WENSINCK AND BOEVÉ (1957) and from recent investigations on the release of ^{32}P labelled compounds normally found inside the cell and the mechanical disintegration or 'slicing up' of Gram-positive bacteria (SALTON, 1963; TCHAN, 1963). If the Gram reaction is

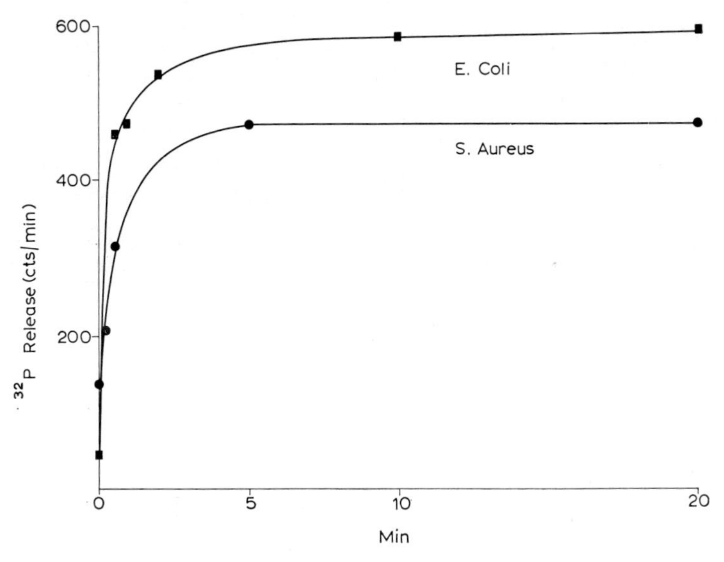

FIGURE 11. The time course for the release of ^{32}P compounds from *Escherichia coli* and *Staphylococcus aureus* suspended in 100% ethanol at room temperature is similar for both organisms (SALTON, 1963).

thus due to the physico-chemical state of the wall during differentiation with 95% ethanol, then many of the conflicting results of earlier investigations can be readily understood. Any breach in the mechanical integrity of the wall either before or after Gram staining would render the crystal violet-iodine complex more accessible to extraction. GERHARDT, VENNES AND BRITT (1956) indeed attempted such a conclusive experiment with *Bacillus megaterium* treated with lysozyme after having been Gram stained as normal smear preparations. Their results were not entirely satisfactory and not as clear cut as the data obtained from washed Gram-stained suspensions. The loss of Gram-positivity on disintegration, autolysis and digestion of heated organisms with lysozyme (WEBB, 1948) and even the effects of ageing, are all readily understandable if an intact rigid wall is needed as a barrier for the retention of the crystal violet-iodine complex. The presence of cell-wall degrading enzymes in bacteria has been well established (MITCHELL AND MOYLE, 1957; SALTON, 1956) and could account for the conversion of bacteria to the Gram negative state. These general conclusions about the mechanism of the Gram reaction have been further substantiated by recent experiments reported by SCHERRER (1963), who has also indicated that "chemical integrity of the cell wall is a prerequisite for Gram positivity".

Thus a re-examination of the problem of the mechanism of the Gram stain suggests that the original hypothesis of BURKE AND BARNES (1929) is correct and that the Gram differentiation is due to a trapping of the crystal violet-iodine complex within the cells of Gram-positive bacteria, probably by a barrier in the form of the dehydrated, mordanted wall. In terms of the anatomy of the bacterial cell, it then seems reasonable to conclude that the cell wall is not the site at which the crystal-violet and iodine form a complex but that it is the response of this structure to treatment with 95% ethanol as the decolorizing agent which results in the retention of the complex within the cell. The solubility of surface lipids of the Gram-negative bacteria may also contribute to the ease with which the differentiating solvent extracts the crystal violet-iodine from the cell. However, as pointed out by SALTON (1963) more information about the physico-chemical structure of microbial cell walls is obviously needed before a clearer picture will emerge of the effects of the differentiation with ethanol which lead to a Gram-positive or Gram-negative reaction. The role of iodine and the possible groups in the wall with which it may be associated must be further investigated and the effect of mordanting with iodine on the penetrability of ethanol across the dehydrated walls of Gram-positive must also be established. It should then be possible to propose a detailed mechanism for the Gram reaction in terms of the chemistry of the surface structures of the bacterial cell.

REFERENCES

AARONSON, S. and H. BAKER, *J. Protozool.*, 8 (1961) 274.
ABRAMS, A. and P. MCNAMARA, *J. Biol. Chem.*, 237 (1962) 170.
ABRAMS, A., P. MCNAMARA and F. B. JOHNSON, *J. Biol. Chem.*, 235 (1960) 3659.
ADAMS, M. H. and B. H. PARK, *Virology*, 2 (1956) 719.

ARMSTRONG, J. J., J. BADDILEY, J. G. BUCHANAN, B. CARSS, and G. R. GREENBERG, *J. Chem. Soc.*, (1958) 4344.
AVERY, O. T. and R. J. DUBOS, *J. Exptl. Med.*, 54 (1931) 73.
BADDILEY, J., in *Immunological Approaches to Problems in Microbiology*, Eds. M. Heidelberger, O. J. Plescia and R. A. Day, Rutgers Univ. Press, N.J. 1961, p. 91.
BADDILEY, J. and A. L. DAVISON, *J. Gen. Microbiol.*, 24 (1961) 295.
BARTHOLOMEW, J. W., T. CROMWELL and H. FINKELSTEIN, *Nature*, 183 (1959) 123.
BARTHOLOMEW, J. W. and T. MITTWER, *Bacteriol. Rev.*, 16 (1952) 1.
BAZELEY, P. L., *Australian Vet. J.*, 16 (1940) 243.
BENIANS, T. H. C., *J. Pathol. Bacteriol.*, 23 (1920) 401.
BRENNER, S., F. A. DARK, P. GERHARDT, M. H. JEYNES, O. KANDLER, E. KELLENBERGER, E. KLIENEBERGER-NOBEL K. MCQUILLEN, M. RUBIO-HUERTOS, M. R. J. SALTON, R. E. STRANGE, J. TOMCSIK and C. WEIBULL, *Nature* 181 (1958) 1713.
BRINTON, C. C., *Nature*, 183 (1959) 782.
BROWN, J. W., *Biochim. Biophys. Acta*, 52 (1961) 368.
BROWN, A. D., D. G. DRUMMOND and R. J. NORTH, *Biochim. Biophys. Acta*, 58 (1962) 514.
BROWN, A. D. and C. D. SHOREY, *Biochim. Biophys. Acta*, 59 (1962) 258.
BROWN, A. D. and C. D. SHOREY, *J. Cell. Biol.*, 18 (1963) 681.
BRUMFITT, W., A. C. WARDLAW and J. T. PARK, *Nature*, 181 (1958) 1783.
BURKE, V. and M. W. BARNES, *J. Bacteriol.*, 18 (1929) 69.
CAMPBELL, J. J. R., L. A. HOGG and G. A. STRASDINE, *J. Bacteriol.*, 83 (1962) 1155.
CANALE-PAROLA, E., R. BORASKY and R. S. WOLFE, *J. Bacteriol.*, 81 (1961) 311.
CHAPMAN, J. A., R. G. E. MURRAY and M. R. J. SALTON, *Proc. Roy. Soc. (London)*, B (1963). In the press.
CHAPMAN, J. A. and M. R. J. SALTON, *Proc. Fifth Intern. Congr. Elect. Mic. (Philadelphia)* UU-2, Academic Press, New York, 1962.
CHELTON, E. T. J. and A. S. JONES, *J. Gen. Microbiol.*, 21 (1959) 652.
CHURCHMAN, J. W. and L. SIEGEL, *Stain Technol.*, 3 (1928) 73.
CLARKE, P. H. and M. D. LILLY, *Nature*, 195 (1962) 516.
COLOBERT, L., *Ann. Inst. Pasteur*, 92 (1957) 74.
COLOBERT, L. and G. DIRHEIMER, *Biochim. Biophys. Acta*, 54 (1961) 455.
DE LEY, J. and V. VERVLDET, *Biochim. Biophys. Acta*, 50 (1961) 1.
DEUSSEN, E., *Z. Hyg.*, 93 (1921) 512.
DOUGLAS, H. W., *J. Appl. Bacteriol.*, 20 (1957) 390.
DOUGLAS, H. W. and F. PARKER, *Biochem. J.*, 68 (1958) 99.
DREWS, G., *Archiv. Mikrobiol.*, 36 (1960) 99.
DUBOS, R. J. and C. M. MACLEOD, *J. Exptl. Med.*, 67 (1938) 791.
DUGUID, J. P., *Edinburgh Med. J.*, 53 (1946) 401.
DUGUID, J. P., *J. Pathol. Bacteriol.*, 63 (1951) 673.
DUGUID, J. P., I. W. SMITH, G. DEMPSTER and P. N. EDMUNDS, *J. Pathol. Bacteriol.*, 70 (1955) 335.
DUGUID, J. P. and J. F. WILKINSON, *J. Gen. Microbiol.*, 9 (1953) 174.
DUGUID, J. P. and J. F. WILKINSON, *J. Gen. Microbiol.*, 11 (1954) 71.
DUGUID, J. P. and J. F. WILKINSON, in *Soc. Gen. Microbiol. Symposium*, No. 11, 1961, p. 69.
EISENBERG, P., *Zentr. Bakteriol. Parasitenk.* (Orig.), 56 (1910) 193.
FEW, A. V., *J. Gen. Microbiol.*, 10 (1954) 304.
FITZ-JAMES, P. C., *J. Biophys. Biochem. Cytol.*, 4 (1958) 257.
FITZ-JAMES, P. C., *J. Biophys. Biochem. Cytol.*, 8 (1960) 507.
FOGG, G. E., *The Metabolism of Algae*, Methuen, London, 1953.
FREIMER, E. H., *J. Exptl. Med.*, 117 (1963) 377.
FREIMER, E. H., R. M. KRAUSE and M. MCCARTY, J. *Exptl. Med.*, 110 (1959) 853.

FUKASAWA, T. and H. NIKAIDO, *Nature*, 183 (1959) 1131.
GAUDY, E. and R. S. WOLFE, *Appl. Microbiol.*, 9 (1961) 580.
GAUDY, E. and R. S. WOLFE, *Appl. Microbiol.*, 10 (1962) 200.
GEBICKI, J. M. and A. M. JAMES, *Biochim. Biophys. Acta*, 59 (1962) 167.
GEORGI, C. E., W. E. MILITZER and T. S. DECKER, *J. Bacteriol.*, 70 (1955) 716.
GERHARDT, P., J. W. VENNES and E. M. BRITT, *J. Bacteriol.*, 72 (1956) 721.
GIESBRECHT, P. and G. DREWS, *Archiv. Mikrobiol.*, 43 (1962) 152.
GILBY, A. R., A. V. FEW and K. MCQUILLEN, *Biochim. Biophys. Acta*, 29 (1958) 21.
GLAUERT, A. M., *Brit. Med. Bull.*, 18 (1962) 245.
GLAUERT, A. M. and D. A. HOPWOOD, *J. Biophys. Biochem. Cytol.*, 7 (1960) 479.
GRAM, C., *Fortschr. Med.*, 2 (1884) 185.
GUEX-HOLZER, S and J. TOMCSIK, *J. Gen. Microbiol.*, 14 (1956) 14.
GUNSALUS, I. C. and R. Y. STANIER, *The Bacteria*, Vol. I, Academic Press, Inc., New York, 1960.
HEIDELBERGER, M., *Lectures in Immunochemistry*, Academic Press, Inc., New York, 1956.
HENRY, H., M. STACEY and E. G. TEECE, *Nature*, 156 (1945) 720.
HENRY, H. and M. STACEY, *Proc. Roy. Soc. (London)* B, 133 (1946) 391.
HERBST, E. J., R. H. WEAVER and D. L. KEISTER, *Arch. Biochem. Biophys.*, 75 (1958) 171.
HILL, P. B., *Biochim. Biophys. Acta.*, 57 (1962) 386.
HOFSCHNEIDER, P. H., *Proc. Europ. Conf. Elect. Mic.*, (Delft) 1928, De Nederlandse Vereniging Voor Electronmicroscopie Delft (1960).
HOLME, T., A. S. MALMBORG and E. COTA-ROBLES, *Nature*, 185 (1960) 57.
HOPWOOD, D. A. and A. M. GLAUERT, *J. Biophys. Biochem. Cytol.*, 8 (1960) 813.
HOUWINK, A. L., *Biochim. Biophys. Acta*, 10 (1953) 360.
HOUWINK, A. L., *J. Gen. Microbiol.*, 15 (1956) 146.
HUGHES, D. E., *J. Gen. Microbiol.*, 29 (1962) 39.
HUNT, A. L., A. RODGERS and D. E. HUGHES, *Biochim. Biophys. Acta*, 34 (1959) 354.
IVANOVICS, G. and S. HORVATH, *Acta Physiol. Acad. Sci. Hung.*, 4 (1953) 175.
JEYNES, M. H., *Exptl. Cell Res.*, 24 (1961) 255.
JONES, A. S., S. B. H. RIZVI and M. STACEY, *J. Gen. Microbiol.*, 18 (1958) 597.
KANDLER, O. and C. ZEHENDER, *Z. Naturforsch.*, 12b (1957) 725.
KAPLAN, M. L. and L. KAPLAN, *J. Bacteriol.*, 25 (1933) 309.
KASS, E. H. and C. V. SEASTONE, *J. Exptl. Med.*, 79 (1944) 319.
KELLENBERGER, E., in *Soc. Gen. Microbiol. Symposium*, No. 10, 1960, p. 39.
KELLENBERGER, E. and A. RYTER, *J. Biophys. Biochem. Cytol.*, 4 (1958) 323.
KELLENBERGER, E., A. RYTER and J. SECHAUD, *J. Biophys. Biochem. Cytol.*, 4 (1958) 671.
KERRIDGE, D., in *Soc. Gen. Microbiol. Symposium*, No. 11, 1961, p. 41.
KESSLER, G. and W. J. NICKERSON, *J. Biol. Chem.*, 234 (1959) 2281.
KLIENEBERGER-NOBEL, E., in *The Bacteria*, Eds. I.C. Gunsalus and R.Y. Stanier, Vol. I, Academic Press, Inc., New York, 1960, p. 361.
KLIENEBERGER-NOBEL, E., *Pleuropneumonia-like Organisms (PPLO) Mycoplasmataceae*, Academic Press, Inc., New York, 1962.
KNAYSI, G., *Elements of Bacteriol Cytology*, 2nd ed., Comstock Publishing Company, Inc., Ithaca, N.Y., 1951.
KOIKE, M. and K. TAKEYA, *J. Biophys. Biochem. Cytol.*, 9 (1961) 597.
LABAW, W. and V. M. MOSLEY, *J. Bacteriol.*, 67 (1954) 576.
LANCEFIELD, R. C., *J. Exptl. Med.*, 78 (1943) 465.
LARK, K. G., *Canad. J. Microbiol.*, 4 (1958) 165.
LEDERBERG, J., *Proc. Nat. Acad. Sci., U.S.*, 42 (1956) 574.
LEDERBERG, J. and J. ST. CLAIR, *J. Bacteriol.*, 75 (1958) 143.

LINNANE, A. W., E. VITOLS and P. G. NOWLAND, *J. Cell Biol.*, 13 (1962) 345.
LOMINSKI, I., J. CAMERON and G. WYLLIE, *Nature*, 181 (1958) 1477.
LOMINSKI, I. and S. GRAY, *Nature*, 192 (1961) 683.
LUKIOANOVA, M. A., N. S. GELMAN and V. I. BIRIUSOVA, *Biochimia*, 26 (1961) 916.
MACFARLANE, M. G., *Biochem. J.*, 80 (1961) 45P.
MALAMY, M. and B. L. HORECKER, *Biochim. Biophys. Res. Comm.*, 5 (1961) 104.
MARINETTI, G. V., J. ERBLAND and E. STOTZ, *J. Biol. Chem.*, 233 (1958) 562.
MARKOVITZ, A. and A. DORFMAN, *J. Biol. Chem.*, 237 (1962) 273.
MARR, A. G. (a) in *The Bacteria*, Eds. I.C. Gunsalus and R. Y. Stanier, Vol. I, Academic Press, Inc., New York, 1960, p. 443.
MARR, A. G. (b) *Ann. Rev. Microbiol.*, 14 (1960) 241.
MARR, A. G. and E. H. COTA-ROBLES, *J. Bacteriol.*, 74 (1957) 79.
MARTIN, H. H. and H. FRANK, *Zentr. Bakt. Parasitenk.* (Orig.), 184 (1962) 306.
MAYALL, B. H. and C. ROBINOW, *J. Appl. Bacteriol.*, 20 (1957) 333.
MCQUILLEN, K. (a) *Biochim. Biophys. Acta*, 6 (1951) 534.
MCQUILLEN, K. (b) *Biochim. Biophys. Acta*, 7 (1951) 54.
MCQUILLEN, K., in *Soc. Gen. Microbiol. Symposium*, No. 6, 1956, p. 127.
MCQUILLEN, K., in *The Bacteria*, Eds. I. C. Gunsalus and R. Y. Stanier, Vol. I, Academic Press, Inc., New York, 1960, p. 250.
MCQUILLEN, K., in *Progress in Biophysics and Biophysical Chemistry*, Vol. 12, Pergamon Press, London, 1961, p. 67.
MITCHELL, P. and J. MOYLE, *Nature*, 166 (1950) 218.
MITCHELL, P. and J. MOYLE, *J. Gen. Microbiol.*, 10 (1954) 533.
MITCHELL, P. and J. MOYLE, *Biochem. J.*, 64 (1956) 19P.
MITCHELL, P. and J. MOYLE, *J. Gen. Microbiol.*, 16 (1957) 184.
MITCHELL, P. and J. MOYLE, *Proc. Roy. Physics Soc. (Edinburgh)*, 27 (1958) 79.
MORRISON, T. H. and C. WEIBULL, *Acta. Pathol Microbiol. Scand.*, 55 (1962) 475.
MURRAY, R. G. E., in *Soc. Gen. Microbiol. Symposium*, No. 12, 1962, p. 119.
MURRAY, R. G. E., (a) *Canad. J. Microbiol.* 9 (1963) 593.
MURRAY, R. G. E., (b) *Canad. J. Microbiol.* 9 (1963) 381.
NEIDE, E., *Zentr. Bakteriol. Parasitenk. (Orig.)*, 35 (1904) 508.
NORTHCOTE, D. H. and R. W. HORNE, *Biochem. J.*, 51 (1952) 232.
PANGBORN, J., A. G. MARR and S. A. ROBRISH, *J. Bacteriol.*, 84 (1962) 669.
PLACKETT, P., *Biochim. Biophys. Acta*, 35 (1959) 260.
PLACKETT, P., *Nature*, 189 (1961) 125.
PRINGSHEIM, E. G., *J. Gen. Microbiol.*, 13 (1955) 285.
RAZIN, S. and M. ARGAMAN, *Nature*, 193 (1962) 502.
REBERS, P. A. and M. HEIDELBERGER, *J. Am. Chem. Soc.*, 81 (1959) 2415.
REBERS, P. A. and M. HEIDELBERGER, *J. Am. Chem. Soc.*, 83 (1961) 3056.
REITER, B. and J. D. ORAM, *Nature*, 193 (1962) 651.
RIS, H. and R. N. SINGH, *J. Biophys. Biochem. Cytol.*, 9 (1961) 63.
ROBERTSON, J. D., *Biochem. Soc. Symposia*, Cambridge, Engl., 16 (1959) 3.
ROBINOW, C. F., *Canad. J. Microbiol.*, 3 (1957) 771.
RODWELL, A. W. and A. ABBOT, *J. Gen. Microbiol.*, 25 (1961) 201.
RYTER, A. and E. KELLENBERGER, *Z. Naturforsch.*, 13b (1958) 597.
SALTON, M. R. J., *Biochim. Biophys. Acta*, 10 (1953) 512.
SALTON, M. R. J., *Proc. Intern. Congr. Biochem.*, *3rd Congr.* (Brussels), 1955, p. 404.
SALTON, M. R. J. in *Soc. Gen. Microbiol.*, No. 6, 1956, p. 81.

SALTON, M. R. J., *Bacteriol. Revs.*, 21 (1957) 82.
SALTON, M. R. J., *Polysaccharides in Biology, Fifth. Confr. Josiah Macy, Jr. Foundation*, 1959, p. 71.
SALTON, M. R. J., (a) in *The Bacteria*, Eds. I. C. Gunsalus and R. Y. Stanier, Vol. I, p. 97, Academic Press, Inc., New York, 1960.
SALTON, M. R. J. (b) *Microbial Cell Walls*, John Wiley and Sons, New York, 1960.
SALTON, M. R. J., *Bacteriol. Revs.*, 25 (1961) 77.
SALTON, M. R. J., *J. Gen. Microbiol.*, 30 (1963).
SALTON, M. R. J. and J. A. CHAPMAN, *J. Ultrastructure Res.*, 6 (1962) 489.
SALTON, M. R. J. and F. SHAFA, *Nature*, 181 (1958) 1321.
SALTON, M. R. J. and R. C. WILLIAMS, *Biochim. Biophys. Acta*, 14 (1954) 455.
SCHERRER, R., *J. Gen. Microbiol.*, 31 (1963) 135.
SCHUMACHER, J., *Zentr. Bakteriol. Parasitenk.* (Orig.), 109 (1928) 181.
SHAFA, F., *A study of the surface structure of some bacteria*, Ph. D. Thesis, University of Manchester, 1958.
SHARP, J. T., W. HIJMANS and L. DIENES, *J. Exptl. Med.*, 105 (1957) 153
SHINOHARA, C., K. FUKUSHI and J. SUZUKI, *J. Bacteriol.*, 74 (1957) 413.
SHOCKMAN, G. D., J. J. KOLB, B. BAKAY, M. J. CONOVER and G. TOENNIES, *J. Bacteriol.*, 85 (1963) 168.
SHUGAR, D. and J. BARANOWSKA, *Biochim. Biophys. Acta*, 23 (1957) 227.
SHUGAR, D. and J. BARANOWSKA, *Nature*, 181 (1958) 357.
SJÖSTRAND, F. S., *Radiation Res.*, Suppl. 2, (1960) 349.
SLADE, H. D., *J. Gen. Physiol.*, 41 (1957) 63.
SMITHIES, W. R. and N. E. GIBBONS, *Canad. J. Microbiol.*, 1 (1955) 614.
STACEY, M. and S. A. BARKER, *Polysaccharides of Micro-organisms*, Clarendon Press, Oxford, 1960.
STEARN, E. W. and A. E. STEARN, *J. Bacteriol.*, 9 (1924) 479.
STEARN E. W. and A. E. STEARN, *J. Bacteriol.*, 20 (1930) 287.
STOCKER, B. A. D., in *Soc. Gen. Microbiol. Symposium*, No. 6, 1956, p. 19.
STORCK, R. and J. T. WACHSMAN, *J. Bacteriol.*, 73 (1957) 784.
TAKAHASHI, I. and N. E. GIBBONS, *Canad. J. Microbiol.*, 3 (1957) 687.
TAUBENECK, U., H. BÖHME and E. SCHUHMANN, *Biologisches Zentr.*, 77 (1958) 663.
THORSSON, K. G. and C. WEIBULL, *J. Ultrastructure Res.*, 1 (1958) 412.
TOMCSIK, J., *Experientia*, 7 (1951) 459.
TOMCSIK, J., in *Soc. Gen. Microbiol. Symposium*, No. 6, 1956, p. 41.
TOMCSIK, J. and S. GUEX-HOLZER, *Schweiz. Z. Allgem. Pathol. Bakteriol.*, 14 (1951) 515.
TOMCSIK, J. and S. GUEX-HOLZER, *Experientia*, 12 (1954) 484.
TORII, M., *Med. J. Osaka Univ.*, 6 (1955) 725.
VANDERWINKEL, E. and R. G. E. MURRAY, *J. Ultrastructure Res.*, 7 (1962) 185.
VAN ITERSON, W., *Gallionella ferruginea Ehrenberg in a different light*, N.V. Noord-Hollandsche Uitg. Mij., Amsterdam, 1958.
VAN ITERSON, W., *J. Biophys. Biochem. Cytol.*, 9 (1961) 183.
VAN ITERSON, W. and A. C. RUYS, *J. Ultrastructure Res.*, 3 (1960) 282.
VENNES, J. W. and P. GERHARDT, *Science*, 124 (1956) 535.
WARTH, A. D., D. F. OHYE and W. G. MURRELL, *J. Cell. Biol.*, (1963). In the press.
WEBB, M., *J. Gen. Microbiol.*, 2 (1948) 260.
WEIBULL, C., *Biochim. Biophys. Acta*, 2 (1948) 351.
WEIBULL, C., (a) *J. Bacteriol.*, 66 (1953) 688.
WEIBULL, C., (b) *J. Bacteriol.*, 66 (1953) 696.
WEIBULL, C., *Exptl. Cell Res.*, 9 (1955) 139.
WEIBULL, C., *Ann. Rev. Microbiol.*, 12 (1958) 1.
WEIBULL, C., *Acta Pathol. Microbiol. Scand.*, 42 (1958) 324.

WEIBULL, C., in *The Bacteria*, Eds. I. C. Gunsalus and R. Y. Stanier, Vol. I, Academic Press, Inc., 1960, p. 153.
WEIBULL, C. and L. BERGSTRÖM, *Biochim. Biophys. Acta*, 30 (1958) 340.
WEIBULL, C., B. ZACHARIAS and H. BECKMAN, *Nature*, 184 (1959) 1744.
WEIDEL, W., H. FRANK and H. H. MARTIN, *J. Gen. Microbiol.*, 22 (1960) 158.
WELSCH, M., *Schweiz. Allgem. Pathol. Bakteriol.*, 21 (1958) 741.
WELSHIMER, H. J., *J. Bacteriol.*, 66 (1953) 112.
WENSINCK, F. and J. J. BOEVÉ, *J. Gen. Microbiol.*, 17 (1957) 401.
WESTPHAL, O., O. LÜDERITZ and F. BISTER, *Z. Naturforsch.*, 7b (1952) 148.
WILEY, B. B. and J. C. WONNACOTT, *J. Bacteriol.*, 83 (1962) 1169.
WILKINSON, J. F., *Bacteriol. Revs.*, 22 (1958) 46.
WILKINSON, J. F. and J. P. DUGUID, in *Intern. Rev. Cytol.*, Vol. 9, Academic Press, New York, 1960, p. 1.
YAMAGUCHI, J., *Science Repts. Res. Inst. Tohoku Univ.*, C9 (1960) 125.
ZINDER, N. D. and W. F. ARNDT, *Proc. Nat. Acad. Sci., U.S.*, 42 (1956) 586.

CHAPTER 2

Isolation of bacterial cell walls

One of the first attempts to isolate the rigid cell-wall structure of a bacterium was made by VINCENZI (1887). The material he isolated represented the alkali resistant fraction of *Bacillus subtilis* and still possessed the rod-shaped appearance of the organism from which the 'wall' had been isolated. As with plants and fungi, extraction with alkali became a standard procedure for obtaining what was then regarded as the wall of the bacterial cell. Apart from microscopic examination indicating preservation of the general shape of the cell and the reduced stainability of these alkali-resistant residues, there were few tests available at that time to indicate the homogeneity of the preparations. The limits of resolution of the light microscope imposed severe restrictions on any attempt to fractionate cells into the various morphological entities and substructures we now recognize so readily under the electron microscope. Thus when the electron microscope was developed the structural and biochemical fractionation of cells could be placed on a sound basis with suitable methods for critically assessing the homogeneity of the preparations. Moreover, the larger structures of the cell that were visible by light microscopy could now be examined in the electron microscope for the presence of cellular particles as small as 100–200 Å in diameter.

The stimulus to re-examine the problem of isolating the bacterial cell wall by mechanical rupture of cells came from several of the early studies of the electron microscopy of bacteria. MUDD AND LACKMAN (1941) and MUDD, POLEVITSKY, ANDERSON AND CHAMBERS (1941) showed that the rigid cell-wall structures of several *Bacillus* species and *Streptococci* were not disintegrated when cells had been disrupted in a sonic oscillator. The electron-dense cytoplasm was lost from the cells, leaving broken fragments of the electron transparent wall. Little further was done until DAWSON (1949) showed that when *Staphylococcus aureus* cells were disintegrated with minute glass beads (using conditions from a study made by KING AND ALEXANDER, 1948), the debris contained empty cell walls, a few residual intact cells and small electron-dense particles. It became apparent to several of us (MITCHELL AND MOYLE, 1951; SALTON AND HORNE, 1951b; NORTHCOTE AND HORNE, 1952) that mechanical disintegration of microbial cells, combined with differential centrifugation and electron microscopy of the preparations, would provide the basic requirements for satisfactory methods of cell-wall isolation. Since the development of these methods more than ten years ago, many procedures for isolating bacterial cell walls have been used and these can be reviewed in some detail. All of the methods are essentially similar and are dependent on the removal of cellular cytoplasmic material either by disintegration of the bacteria or by autolysis and digestion of intracellular material. Because of the greater mechanical strength of the walls and compound membranes they survive disintegration whereas the cell cytoplasm and intracellular mem-

The tables are printed together at the end of the book.

branes are broken up and dispersed. Although we now know that the cell walls of many bacteria (especially those of Gram-positive organisms) are resistant to many chemical agents and are insoluble in a variety of organic solvents (SALTON AND HORNE, 1951b), an examination of the older methods of extracting bacteria with alkali, clearly showed that the alkali-resistant residues were morphologically heterogeneous and did not correspond to the cell-wall structures obtained by mechanical disintegration (SALTON AND HORNE, 1951b). Subsequent chemical investigations of microbial cell walls have also reinforced the view that chemical methods of isolating the wall structures would lead to a loss of certain components present in the native wall (e.g. O-acetyl groups, O-alanyl groups removed on extraction with alkali and removal of some polysaccharide material).

The older methods of chemical extraction have therefore been largely replaced by procedures involving mechanical disintegration of cells under controlled conditions at low temperature (c. 0°) followed by differential centrifugation and washing on the centrifuge. Such methods obviously offer the best opportunity of isolating 'native' structures as they exist in the intact cell.

METHODS OF CELL DISINTEGRATION

A great variety of cell disintegration procedures are now available, many having been designed for the extraction or release of intracellular enzymes for biochemical studies. HUGO (1954) has reviewed in some detail a number of the techniques used for cell disruption. For the isolation of microbial cell walls methods for cell rupture and disintegration can be listed under the following general headings:

1. Autolysis
2. Osmotic lysis
3. Heat-treatment rupture
4. Mechanical disintegration

1. Autolysis

WEIDEL (1951) was the first to use autolytic methods for the isolation of the cell walls of *Escherichia coli*. Cells were autolysed under toluene for several days and then digested with trypsin. This procedure obviously removed much of the cytoplasmic material which was probably degraded to relatively small molecular weight substances. However, electron micrographs of the wall fractions obtained in this way showed electron dense material trapped inside the wall structure (WEIDEL, 1951). Although the preparations behaved in a homogeneous manner in the ultracentrifuge, such material is of course structurally heterogeneous and the autolytic method may thus have disadvantages when applied to some bacteria. It is unfortunate that autolytic methods cannot be recommended without reservation as they would obviously have the great advantage of being applicable to large scale isolation procedures without requiring special equipment for cell disintegration. Isolation techniques utilizing autolysis have the additional serious disadvantage of allowing enzymic modification of the wall; it is thus conceivable that the structures isolated by these methods have suffered considerable degradation by enzymes active at the time of autolysis. Lysozyme-like enzymes occur widely in bacteria and these may be so active as to degrade the glycosaminopeptide part of the wall.

NORRIS (1957) did, however, find that the autolytic enzymes produced by several Bacillus species could be used to give very clean cell-wall structures from *Bacillus cereus*. Whether the walls obtained by this method had lost any of their chemical constituents cannot be stated, but at least the structures were completely devoid of electron-dense intracellular components and possessed a very similar appearance and thickness to those isolated by mechanical procedures.

Another autolytic system potentially useful for isolating wall structures of *Staphylococcus aureus* was used by MITCHELL AND MOYLE (1957) for the preparation of protoplasts of this organism. Autolysis of staphylococcal cells resulted in the separation of the walls into two hemispherical shells, thus permitting the escape of intact protoplasts when this was allowed to take place in a buffered sucrose medium. The hemispherical wall fragments separated under these autolytic conditions have apparently undergone a minimum of enzymic modification but prolonged autolysis may lead to further degradation of the staphylococcal wall. That *Staphylococcus aureus* cells possess enzymes capable of degrading isolated walls was first indicated by SALTON (1956) when washed isolated cell walls underwent almost complete dissolution on storage at 4° for several weeks in the presence of chloroform.

BROWN (1961) isolated cell-wall fractions from a marine pseudomonad after mechanical disintegration of the cells. A lytic enzyme was associated with the cell envelope fraction and it was active on the outer component of the double membrane system, accounting for a solubilization of anything from 0–50% of the 'whole' cell-wall fraction according to the medium on which the organism was grown. The activity of such autolytic enzymes has proved useful in the structural analysis of the envelopes but for routine isolation of 'walls' these enzymes would have to be inhibited.

In the absence of suitable equipment for cell disintegration autolytic methods may find a limited application for cell disintegration as the first step for wall isolation. However, because of the difficulty of controlling the simultaneous action of wall degrading enzymes the method has not found wide application.

2. *Osmotic lysis*

The 'envelope' structures of certain bacteria, especially halophilic organisms should be amenable to isolation by inducing osmotic lysis of the cells. Simple dilution of the suspending medium was sufficient to cause rapid lysis of a number of halophilic organisms studied by CHRISTIAN AND INGRAM (1959). SMITHIES, GIBBONS AND BAYLEY (1955) did indeed find that they could produce relatively undamaged cell walls or envelopes of *Vibrio costicolus* and *Micrococcus halodenitrificans* on disrupting the cells by shaking in water and adding deoxyribonuclease to reduce the viscosity of the suspension. However, with another halophilic organism, *Pseudomonas salinaria*, the cell substance was completely dispersed and no particulate wall fraction could be recovered.

BROWN (1960, 1961) also utilized sensitivity to osmotic lysis in preparing walls of a marine pseudomonad, although greater reproducibility was obtained by mechanical disruption.

The reasons for the complete absence of a particulate wall fraction on shaking *Pseudomonas salinaria* in water (SMITHIES, GIBBONS AND BAYLEY, 1955) became apparent when BROWN AND SHOREY (1962) reported the rapid dissolution of isolated cell envelopes of *Halobacterium halobium* and *Halobacterium salinarium* on suspension in aqueous solutions of relatively low ionic strength. The stability of the envelopes of these extreme halophiles is dependent upon di-valent cations and BROWN (1962) has suggested that the complete disaggregation of these structures in distilled water may be a non-enzymic process. MOHR (personal communication) has also found that the 'wall' preparations of *Halobacterium halobium* were sensitive to dissolution by protein denaturing agents. Thus osmotic lysis as a method of wall or envelope isolation may be applicable only to those organisms possessing structures that are stable at low ionic strengths and in the absence of di-valent cations.

Another method of disrupting cells by osmotic shock was reported by ROBRISH AND MARR (1957). They increased the solute concentration of the cytoplasm with glycerol and by this means they were able to disrupt *Azotobacter agilis*, *Rhodospirillum rubrum* and *Serratia plymuthica*. However, other bacteria were not sensitive to this glycerol induced osmotic lysis (ROBRISH AND MARR, 1957).

3. Heat-treatment rupture

During the course of observations on the effects of rapidly introducing thick bacterial suspension into larger volumes of distilled water at various temperatures, SALTON AND HORNE (1951a) noticed that the walls of certain Gram-negative bacteria were ruptured at temperatures above 70° C. By rapidly pipetting suspensions of *Escherichia coli* into distilled water at 100° and holding at that temperature for 5 minutes, virtually all of the walls had been ruptured, leaving a naked coagulated protoplast. This formed the basis of a method developed by SALTON AND HORNE (1951b) for isolating cell-wall material from *Escherichia coli* and *Salmonella pullorum*. The appearance of the cell wall fragments isolated from *Escherichia coli* by the heat-treatment rupture method is shown in Fig. 12. As pointed out by SALTON (1956) this method gave only small yields of material with Gram-negative bacteria and was not applicable to Gram-positive organisms. These factors together with the probability that some of the chemical components of the walls are lost by this method led SALTON (1953a, 1956) to indicate that this procedure was unsatisfactory for the routine quantitative isolation of walls from all types of bacteria.

In a study of bacteriophage action and receptor activity BROWN AND KOZLOFF (1957) and KOZLOFF AND LUTE (1957) compared walls of *Escherichia coli* prepared by the heat-treatment rupture method with those isolated by mechanical disintegration of cells. Apart from the satisfactory behaviour of the material isolated by heat-treatment rupture in the experiments of KOZLOFF and his colleagues the method has found little application.

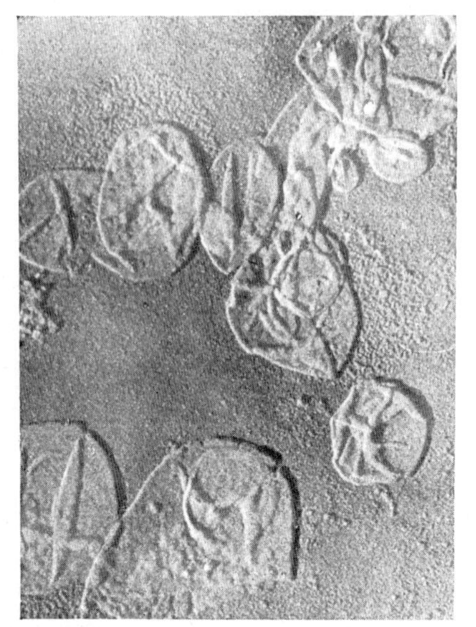
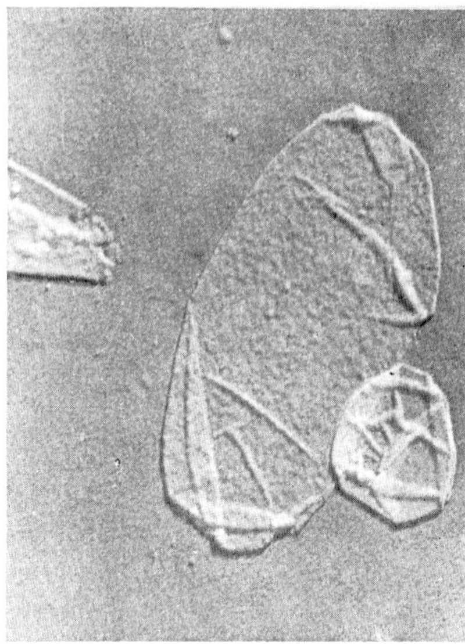

FIGURE 12. Cell-wall fragments isolated from heat-ruptured cells of *Escherichia coli* (a) $\times$ 10,000; (b) $\times$ 20,750. (SALTON AND HORNE, 1951).

4. Mechanical disintegration

Mechanical methods of cell disintegration have provided the most successful techniques universally applicable to the isolation of bacterial structures, enzymes and intracellular particles. Many of the mechanical methods of cell rupture involve agitation with an abrasive such as sand or glass particles. Such techniques have been in use for a very long time for MACFADYEN AND ROWLAND (1901) devised suitable equipment for disintegrating bacteria and yeasts. Indeed their photomicrographs of ruptured yeast cells showed the robust empty walls devoid of intracellular cytoplasmic material. The various methods broadly grouped as 'mechanical' can be classified under the following general headings:

(a) mechanical disintegration by grinding, violent agitation or compression with abrasives such as sand, glass beads or steel balls and alumina.
(b) sonic and ultrasonic disintegration.
(c) decompression rupture.
(d) pressure cell disintegrator.

(a) Mechanical disintegration by agitation with abrasives

A variety of techniques involving disintegration of bacterial cells by grinding or vigorous agitation with suitable materials have now been used in isolating cell walls. With some bacteria such as *Bacillus megaterium* almost quantitative disintegration can be achieved by simple grinding in a mortar with Ballotini beads of about 0.13 mm in diameter. However, fragmentation of glass and sand particles to give colloidal suspensions can prove troublesome and not all bacteria can be as readily disrupted as this organism. More satisfactory results are obtained when cell disintegration is performed in some sort of a shaking device.

Disintegration and death of bacteria on violent agitation with glass particles was first studied in some detail by CURRAN AND EVANS (1942). More extensive investigations were reported by KING AND ALEXANDER (1948) who introduced the use of small smooth glass beads. Several grades of these Ballotini beads were examined and those beads with diameters of about 0.13–0.26 mm were the most effective in killing a variety of bacteria when suspensions of cells were shaken at speeds of 300–500 strokes/minute with an amplitude of $2\frac{1}{2}$ inches.

Careful quantitative studies of cell rupture of *Staphylococcus aureus* were made by COOPER (1953). Studies were performed with a 'microid' flask shaker using Ballotini grade '12' beads (0.1–0.2 mm in diameter). Loss of turbidity, viable counts and the ratios of whole bacteria to cell walls were determined from electron micrographs, at frequencies of shaking varying from 28 to 53.5 cycles/sec. The time taken to reach the limiting turbidity on disrupting *Staphylococcus aureus* at 2 mg dry weight/ml in distilled water at 28 and 49 cycles/sec from COOPER's data are presented in Fig. 13. The effect of varying the cell concentration on the rate of disruption in distilled water was also studied and the influence of this factor is clearly seen in Fig. 14.

DAWSON (1949) was the first to show that rupture of *Staphylococcus aureus* cells with Ballotini beads in the MICKLE (1948) tissue disintegrator for one hour gave fractions containing cell walls freed from their cytoplasmic constituents. Since then the Mickle apparatus has been widely used as the first step in the preparation of cell walls of a variety of microorganisms

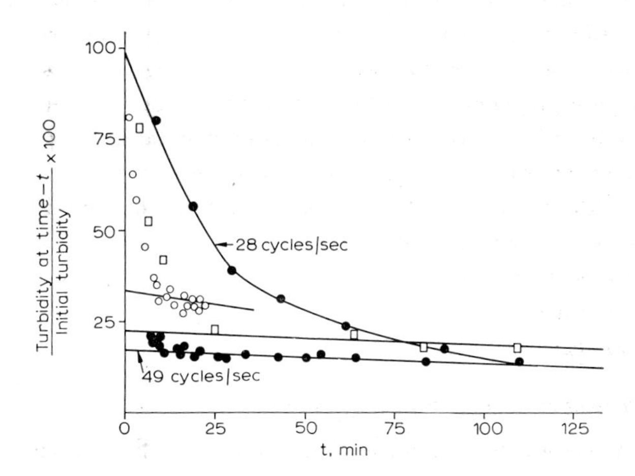

FIGURE 13. Factors affecting the disruption of *Staphylococcus aureus*. Influence of the frequency and time of shaking on the disruption of *Staphylococcus aureus* cells at 2 mg dry weight/ml in distilled water and media reducing enzymic activities; ●——● water; 'Medium B' □——□ 49 cycles/sec; 'Medium A' o——o, 53.5 cycles/sec (COOPER, 1953).

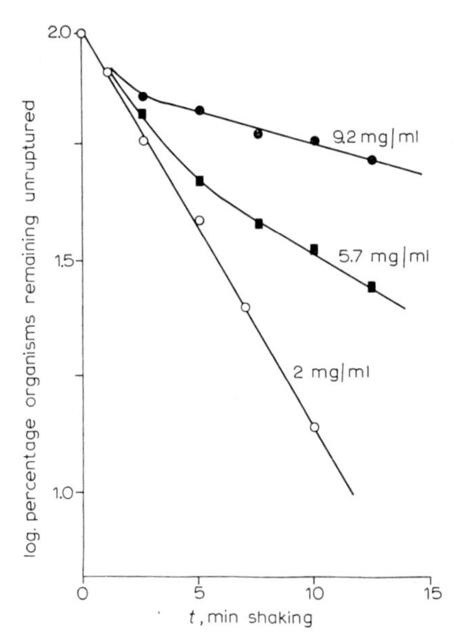

FIGURE 14. Effect of varying the concentration of *Staphylococcus aureus* cells on the rate of disruption in distilled water calculated from turbidity measurements. Shaking frequency, 40 cycles/sec (COOPER, 1953).

(SALTON AND HORNE, 1951b; MITCHELL AND MOYLE, 1951; NORTHCOTE AND HORNE, 1952). This device shakes the cups containing cell suspensions and beads at a frequency of 50 cycles/sec with a maximum amplitude of about 5 cm. For disintegration of bacterial cells 10 ml of the cell suspension (containing 10–20 mg dry weight bacteria/ml) is mixed with an equal volume of Ballotini beads of 0.1–0.2 mm diameter (SALTON AND HORNE, 1951b) in plain pyrex glass cups. The cups normally supplied with this apparatus have two series of indentations but the author found that it was virtually impossible to obtain satisfactory cell rupture when these were used. The indentations in the cups so seriously impaired cell disintegration by Ballotini beads that the cups could not be used for wall preparation by this method and plain glass cups have invariably been employed in the author's laboratory.

Over a period of years, the author has found that the Mickle apparatus can be used to disintegrate a wide variety of Gram-negative and Gram-positive bacteria. With distilled water suspensions containing 10–20 mg dry weight bacteria/ml cell rupture is usually complete within 10 minutes when the conditions described above (and elsewhere: SALTON AND HORNE, 1951b) are used. Bacterial spores and filamentous, clumped and aggregated organisms such as Streptomyces species, Mycobacteria and fungi (e.g. *Neurospora crassa*) may take longer periods of shaking to bring about complete disintegration. CUMMINS AND HARRIS (1956, 1958) have also found that disintegration in the Mickle apparatus was of wide applicability to the extensive range of bacterial species used in their cell wall studies.

As shown in the careful experiments performed by COOPER (1953) there is a decrease in turbidity on disintegration of *Staphylococcus aureus* in the Mickle apparatus. With some organisms the progress of disruption can be roughly gauged by observing the change in turbidity of the suspension in the cups. However, with organisms containing large amounts of the β- hydroxybutyrate polymer granules, the changes in turbidity are less conspicuous and may give the impression that the cells are very resistant to disintegration. Thus a more accurate control will involve microscopic examination and this may be followed conveniently in the phase contrast microscope or by preparation of Gram stained smears or smears stained with only crystal violet or even negative staining with nigrosin. All of these methods have given useful information but the ultimate question of the homogeneity of the preparations can be answered by examination in the electron microscope and sometimes only with the additional aid of other criteria.

Although disintegration of bacteria in the Mickle apparatus has been widely used in the past as the first step in cell wall preparation, this procedure has certain disadvantages. When the Mickle apparatus is used at room temperature the contents of the cups may rise to as much as 40° during a 10–20 minute disintegration period. The possible deleterious effects of this can be obviated by carrying out the complete process in a cold room, and if the temperature is to be maintained below 10° only short periods of several minutes can be used and the cups chilled to 0° again. The second principal disadvantage of this method of cell rupture is its rather limited capacity. The maximum capacity under optimal working conditions is about 1–2 g dry weight bacteria/hour. Thus for large scale preparations of cell walls, disintegration in the Mickle apparatus cannot be recommended and other devices with larger capacities and more accurate temperature control such as those described by NOSSAL (1953), LAMANNA AND MALLETTE (1954), SHOCKMAN, KOLB AND TOENNIES (1957) and RIBI, PERRINE, LIST, BROWN AND GOODE (1959).

One of the most useful pieces of equipment introduced for cell disruption is the high speed shaker head illustrated in Fig. 15 and described in detail by SHOCKMAN, KOLB AND TOENNIES (1957). This apparatus is fixed to the refrigerated International Centrifuge (Model PR2) and at a speed of 1900–2000 r.p.m the capsules mounted on the head are shaken vertically as well as with a slight horizontal movement. The stroke of the capsule is $1\frac{3}{8}''$. This method of cell disintegration with the shake head has the two-fold advantages of operation at temperatures near 0° when the device is fitted in a refrigerated centrifuge and an excellent capacity of about 6 g dry weight of bactria for a single run. As shown in Fig. 16 there is a rapid loss of soluble nitrogenous material from the cells during disintegration in this shaker. Even if the operation time required for complete disintegration of the cell suspensions is as much as one hour, the capacity is considerably greater than that of the Mickle apparatus and has the added advantage of a reduced number of manipulations. Some modifications designed to reduce vibration and heating of the centrifuge spindle have been recently introduced by SHOCKMAN (1962). Essentially complete disruption of 500 mg dry weight of *Streptococcus faecalis* after 15 minutes shaking at 1900–200 rev/min in 25 ml water with 25 g of 0.2 mm diameter glass beads was achieved.

FIGURE 15. The high-speed shaker head for attachment to the International Centrifuge as described by SHOCKMAN *et al.*, 1957, for the disintegration of microorganisms.

More recently, a continuous-flow mechanical cell disintegrator has been described by ROSS (1963). This cell disruption equipment has a capacity ranging from 840 ml/hour to 4,470 ml/hour, can be kept at 0° and can effect complete rupture of cells within a very short time. This disintegrator appears to be very suitable for cell wall isolation procedures.

Glass beads of about 0.2 mm diameter have been used for cell disintegration in the Waring blendor (LAMANNA, 1954) and a special device for compressing cells and beads packed into a circle of plastic tubing has been used by LAMANNA, CHATIGNY AND COLLEDGE (1959). The former method has been tried by the author but rather long periods were required to disintegrate some bacteria and considerable heating of the contents occurred. The disintegration of bacteria within the sealed plastic tubes is extremely advantageous for handling pathogenic organisms, but so far as the author is aware, the method has not been specifically adopted for cell-wall isolation.

Thus small Ballotini beads have been used most successfully in a variety of devices for mechanical rupture of bacteria. The beads are used once only and then they may be recovered and cleaned by washing away adsorbed protein and nucleic acid and cell constituents and treated with cleaning fluid or nitric acid and thoroughly washed. If the beads are used a number of times without washing, cell disintegration becomes progressively less effective.

Only one disadvantage of their use became apparent when KOLB (1960) reported that certain brands of beads could give rise to a substantial generation of alkali during the shaking process, the supernatant liquid rising to pH 10.2. Even washing the beads by heating with HCl or HNO_3 was ineffective in preventing the suspending medium from becoming alkaline.

For rapid disintegration of bacterial cells Ballotini beads have proved effective when used in conjunction with mechanical shaking devices. However, cell disintegration may also be carried out with larger steel balls in a ball-mill. This latter method was used by MCCARTY (1952) for isolation of walls of group A streptococci. Cell disruption was achieved in a ball-mill containing $\frac{1}{4}$ inch diameter stainless steel balls and by rotating at a rate of 50–60 rev/min for 1 to 2 hours at 4°. Acetone dried streptococcal cells were used and after disintegration the material was washed from the mill with 0.85% NaCl (MCCARTY, 1952).

The Hughes press (HUGHES, 1951) has perhaps been used more widely to disintegrate microorganisms for enzyme preparation rather than for cell wall isolation. Many of the above methods involve vigorous shaking and the frothing and heating resulting in denaturation of the proteins released from the cells. Rupture in the Hughes press by compression of a frozen bacterial paste (with or without an abrasive) enables a low temperature to be maintained during the manipulations. Whether the cell fractions obtained by disintegration in the Hughes press give final cell-wall preparations identical to those obtained by more vigorous methods has not so far been fully established. There is however, good evidence from the studies of HUNT, RODGERS AND HUGHES (1959) and HUGHES (1962) that cell disintegration in the Hughes press yields a wall-membrane ('envelope') fraction rather than a simple wall preparation. The

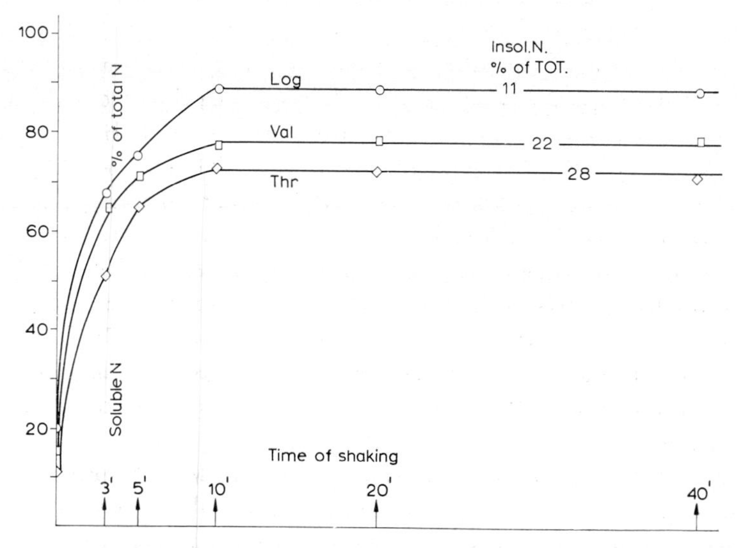

FIGURE 16. The course of mechanical disruption of exponential cells and valine and threonine-limited post-exponential cells in the shaker-head equipment, followed by release of soluble N. Data from TOENNIES AND SHOCKMAN, 1959.

Hughes press will undoubtedly find wider application for investigations of the distribution of enzymes in the wall and membrane fractions of bacteria. EATON (1962) has recently reported the design of a new block to overcome some of the disadvantages of that used in the Hughes press. The modified press makes quantitative recovery of the material easier and facilitates the separation of the blocks.

(b) Sonic and ultra-sonic disintegration

One of the most widely used methods of disrupting cells, especially for extraction of enzymes, has been provided by the sonic and ultra-sonic oscillators. This method has also been used for cell rupture to permit wall isolation by SALTON (1953b); BOSCO (1956); IKAWA AND SNELL (1960). The virtues of these procedures for cell disintegration for enzyme studies are undisputed and they have been discussed in detail by HUGO (1954) and MARR (1960). The method has certain advantages over some of the mechanical procedures for cell-wall or membrane isolation discussed above. The capacity may be greater and the temperature can be kept low during disintegration. The Raytheon magnetostriction oscillators (9 and 10 Kc) have been satisfactorily used for wall isolation. There are however, disadvantages in the use of sonic disintegrators for some organisms such as staphylococci are quite resistant to disruption and have to be exposed for long periods to achieve quantitative disintegration. Prolonged exposure of even the robust cell walls of *Staphylococcus aureus* in the 10 Kc Raytheon results in a progressive loss of sedimentable wall material.

SLADE AND VETTER (1956) firmly established the release or 'solubilization' of cell-wall material from group A streptococci during sonic disintegration of the cells. Rhamnose, one of the typical constituents of the streptococcal wall, was found in the suspending medium after sedimentation of the walls. From the information available it is not possible to say whether the rhamnose was in a soluble product from enzymic action on the walls or attached to colloidal fragments of poor sedimentability. Judging from the author's experience with *Staphylococcus aureus* and *Rhodospirillum rubrum* walls it would seem likely that in some instances the 'solubilized' material is in fact of particulate nature requiring higher gravitational fields for complete sedimentation. Cell walls of *Escherichia coli* isolated by the Mickle procedure (SALTON AND HORNE, 1951b) were broken down into a small particle fraction by exposure to the 9 and 10 Kc Raytheons for 15–30 minutes. No spiral shaped walls were ever obtained from *Rhodospirillum rubrum* when cell disintegration was carried out for up to 10 minutes in the 9 Kc Raytheon (SALTON AND WILLIAMS, 1954). The suggestion that the solubilized material can be sedimented at higher gravitational fields has been confirmed (ROBERSON AND SCHWAB, 1960). They found that after sonic disintegration of group A streptococci for 1 h in the absence of glass beads it was necessary to carry out centrifugation for 1 h at 144,000 × g to sediment an amount of cell-wall rhamnose comparable to that found in deposits (30 minutes at 11,500 × g) from cells sonically vibrated in the presence of glass beads.

The extent to which enzymes are involved in cell disintegration by sonic oscillation is not known but the study made by ROTMAN (1956) has shown that bacteria such as *Escherichia coli* and *Azotobacter vinelandii* do not lyse on sonic vibration when suspended in media of low ionic strength or at a low pH. Thus ROTMAN (1956) established that although structural damage occurs on sonic vibration the leakage of intracellular material occurred only in the presence of neutral buffer. Such a lytic effect could be due to either enzymic lysis or to solvation of the cytoplasm. MARR AND COTA-ROBLES (1957) carefully studied the disintegration of *Azotobacter vinelandii* by sonic disruption and they concluded that concomitant with the release of intracellular ribosomes there was a fragmentation of the envelope structure. The simultaneous fragmentation of wall or envelope structures and cell rupture makes this method less attractive for large scale quantitative isolation of cell walls and undoubtedly accounts for the low yields reported by several investigators (BROOKES, CRATHORN AND HUNTER, 1959; IKAWA AND SNELL, 1960).

(c) Decompression rupture

FRASER (1951) showed that *Escherichia coli* cells could be ruptured by suddenly releasing gas pressure applied in a 'bomb'. Bacterial suspensions containing up to 10^8 cells/ml were subjected to pressures ranging from 500–900 lb/in^2 using inert gases and CO_2. This method as originally described has not been adopted for bulk preparation of walls but it gave material suitable for studies of bacteriophage attachment (WILLIAMS AND FRASER, 1956).

FOSTER, COWAN AND MAAG (1962) described an apparatus for instantaneous rupture of bacteria and other cells in a closed system with controlled conditions of explosive decompression. Bacterial suspensions were ruptured after saturation with nitrogen at 1740 psi. The pressure and duration of saturation of cells with gas affected the efficiency of rupture. With this device 31–59% of *Serratia marcescens* (suspension dry weight ranging up to 20 mg) and 10–25% *Brucella abortus* and *Staphylococcus aureus* cells were ruptured. Clean cell walls were isolated by centrifugation at 5,480 $\times$ g for 30 minutes in a linear glycerol gradient.

(d) Pressure cell disintegrator

MILNER, LAWRENCE AND FRENCH (1950) constructed a dispersion unit to disintegrate chloroplasts by forcing the suspension through a small orifice at high pressures of 20,000 psi. Best results were obtained when the needle valve was adjusted for a flow rate between 9–10 ml/minute. Under these conditions a 20% breakage of yeast cells in a dilute suspension was observed and with thick suspensions of *Escherichia coli* cell disruption was indicated by a several fold increase in glutamic acid decarboxylase activity.

An extension of this method of cell disintegration was made when RIBI, PERRINE, LIST, BROWN AND GOODE (1959) designed a new type of pressure cell in which the bacterial suspensions contained in a steel cylinder under a total load of 35,000– 40,000 lb were bled through a needle valve cooled to 2° with a controlled stream of CO_2. The bacterial products 'bled' from the pressure cell did not rise above 15°.

This method has the advantages of a good capacity (100 ml suspension containing 3 mg bacteria/ml) and control of temperature during cell rupture. For the isolation of the walls of

Mycobacteria, Ribi, Perrine, List, Brown and Goode (1959) also found that non-dispersable aggregates formed in the Mickle apparatus were not encountered when the pressure cell was used and the final yields of cell walls were much larger. The final yield of purified cell wall

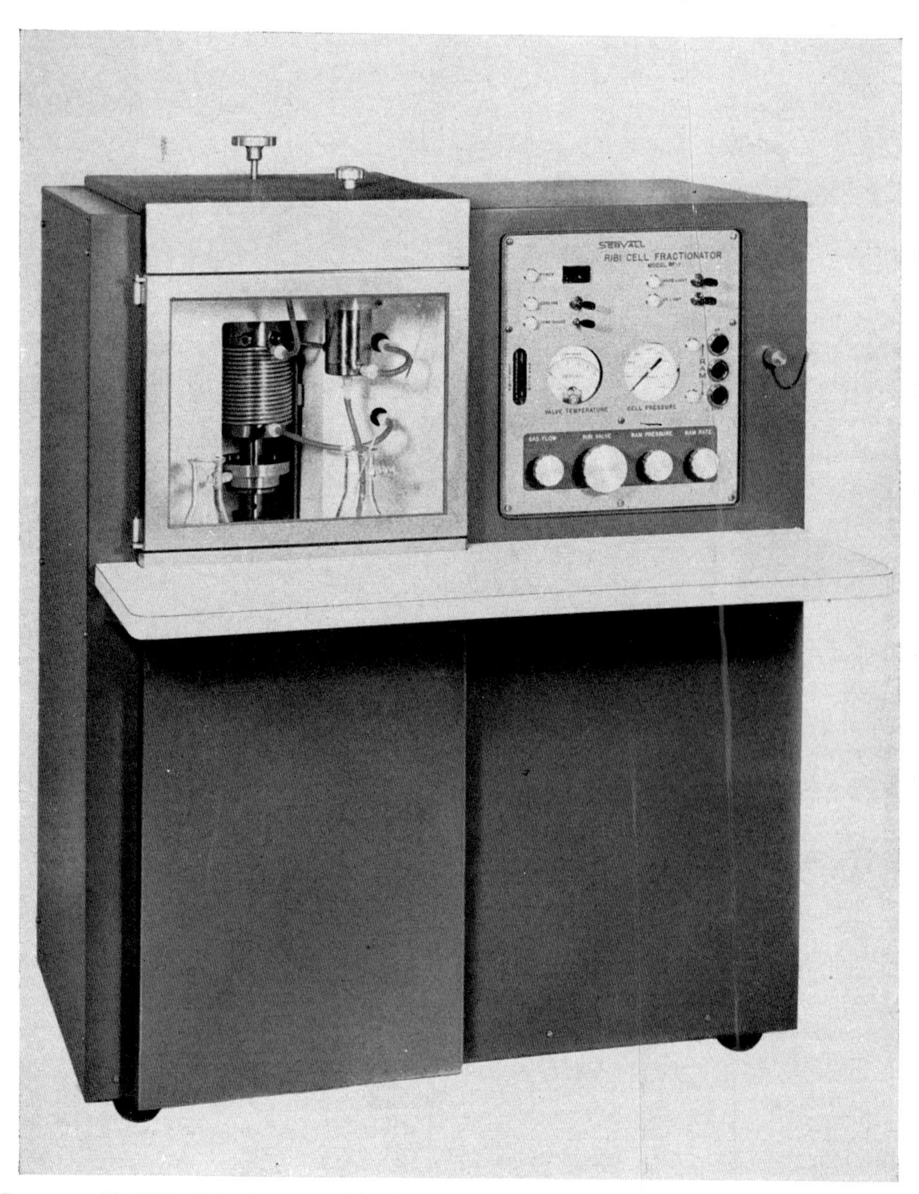

Figure 17. The Ribi cell fractionator used for the disruption of microorganisms under the conditions described by Ribi *et al.*, 1959.

from 1 g dry weight starting material was about 60 mg. If a bacterial suspension density of 3 mg/ml represents the maximum that can be used in a single 'run' with the pressure cell then its total capacity is not as great as the shaker-head device used by SHOCKMAN (1962). The RIBI pressure cell disintegrator is illustrated in Fig. 17.

CELL WALL ISOLATION PROCEDURES

Pretreatment of bacteria

In general untreated washed bacterial suspensions are used for cell wall isolation; the cells being washed and suspended in distilled water, saline, buffer or salt solutions depending on the organisms to be disrupted. For the isolation of walls from certain pathogenic bacteria there is some advantage in working with killed cells since with the exception of the sealed tubes used by LAMANNA, CHATIGNY AND COLLEDGE (1959) most of the methods of cell disintegration discussed above readily produce aerosols. CUMMINS AND HARRIS (1956, 1958) in their extensive surveys of cell-wall composition have widely adopted the practice of killing the bacteria with 2% formaldehyde prior to harvesting and disintegration.

The normal tendency for Mycobacteria and *Streptomyces* spp. to aggregate and clump even after disintegration in the Mickle apparatus has made it difficult to isolate satisfactory wall preparations by the usual procedures. This problem was overcome with several *Mycobacterium* spp. by washing the cells in 0.5% Tween solution prior to cell rupture and carrying out disintegration in the Tween as a suspending medium RIBI, LARSON, LIST AND WICHT (1958). CUMMINS AND HARRIS (1958) pretreated *Nocardia* and *Mycobacterium* spp. in 0.5% w/v KOH in ethanol for 48 hours at 37°. After washing with neutral ethanol and distilled water the cells could be disintegrated in the usual way.

Preliminary dispersion of aggregates in high-speed blending equipment will also reduce the size of clumps sufficiently for satisfactory cell disintegration in the Mickle apparatus or other shaking devices using small glass beads of about 0.1–0.2 mm diameter.

In bacteria containing large amounts of the β-hydroxybutyrate polymer which may frequently be rather difficult to separate from the walls, the solubility of the polymer in chloroform can be utilized to simplify the wall isolation procedure. At the suggestion of the author, VINCENT, HUMPHREY AND NORTH (1962) and HUMPHREY AND VINCENT (1962) found that the polymer could be removed by prior extraction of lyophilized cells with chloroform for 24 hours at room temperature and this gave satisfactory cell-wall preparations by the Mickle disintegration procedure.

Procedures for isolation of walls after cell disintegration

Bacterial cells may be disrupted by any of the methods outlined above. When glass beads have been used they are separated from the disintegrated bacteria by filtration on a coarse sintered ('frittered') glass filter (porosity grade 2). The beads are generally washed and the combined filtrates are then subjected to differential centrifugation by procedures designed to separate the walls from any intact cells, intracellular granules such as β-OH-butyrate granules or polysaccharide or polymetaphosphate granules from the soluble and particulate materials derived from membranes, mesosomes, ribosomes and cell cytoplasm. The presence of the various structures, particles and macromolecular aggregates of varying size can be demonstrated by centrifuging the disintegrated cell 'mush' at about 10,000–20.000 $\times$ g. The typical layered appearance of the pellet so obtained is illustrated diagramatically in Fig. 18 and shows the presence of a series of layers consisting of intact cells, dense granules, walls and particles obtained from the comminuted plasma and intracellular membranes sedimenting one upon the other.

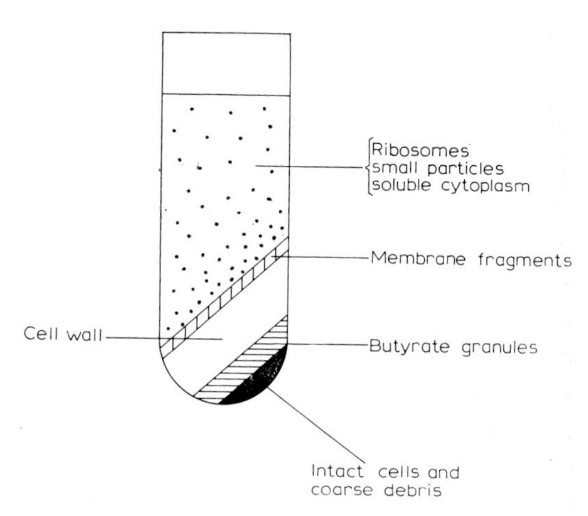

FIGURE 18. A diagrammatic representation of the appearance of the pellet obtained by centrifugation of the cell 'mush' at c. 5,000–10,000 $\times$ g after disruption of a Gram-positive organism such as *Bacillus megaterium*.

If there is a substantial percentage (1%) of unbroken cells then centrifugation at about 3,000–4,000 $\times$ g for 10 minutes will deposit these together with any coarse debris. The bulk of the cell walls will then be recovered in the supernatant. A preliminary low speed centrifugation is not always necessary except perhaps in the final stages of washing providing that the cell breakage is really efficient. A low-speed centrifugation is often essential if the organism contains appreciable amounts of intracellular particles such as the β-hydroxybutyrate particles ('lipid granules') and these can be deposited as an opaque, chalky-white pellet by low-speed centrifugation, leaving the walls in the supernatant. With particles soluble in lipid solvents, prior extraction, as already mentioned, may give very satisfactory wall preparations thus avoiding a series of differential centrifugations.

The cell walls of most bacteria can be sedimented by centrifuging at approximately 10,000 $\times$ g for 15–30 minutes. On removal of the bacterial 'protoplasm' and salts by washing with distilled water, the walls of Gram-negative species in particular require rather higher gravitational fields for complete sedimentation (e.g. 10,000–20,000 g for 15–30 minutes). The first crude cell-wall deposit undoubtedly contains contaminating particulate materials, insoluble denatured protein as well as adsorbed soluble components. Although continued washing with distilled water is sufficient to remove these materials from the walls of many Gram-positive bacteria, we have found that several washes with *M* NaCl or buffer are quite effective in cleaning up the cell-wall fractions (SALTON AND HORNE, 1951b; SALTON, 1953a) from both Gram-positive and Gram-negative bacteria.

Many bacteria contain enzymes capable of degrading their own cell walls (SALTON, 1956; MITCHELL AND MOYLE, 1957) and such enzymes may be especially active after cell disintegration. STRANGE AND DARK (1957) experienced considerable difficulty in isolating stable wall preparations from *Bacillus cereus* for subsequent studies of the effects of enzymes on these structures. Heating the wall suspensions at 100° for 15 minutes was sufficient to prevent the release of bound hexosamine. Where such enzymes are sensitive to heat, it may be an advantage to inactivate cell-wall degrading systems by heating prior to carrying out the final stages of wall isolation. MANDELSTAM AND ROGERS (1959) used this procedure as a precaution against loss of mucopeptides due to enzymic digestion of walls during their studies of wall biosynthesis. The crude wall deposits were heated prior to digestion with proteolytic enzymes.

When the methods of cell-wall isolation were first investigated any enzymic treatments which could have modified the cell wall structure were avoided for it was not known what wall components may be lost by exposing preparations to enzymic attack. However, the retention of the M-protein antigen on the cell wall of *Streptococcus pyogenes* (SALTON, 1953a) and its removal on digestion with trypsin indicated that proteolytic enzymes could be used in wall fractionation without degrading the rigid cell wall. Cell walls prepared by mechanical disintegration and washing with distilled water, *M* NaCl or buffers, were shown to be completely resistant to crude and crystalline trypsin (SALTON, 1953b). Thus for routine preparations of walls, MCCARTY (1952), CUMMINS AND HARRIS (1956, 1958) have used digestion with proteolytic enzymes, trypsin and pepsin and ribonuclease. Digestion of the crude cell-wall fractions with trypsin can therefore be used to considerable advantage and shortens the procedure of wall isolation by avoiding the prolonged washing steps with distilled water and *M* NaCl which are sometimes necessary with walls from certain bacteria. Whether all bacterial cell walls will resist digestion with proteolytic enzymes has not been established and it is well to bear in mind the possible effects of these enzymes on the cell-wall structures of Gram-negative bacteria. None of the constituents of the glycosaminopeptide components of walls have been detected in the soluble products obtained from trypsin-digested walls (SALTON, unpublished observations).

The difficulty encountered by CUMMINS AND HARRIS (1958) in obtaining clean wall preparations of *Mycobacteria* spp. by the usual techniques has already been mentioned. The problem was overcome by pretreating the cells with KOH in ethanol. They were also able to isolate similar cell-wall material by treating the crude isolated cell-wall deposits with alcoholic KOH after initially disintegrating untreated cells. This procedure apparently removes material which caused the wall fragments to flocculate and adsorb cytoplasmic debris. To what extent other components not co-valently bound to the rigid wall structure are removed cannot be said at present, but it is conceivable that some of the lipid-polysaccharide complexes could have been released from the wall surface.

A general scheme for the isolation of cell walls by fractionation of disintegrated bacteria is presented in Fig. 19.

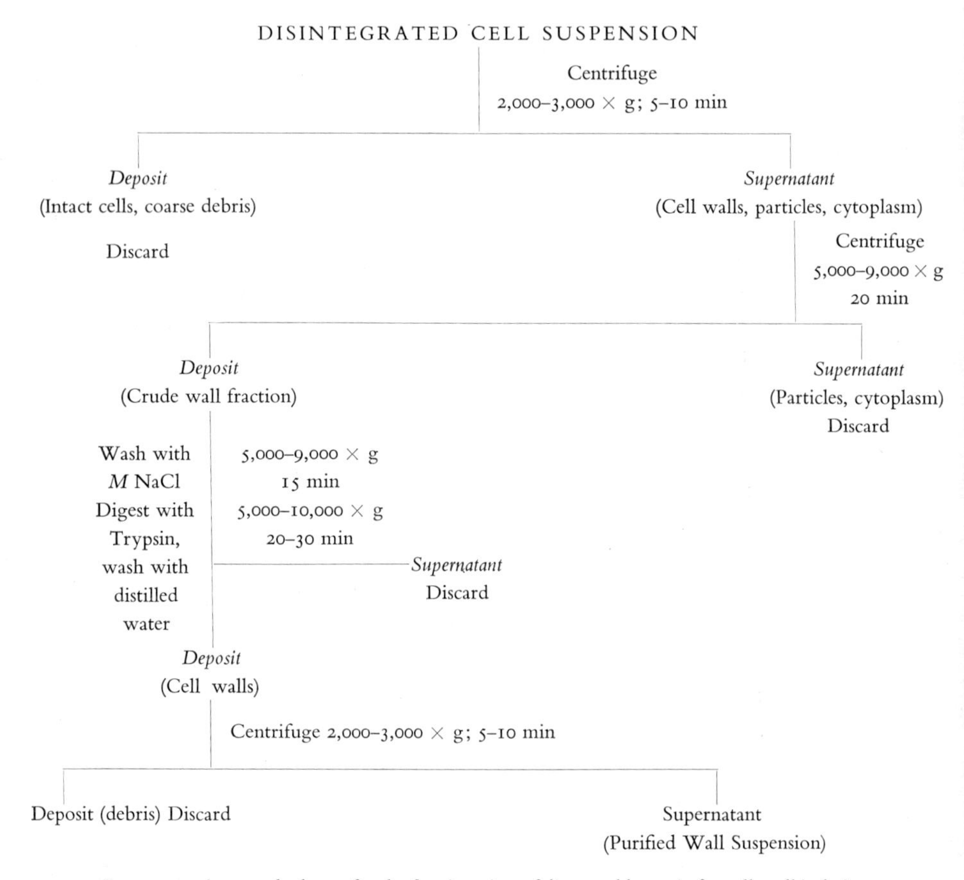

FIGURE 19. A general scheme for the fractionation of disrupted bacteria for cell-wall isolation.

Separation in two-phase polymer systems and in sucrose density gradients

The separation of cellular particles and structures has been improved by more widespread use of isolation in sucrose density gradients and by the introduction of a new technique for separation of biological materials in liquid two-phase polymer systems by ALBERTSSON (1958).

The behaviour of various particles in phosphate-polyethylene glycols (PEG) and dextrin – PEG systems together with other liquid two phase systems was studied by ALBERTSSON (1958) and by selecting suitable conditons, a variety of cellular particles were isolated as homogeneous fractions after separation in the two-phase systems. Cell walls of *Aerobacter* and *Chlorella* were isolated from cells broken in the Mickle apparatus and fractionated in a phosphate – PEG (molecular weight 4,000) system. As suggested by ALBERTSSON (1958) the method serves as a useful complement to centrifugation techniques.

Improved separation of cell walls from electron-dense 'granules' and unbroken cells of group A streptococci by sucrose zone centrifugation, had been reported by ROBERSON AND SCHWAB (1960). The group A streptococci were disrupted in a sonic vibrator in the presence of glass beads and the cell walls separated by sucrose zone centrifugation displayed a high degree of homogeneity as indicated by electron microscopy and moving boundary electrophoresis. The walls were centrifuged through a sucrose gradient varying in density from 1.025 to 1.30 at the bottom of the centrifuge bottle. This method gave products of greater reproducibility and homogeneity than those obtained by differential centrifugation. Immunological studies showed that the cell-wall preparations were antigenically complex though not necessarily contaminated with non-cell wall antigens (ROBERSON AND SCHWAB, 1960).

SALTON (unpublished experiments) also found that the walls of the photosynthetic bacteria *Rhodospirillum rubrum* and *Rhodopseudomonas spheroides* could be separated more cleanly from the chromatophores by centrifugation in a sucrose gradient.

Cell wall isolation for biochemical studies

The large number of manipulations involved in the isolation of bacterial cell walls by all of the methods outlined above impose a severe limitation on the number of preparations that can be conveniently handled in biochemical experiments. For investigations of the kinetics of biochemical reactions a simple method of quantitative wall isolation from a large number of bacterial samples would have much to offer. In studying the incorporation of ^{14}C amino acids into cell-wall mucopeptide of *Staphylococcus aureus* HANCOCK AND PARK (1958) and PARK AND HANCOCK (1960) described a method for isolating wall material by extracting the cells with 5% trichloroacetic acid (TCA) at 90°C or 100°C for 5–10 minutes and subsequent digestion with trypsin. The hot-TCA, trypsin-resistant residue was largely cell-wall mucopeptide contaminated by less than 5% cytoplasmic protein. The teichoic acids are of course solubilized by the hot TCA extraction.

This method unfortunately is not suitable for the isolation of intact wall or even mucopeptide from walls of other Gram-positive bacteria such as *Micrococcus lysodeikticus* and *Bacillus subtilis* as only a small yield of hot-TCA insoluble material was obtained (SALTON AND PAVLIK, 1960). The technique may find some application to bacteria other than *Staphylococcus aureus* but it is obvious that conditions would have to be established for each individual organism.

BROWN AND SALTON (unpublished observations) found satisfactory quantitative recovery of walls labelled with radioactive tracers by small scale disintegration in the Mickle using 1 ml samples of bacterial suspensions and Ballotini beads and performing the whole operation in heavy wall centrifuge tubes. The method does suffer from the limited number of preparations that can be handled.

Criteria for homogeneity of isolated cell walls

Criteria used for assessing the homogeneity of isolated cell structures include direct microscopic examination in the phase contrast and electron microscopes, behaviour on ultra-centrifugal and electrophoretic analysis and at the chemical level tests for detection of cellular components known to be present in specific structures. In addition to these methods separation in sucrose density gradients and immunochemical studies can provide further valuable information about the homogeneity of cell fractions.

Defining the homogeneity of structures as large as bacterial cell walls or compound membranes presents some difficulties. This question could hardly be approached until the resolution of the electron microscope facilitated the detection of other cellular subunits and particles. As emphasized by the author (SALTON, 1956) examination of isolated cell-wall preparations in the electron microscope is at present the most useful single guide to the homogeneity of the material. Thus in shadow-cast preparations it is possible to detect contamination with electron-dense particles as small as 100–200 Å in diameter. Any coarser contaminating cytoplasmic material, flagella fragments, β-hydroxybutyrate polymer and other storage granules can be readily seen in the electron microscope. Morover, contamination with viscous mucoid material can frequently be detected as 'background' material on the shadowed supporting films. Examination in the electron microscope by the usual techniques does not give any quantitative data about the homogeneity of wall fractions and the information has to be supplemented with other criteria.

Although in the initial stages of wall preparation, staining with crystal violet or examination under the phase contrast microscope gives useful information about the efficiency of cell rupture these techniques can not be relied upon to give a final assessment of homogeneity.

Direct microscopic examination can be supplemented with other valuable criteria of physical behaviour of cell-wall preparations. Ultra-centrifugal examination of wall preparations by WEIDEL (1951) indicated their homogeneity although it should be pointed out that in the electron microscope, residual electron-dense material remained inside the walls. By the latter criterion the preparation was structurally heterogeneous. However, with walls produced by cell disintegration it is unlikely that this situation would have arisen. ROBERSON AND SCHWAB (1960) also used physical criteria in determining the homogeneity of streptococcal walls. They examined the electrophoretic behaviour of the preparations as well as carrying out separation by sucrose zone centrifugation. Such methods combined with electron microscopy of the fractions give the best available information for estimation of homogeneity.

Where the isolated cell walls are susceptible to complete digestion with an enzyme such as lysozyme, an estimate of the amount of non-wall material can be made by direct weighing on centrifuging down the lysozyme-insoluble matter after cell-wall dissolution. Preparations of *Micrococcus lysodeikticus* walls were contaminated to the extent of 0.8% by weight with non-cell wall material (SALTON, 1956). Similar results were also obtained for wall preparations of *Sarcina lutea* and *Bacillus megaterium*, indicating that contamination with insoluble material can be kept below 1%. It is obvious that such an estimate would give a minimum value for contaminating materials and would not include any soluble materials strongly adsorbed on to the cell walls.

Useful supplementary information about the homogeneity of the wall preparations can be obtained by examining the distribution of specific types of chemical substances such as nucleic acids and cellular pigments in wall and 'cytoplasmic' fractions from disintegrated cells. As shown in Fig. 20 the ultra-violet absorption spectrum of the 'cytoplasmic' fraction from *Streptococcus faecalis* has a sharp peak at 260 mμ corresponding to that of the nucleic acids, whereas the wall gives a smooth scattering curve. Gross contamination with nucleic acid can therefore be readily detected. However, as BARKULIS AND JONES (1957) have pointed out material absorbing at ultra-violet wavelengths may be present in the wall fraction, the absorption being masked by the high scattering properties of the wall. They showed that some u.-v. absorbing material could be extracted from the isolated walls with hot water.

The cell walls of pigmented Gram-positive bacteria so far studied have been devoid of pigments. Thus in the organism, *Micrococcus lysodeikticus*, the carotenoids are localized in the

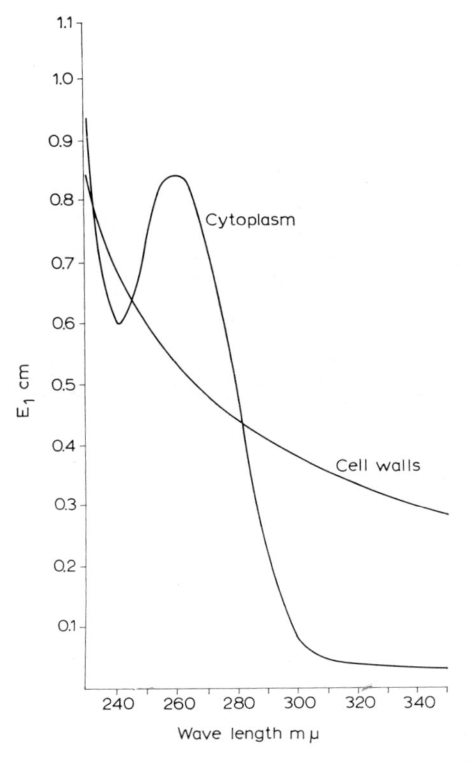

FIGURE 20. The ultra-violet absorption spectra of isolated cell-wall suspension and cytoplasmic supernatant fraction from *Streptococcus faecalis* (SALTON AND HORNE, 1951).

membrane-mesosome particle fraction (SALTON AND CHAPMAN, 1962). A pigmented wall fraction from those Gram-positive bacteria containing carotenoid pigments can thus be taken as an indication of contamination with intracellular particulate material.

However, the cell-wall preparations for Gram-negative bacteria frequently show the presence of pigments which are not readily released by preparative steps such as washing in water, buffers and digestion with trypsin. Whether the pigments become adsorbed to the walls or membranes during isolation or whether they are present as contaminating or normal structures cannot be said at present. It is generally rather more difficult to obtain clean walls or outer compound membranes from Gram-negative bacteria than it is to isolate them from Gram-positive organisms. Pigmentation of 'wall' or 'envelope' fractions may be a useful indicator of inhomogeneity of preparations isolated from some organisms. The walls of the photosynthetic bacterium, *Rhodospirillum rubrum*, show the presence of carotenoids and photosynthetic pigments, the latter in much smaller amounts than found in isolated chromatophore preparations. It is possible that some chromatophore material is strongly adsorbed on the walls and is fractionated along with these structures, or alternatively the chromatophore components may form part of the outer multilayered envelope of these organisms. Carotenoids have been found in the isolated walls of *Myxococcus xanthus* (MASON AND POWELSON, 1958) and also in several blue-green algae (SALTON, 1960). The walls of the latter microorganisms did not contain any of the photosynthetic pigments which separated cleanly from the walls as a particle fraction.

Pigment of *Serratia marcescens* was also found in the envelope fraction of this organism (PURKAYASTHA AND WILLIAMS, 1960). The distribution of pigment and hexosamines were closely parallel in the various particle fractions and PURKAYASTHA AND WILLIAMS (1960) concluded that the pigment was in the cytoplasmic membrane.

Immunological methods are also potentially useful for establishing the homogeneity of subcellular fractions and structures. The excellent experiments on the immunological reactions of intact cells and isolated flagella, walls and cytoplasmic membranes of *Bacillus megaterium* by VENNES AND GERHARDT (1956, 1959) showed that these cell structures were antigenically distinct. Although immunological methods have not been applied to routine examination for testing the homogeneity of cell-wall fractions, the studies of VENNES AND GERHARDT (1956, 1959) offer a model of the types of reactions that would be of great value to this aspect of cell-wall isolation procedures.

Yield of isolated cell wall and contribution to cell mass

It was evident from several early estimates that the cell wall of *Staphylococcus aureus* accounts for a considerable proportion of the cell dry weight. By direct isolation of the walls, COOPER, ROWLEY AND DAWSON (1949) found that they accounted for 20% of the dry weight of the organism. This value was later confirmed by MITCHELL AND MOYLE (1951). The cell-wall fraction isolated from *Corynebacterium diphtheriae* by HOLDSWORTH (1952) formed 45% of the dry weight of the cells. SALTON (1956) also reported that the walls of a number of Gram-positive organisms contribute approximately 20% to the dry weight of the bacterial cell and that in some instances the wall of *Micrococcus lysodeikticus* may account for as much as 30% of the dry bacterial substance. The thinner walls or membranes of Gram-negative organisms probably account for less than 20% of the dry weight of the organisms. FERNELL AND KING (1953) found that the total insoluble fractions isolated from a number of Gram-positive and Gram-negative bacteria represented considerably higher proportions of the dry weights of whole organisms and it seems doubtful that these residues could have been solely cell-wall material. The insoluble residue of *Pseudomonas aeruginosa* accounted for 76–78% of the dry weight of the whole organism. From our present knowledge of bacterial anatomy it seems likely that the unfractionated insoluble residues could include walls, comminuted membranes, reserve granules and particles.

The only investigations indicating the nature of factors affecting the contribution of wall to cell weight are those of SHOCKMAN, KOLB AND TOENNIES (1958) and TOENNIES AND SHOCKMAN (1959). These workers showed that the actual amount of cell wall formed in *Streptococcus faecalis* depended on the phase of growth. Thus in stationary phase cells the wall contributed 38% to the weight of the cells, while the corresponding value in the exponential phase was 27%. Furthermore, the nutritional status of the organism was also shown to affect the amount of wall formed by the cells. Some of the data for *Streptococcus faecalis* are compared with the results for other organisms in Table 10.

The values presented in Table 10 represent quantitative recoveries of cell-wall material and are in accord with estimates based on wall thickness. It would be of interest to determine the amounts of wall present in those organisms with wall thicknesses as much as 800 Å (*Lactobacillus acidophilus* studied by GLAUERT, 1962). The yields of cell-wall material under conditions of bulk preparation do not, however, always approach the expected values. In early experiments on wall isolation where it was of great importance to obtain very clean preparations an overall yield of 5% of the dry weight of the cells of *Streptococcus faecalis* was achieved as final wall material (SALTON, 1952). However, in these early studies much material was sacrificed in an attempt to obtain homogeneous preparations. Under normal conditions of wall isolation now used there is usually no difficulty in obtaining a final cell-wall product accounting for 20% of the weight of the cells of many Gram-positive bacteria, representing a 75–95% yield. Because of the difficulties in obtaining as clean a separation, the yields of cell walls or membranes from Gram-negative bacteria are generally lower.

REFERENCES

ALBERTSSON, P. Å., *Biochim. Biophys. Acta*, 27 (1958) 378.
BARKULIS, S. S. and M. F. JONES, *J. Bacteriol.*, 74 (1957) 207.
BOSCO, G., *J. Infectious Diseases*, 99 (1956) 270.
BROOKES, P., A. R. CRATHORN and G. D. HUNTER, *Biochem. J.*, 73 (1959) 396.
BROWN, A. D., *Biochim. Biophys. Acta*, 44 (1960) 178.
BROWN, A. D., *Biochim. Biophys. Acta*, 48 (1961) 352.

Brown, A. D., *Biochim. Biophys. Acta*, 62 (1962) 132.
Brown, A. D. and C. D. Shorey, *Biochim. Biophys. Acta*, 59 (1962) 258.
Brown, D. D. and L. M. Kozloff, *J. Biol. Chem.*, 225 (1957) 1.
Christian, J. H. B. and M. Ingram, *J. Gen. Microbiol.*, 20 (1959) 32.
Cooper, P. D., *J. Gen. Microbiol.*, 9 (1953) 199.
Cooper, P. D., D. Rowley and I. M. Dawson, *Nature*, 164 (1949) 842
Cummins, C. S. and H. Harris, *J. Gen. Microbiol.*, 14 (1956) 583.
Cummins, C. S. and H. Harris, *J. Gen. Microbiol.*, 18 (1958) 173.
Curran, H. R. and F. R. Evans, *J. Bacteriol.*, 43 (1942) 125.
Dawson, I. M. in *Soc. Gen. Microbiol. Symposium*, No. 1, 1949, p. 119.
Eaton, N. R., *J. Bacteriol.*, 83 (1962) 1359.
Fernell, W. R. and H. K. King, *Biochem. J.*, 55 (1953) 758.
Foster, J. W., R. M. Cowan and T. A. Maag, *J. Bacteriol.*, 83 (1962) 330.
Fraser, D., *Nature*, 167 (1951) 33.
Glauert, A. M., *Brit. Med. Bull.*, 18 (1962) 245.
Hancock, R. and J. T. Park, *Nature*, 181 (1958) 1050.
Holdsworth, E. S., *Biochim. Biophys. Acta*, 9 (1952) 19.
Hughes, D. E., *Brit. J. Exptl. Pathol.*, 32 (1951) 97.
Hughes, D. E., *J. Gen. Microbiol.*, 29 (1962) 39.
Hugo, W. B., *Bacteriol. Revs.*, 18 (1954) 87.
Humphrey, B. and J. M. Vincent, *J. Gen. Microbiol.*, 29 (1962) 557.
Hunt, A. L., A. Rodgers and D. E. Hughes, *Biochim. Biophys. Acta*, 34 (1959) 354.
Ikawa, M. and E. E. Snell, *J. Biol. Chem.*, 235 (1960) 1376.
King, H. K. and H. Alexander, *J. Gen. Microbiol.*, 2 (1948) 315.
Kolb, J. J., *Biochim. Biophys. Acta*, 38 (1960) 373.
Kozloff, L. M. and M. Lute, *J. Biol. Chem.*, 228 (1957) 529.
Lamanna, C., M. A. Chatigny and E. H. Colledge, *J. Bacteriol.*, 77 (1959) 104.
Lamanna, C. and M. F. Mallette, *J. Bacteriol.*, 67 (1954) 503.
Macfadyen, A and S. Rowlands, *Zentr. Bakteriol.*, Abt. 1, 30 (1901) 48.
Mandelstam, J. and H. J. Rogers, *Biochem. J.*, 72 (1959) 654.
Marr, A. G., in '*The Bacteria*', Eds. I. C. Gunsalus and R. Y. Stanier, Vol. I, Academic Press, Inc., New York, 1960, p. 443.
Marr, A. G. and E. H. Cota-Robles, *J. Bacteriol.*, 74 (1957) 79.
Mason, D. J. and D. Powelson, *Biochim. Biophys. Acta*, 29 (1958) 1.
McCarty, M., *J. Exptl. Med.*, 96 (1952) 569.
Mickle, H., *J. Roy. Microscop. Soc.*, 68 (1948) 10.
Milner, H. W., N. S. Lawrence and C. S. French, *Science*, 111 (1950) 633.
Mitchell, P. and J. Moyle, *J. Gen. Microbiol.*, 5 (1951) 981.
Mitchell, P. and J. Moyle, *J. Gen. Microbiol.*, 16 (1957) 184.
Mudd, S. and D. B. Lackman, *J. Bacteriol.*, 41 (1941) 415.
Mudd, S., K. Polevitsky, T. F. Anderson and L. A. Chambers, *J. Bacteriol.*, 42 (1941) 251.
Norris, J. R., *J. Gen. Microbiol.*, 16 (1957) 1.
Northcote, D. H. and R. W. Horne, *Biochem. J.*, 51 (1952) 232.
Nossal, P. M., *Australian J. Exptl. Biol. Med. Sci.*, 31 (1953) 583.
Park, J. T. and R. Hancock, *J. Gen. Microbiol.*, 22 (1960) 249.
Purkayastha, M and R. P. Williams, *Nature*, 187 (1960) 349.
Ribi, E., C. L. Larson, R. List and W. Wicht, *Proc. Soc. Exptl. Biol. Med.*, 98 (1958) 263.
Ribi, E., T. Perrine, R. List, W. Brown and G. Goode, *Proc. Soc. Exptl. Biol. Med.*, 100 (1959) 647.

Roberson, B. S. and J. H. Schwab, *Biochim. Biophys. Acta*, 44 (1960) 436.
Robrish, S. A. and A. G. Marr, *Bacteriol. Proc.*, 70 (1957) 130.
Ross, J. W., *Appl. Microbiol.*, 11 (1963) 33.
Rotman, B., *J. Bacteriol.*, 72 (1956) 827.
Salton, M. R. J., *Biochim. Biophys. Acta*, 8 (1952) 510.
Salton, M. R. J., (a) *Biochim. Biophys. Acta*, 10 (1953) 512.
Salton, M. R. J., (b) *J. Gen. Microbiol.*, 9 (1953) 512.
Salton, M. R. J., in *Soc. Gen. Microbiol. Symposium*, No. 6, 1956, p. 81.
Salton, M. R. J., *Microbial Cell Walls*, John Wiley and Sons, New York, 1960.
Salton, M. R. J. and J. A. Chapman, *J. Ultrastructure Res.*, 6 (1962) 489.
Salton, M. R. J. and R. W. Horne, (a) *Biochim. Biophys. Acta*, 7 (1951) 19.
Salton, M. R. J. and R. W. Horne, (b) *Biochim. Biophys. Acta*, 7 (1951) 177.
Salton, M. R. J. and J. G. Pavlik, *Biochim. Biophys. Acta*, 39 (1960) 398.
Salton, M. R. J. and R. C. Williams, *Biochim. Biophys. Acta*, 14 (1954) 455.
Shockman, G. D., *Biochim. Biophys. Acta*, 59 (1962) 234.
Shockman, G. D., J. J. Kolb and G. Toennies, *Biochim. Biophys. Acta*, 24 (1957) 203.
Shockman, G. D., J. J. Kolb and G. Toennies, *J. Biol. Chem.*, 230 (1958) 961.
Slade, H. D. and J. K. Vetter, *J. Bacteriol.*, 71 (1956) 236.
Smithies, W. R., N. E. Gibbons and S. T. Bayley, *Can. J. Microbiol.*, 1 (1955) 605.
Strange, R. E. and F. A. Dark, *J. Gen. Microbiol.*, 16 (1957) 236.
Toennies, G. and G. D. Shockman, *4th Intern. Congr. Biochem.*, Vol. XIII, Vienna, 1959, p. 365.
Vennes, J. W. and P. Gerhardt, *Science*, 124 (1956) 535.
Vennes, J. W. and P. Gerhardt, *J. Bacteriol.*, 77 (1959) 581.
Vincent, J. M., B. Humphrey and R. J. North, *J. Gen. Microbiol.*, 29 (1962) 551.
Vincenzi, L., *Hoppe-Seyl. Z.*, 11 (1887) 181.
Weidel, W., *Z. Naturforsch.*, 6b (1951) 251.
Williams, R. C. and D. Fraser, *Virology*, 2 (1956) 289.

CHAPTER 3

Electron microscopy of isolated walls

The rigid nature of the cell-wall structure of bacteria and of many other microorganisms has been known for some time. VINCENZI's (1887) wall material from *Bacillus subtilis* was detectable as a rod-shaped residue on staining. WAMOSCHER (1930) showed that on micro-dissection the protoplasm flowed out of the cell leaving a rod-shaped 'ghost' which was interpreted as a rigid cell-wall. Bacterial walls were not seen clearly until organisms were examined in the electron microscope. The cellular residues remaining after sonic disintegration of *Bacillus* species and Streptococci were the first preparations to show the presence of rigid wall structures in the electron microscope (MUDD AND LACKMAN, 1941; MUDD, POLEVITSKY, ANDERSON AND CHAMBERS, 1941 and MUDD, HEINMETS AND ANDERSON, 1943). These investigators concluded that the cells possessed rigid outer membranes or walls which accounted for the characteristic shape and grouping of streptococci into chains and for the cylindrical form of *Bacillus* spp. Rigid cell walls or multilayered envelopes possessing components characteristic of the wall mucopeptides (glycosaminopeptides) have been detected or isolated from a wide variety of bacteria belonging to the various taxonomic groups of the Eubacteria, Actinomycetales, Myxobacteria and Blue-green algae (Myxophyceae). As already indicated, in the PPLO and halophilic species the 'wall' functions are adequately met by a structure with the dimensions and properties of a 'unit' or 'compound' type of membrane. The nature of the surface component surrounding spirochaetes has not been investigated but there is strong evidence from electron microscopy (VAN ITERSON, 1954) and from the penicillin sensitivity of *Treponema pallidum* and the immobilizing action of lysozyme that some members of the group possess similar components in the envelope to those found in other classes of bacteria.

Examination of a variety of isolated cell walls in the electron microscope abundantly confirmed the early impressions that it was indeed the cell-wall structure which was responsible for the typical morphology of a particular organism. Thus walls of *Staphylococcus aureus* were typically spherical (DAWSON, 1949), those of *Streptococcus faecalis* were ellipsoidal (SALTON AND HORNE, 1951), walls of *Bacillus megaterium* and other rod-shaped bacteria were of cylindrical appearance (SALTON, 1953) and even spirilla retained the helicoidal shape in wall or membrane preparations. Moreover, the isolated walls of branching organisms such as *Streptomyces fradiae* and *Nocardia* spp. were also seen to possess branched wall structures (ROMANO AND NICKERSON, 1956; ROMANO AND SOHLER, 1956). The characteristic shapes of shadowed preparations of walls isolated from several bacterial species are shown in the electron micrographs presented in Fig. 21. The walls of the Gram-positive organisms (streptococci, staphylococci and bacilli) have a homogeneous appearance when viewed in the electron microscope and appear to be devoid of any fine structure. However, the walls of streptococci and staphy-

The tables are printed together at the end of the book.

lococci do show thickened bands which correspond to the areas of new cell-wall formation during cell division. As might be expected, these bands representing the division septa are at right angles to the long axis of the cell in *Streptococcus faecalis*. In organisms capable of dividing in two planes (e.g. *Micrococcus lysodeikticus* and *Sarcina lutea*), the thickened bands visible on isolated walls are frequently found at right angles to one another. All of these observations suggest that the wall is laid down in specific areas rather than uniformly over the entire cell-wall surface.

The isolated cell walls are usually viewed for electron miscroscopy by drying down sus-

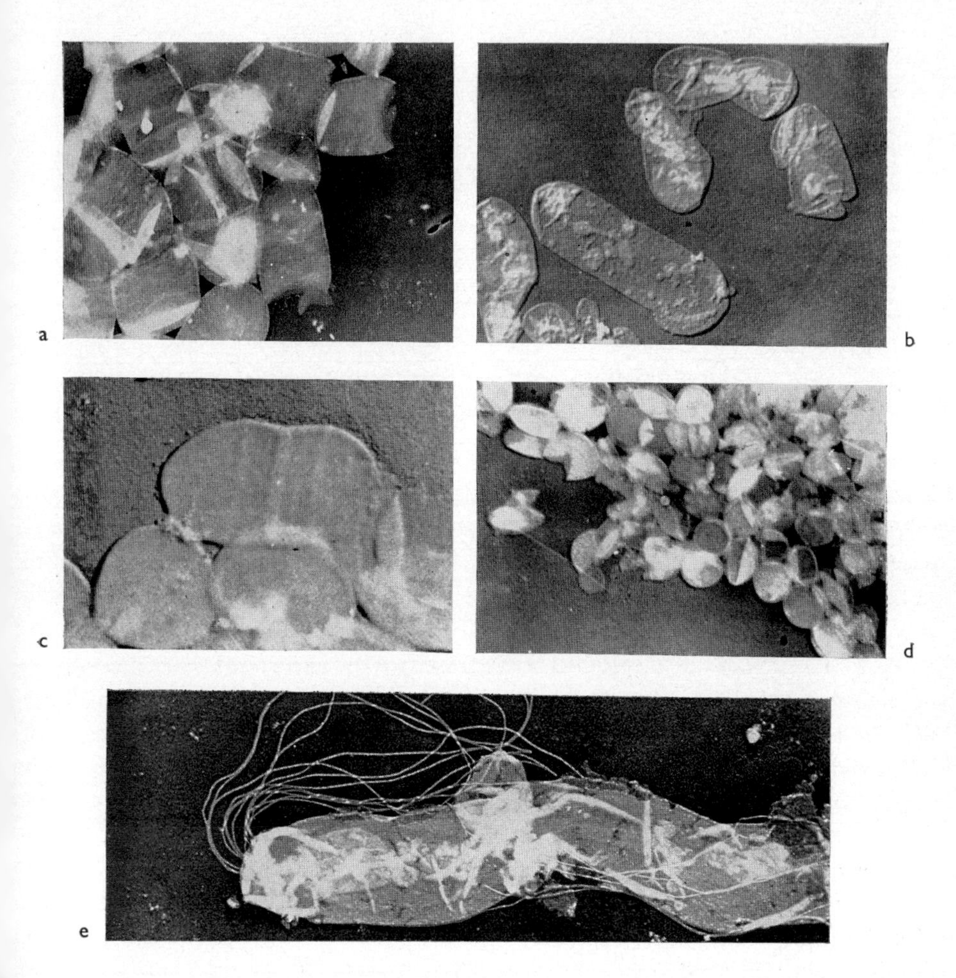

FIGURE 21. Electron micrographs illustrating the characteristic shapes of wall and envelope fractions isolated from bacteria.

(a) *Bacillus megaterium*, (b) *Escherichia coli*; cocci (c) *Streptococcus faecalis*, (d) *Staphylococcus aureus* and (e) the helicoidal-shaped *Spirillum serpens*. Preparations (a) to (d) are wall fractions from disintegrated cells and (e) is an autolysed, trypsin-digested envelope fraction.

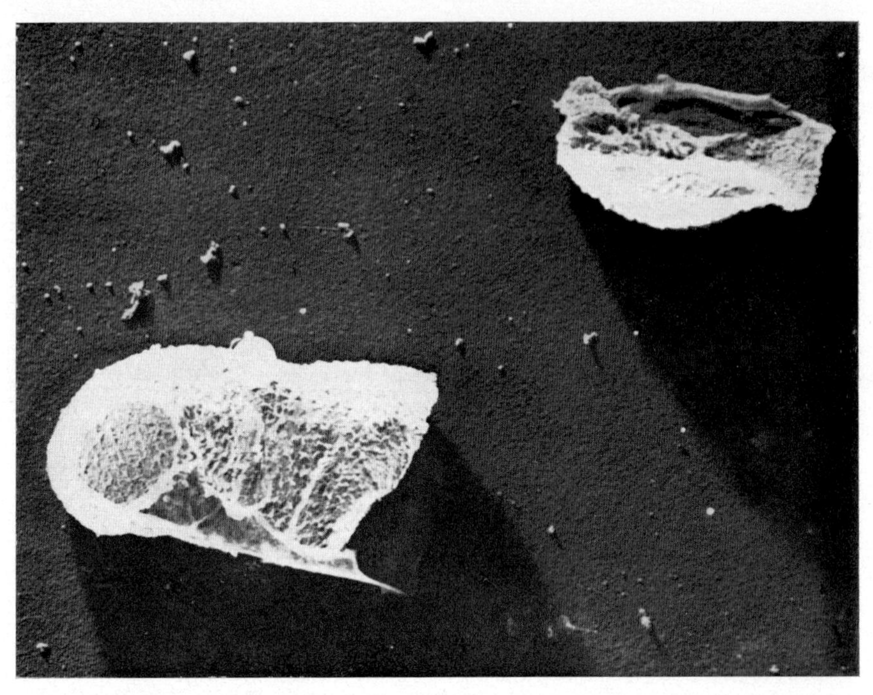

FIGURE 22. Isolated cell walls of *Bacillus megaterium* showing three-dimensional structure when prepared for electron microscopy by freeze-drying, $\times$12,750. (SALTON AND WILLIAMS, 1954).

pensions on the supporting films covering the electron microscope grids. As a result of the surface tension forces exerted during the drying down of droplets of suspension on grids for electron microscopy, the walls appear as flat, collapsed structures. When the walls are prepared for electron microscopy by the method of WILLIAMS (1953) which preserved the three-dimensional structure of biological materials, there is no doubt that the wall is a rigid three-dimensional structure as shown in Fig. 22.

The width and length of the isolated walls as measured from air dried specimens examined in the electron microscope are generally greater than those of whole cells, as would be expected from a certain amount of stretching of the empty walls during drying on the grids. It seems extremely unlikely however, that cell walls possess the elasticity claimed by KNAYSI, HILLIER AND FABRICANT (1950) for the wall of *Mycobacterium tuberculosis*. Examination of isolated walls of strains of *Mycobacterium tuberculosis* by other writers have not confirmed the suggestion that they can be stretched to the extent reported by KNAYSI, HILLIER AND FABRICANT (1950).

Thickness of the cell wall

Until the methods for preparing thin sections of bacteria were perfected, the thickness of the wall was estimated by measuring the length of the shadow cast by the collapsed walls on electron micrographs. DAWSON (1949) estimated the wall thickness of *Staphylococcus aureus* to be 150–200 Å. Similar values (200 Å) were obtained for the thickness of the wall of *Streptococcus faecalis* isolated in the studies of SALTON AND HORNE (1951). The cell wall of *Mycobacterium tuberculosis* on the other hand was estimated to be about 350 Å thick by KNAYSI, HILLIER AND FABRICANT (1950). It is of course obvious that thickness measurements based on the shadow length of walls examined in the electron microscope may be subject to many errors. Factors such as the degree of flatness of the supporting film, thickness of metal shadowing and contaminating materials would contribute to the errors of these measurements.

With the perfection of fixation, embedding and sectioning procedures and their application to bacteria first by CHAPMAN AND HILLIER (1953) and later by KELLENBERGER, RYTER AND SECHAUD (1958), measurements of the thickness of the wall could be made directly on the thin sections. The walls of many species have a fairly uniform thickness except perhaps where the cross-wall or new wall is being formed (e.g. *Micrococcus lysodeikticus*). Thin sectioning has not only provided us with more accurate measurements of wall thickness but also given most valuable information about the multilayered nature of some bacterial walls or surface envelope structures. On the basis of thin sectioning, bacterial cell walls may vary in thickness from about 80 Å for the wall of *Escherichia coli* (KELLENBERGER AND RYTER, 1958) to as much as 800 Å for the wall of *Lactobacillus acidophilus* (GLAUERT, 1962).

A selection of data based on measurements from shadowed preparations of isolated walls and from thin sections is presented in Table 11. Even though the determination of thickness from thin sections may be more accurate than from shadowed wall preparations, it is obvious that the values cannot be regarded as absolute for any given organism as the wall thickness may be appreciably affected by growth conditions and by fixation and embedding procedures.

Fine structure and anatomy of isolated cell walls

The first bacterial cell walls to be examined in the electron microscope as partly fractionated and homogeneous preparations were those isolated respectively from the Gram-positive organisms *Staphylococcus aureus* (DAWSON, 1949) and *Streptococcus faecalis* and from the two Gram-negative bacteria *Escherichia coli* and *Salmonella pullorum* (SALTON AND HORNE, 1951). None of these isolated walls showed any fine structure when examined in the electron microscope and apart from the thickened equatorial bands at the zone of division already mentioned above, the walls possessed a homogeneous, amorphous appearance.

HOUWINK (1953) was the first to demonstrate the presence in the wall of a *Spirillum* sp. a type of fine structure which had not been encountered in walls of other microorganisms. Cell walls of yeast and fungi generally possess microfibrillar components in their walls and the typical appearance of such a structure is illustrated in Fig. 23 being the shadowed preparation of isolated mycelial walls of *Neurospora crassa* examined by CHAPMAN AND SALTON (see SALTON, 1960). The microfibrils are frequently embedded in an amorphous matrix in the walls of yeasts, fungi and algae and in order to reveal the fine structure it is often necessary to subject the walls to chemical treatment. However, no prior chemical treatment was necessary

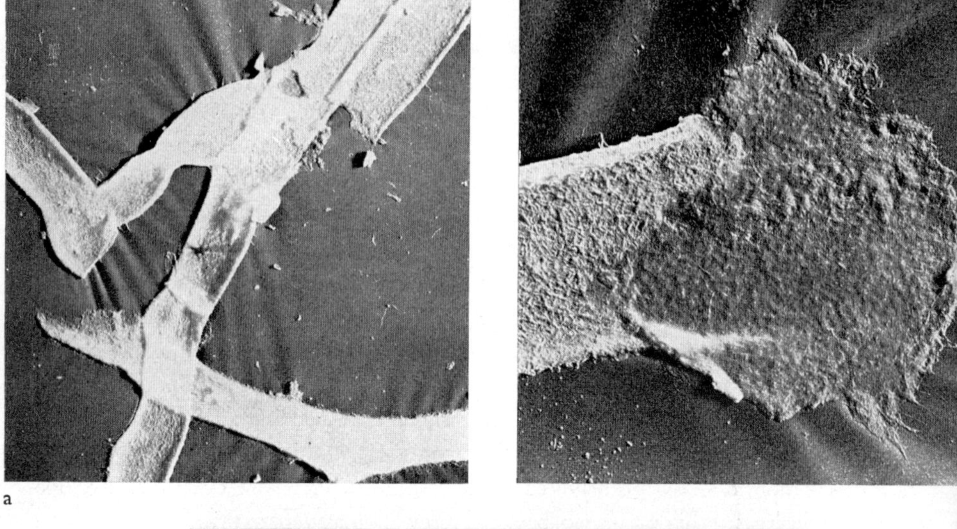

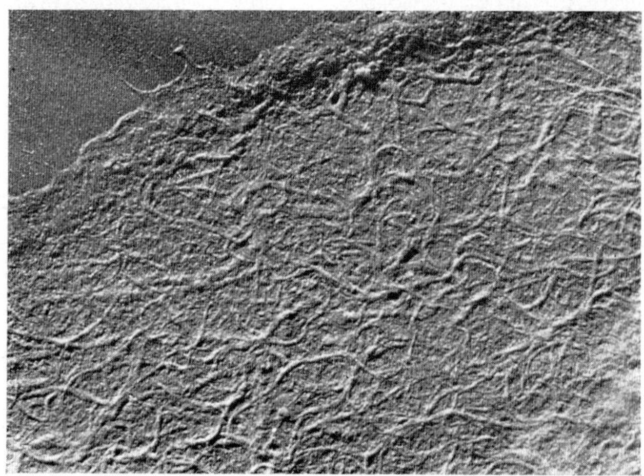

FIGURE 23. Electron micrographs of *Neurospora crassa* walls (a) × 3,000; the microfibrillar structure and difference in texture of outer and inner wall surfaces being clearly shown in (b) × 9,000. The fibrillar structure is further revealed by treatment with 0.5% NaOH for 5 minutes (c) × 18,750. (CHAPMAN AND SALTON, unpublished observations).

→

FIGURE 24. Cell-wall fragment isolated from *Spirillum* sp. showing a layer of hexagonally packed spherical subunits, × 40,000. (HOUWINK, 1953).

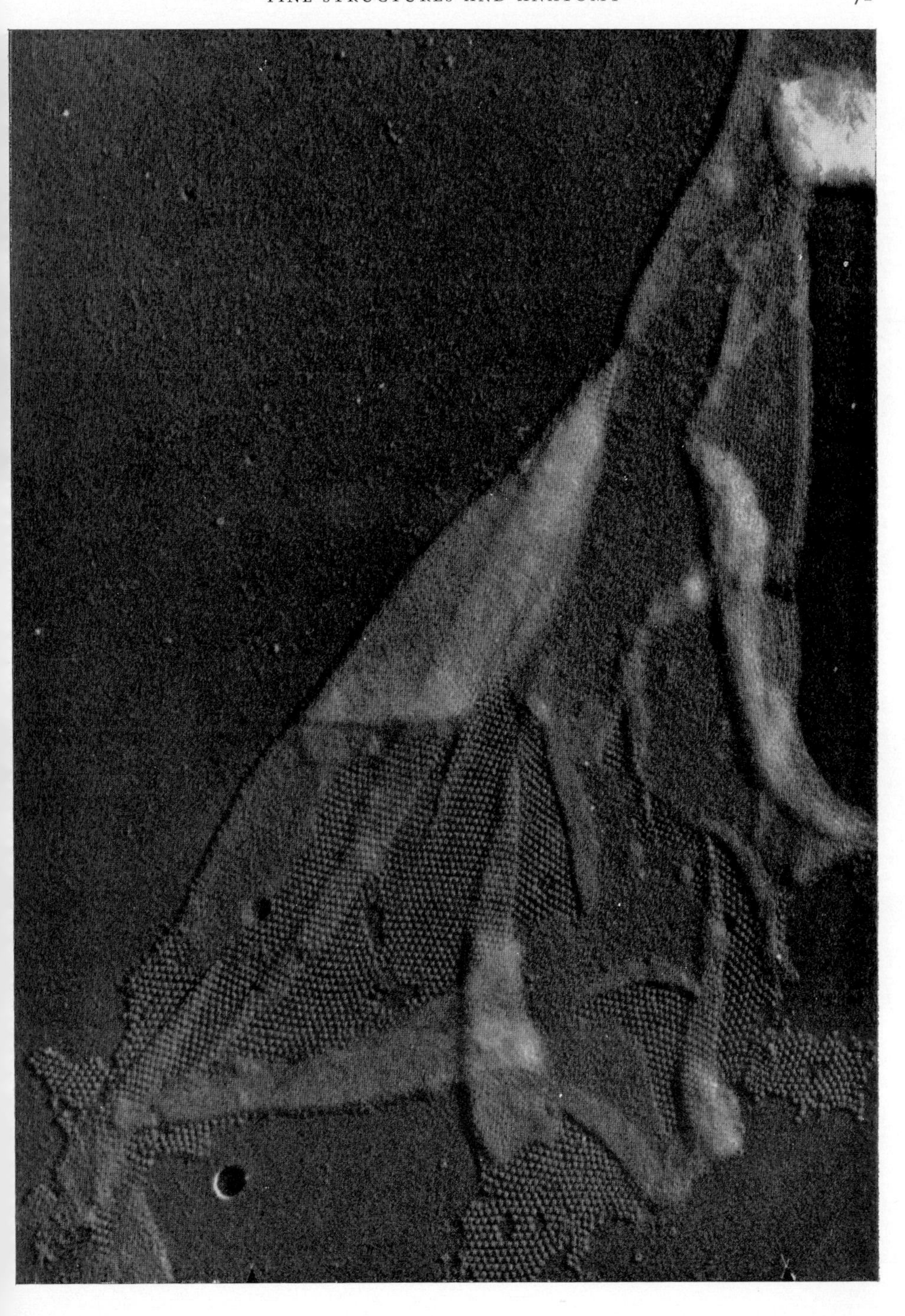

for the detection of the characteristic microfibrillar structure in the walls of *Neurospora crassa* shown in Fig. 23. The wall of the *Spirillum* sp. examined by HOUWINK (1953) was a multilayered structure, one of the layers of which was composed of hexagonally packed, spherical macromolecules measuring about 120 Å in diameter. The microfibrillar type of fine structure is contrasted with that found in the *Spirillum* sp. as shown in Fig. 24.

The fine structure in the walls possessing spherical macromolecules could be seen in the intact cells of the *Spirillum* sp. and also *Halobacterium halobium* (HOUWINK, 1956). As shown in Fig. 25 the macromolecular structure of the wall is readily seen in intact cells of *Halobacterium halobium* and *Selenomonas palpitans* (VAN ITERSON, 1954) examined in the electron microscope.

Hexagonally packed macromolecules of sizes smaller than those observed in the *Spirillum* sp. examined by HOUWINK (1953) have also been seen in related organisms. The macromolecules in the wall of *Spirillum serpens* were about 80 Å in diameter (SALTON, 1956) while those of *Rhodospirillum rubrum* were 100 Å (SALTON AND WILLIAMS, 1954). The walls or envelopes of these organisms were apparently multilayered. The fine structure in the isolated wall fractions from autolysed, trypsin treated *Spirillum serpens* was detected in unshadowed preparations as shown in Fig. 26a. This preparation was examined by HORNE AND SALTON prior to the introduction of the negative staining technique with phosphotungstic acid and is contrasted with the complex layers now revealed by the latter method in recent studies by MURRAY (1963a) in Fig. 26b, c. The spherical macromolecules in the wall of *Rhodospirillum rubrum* are illustrated in the shadowed preparation in Fig. 27. It is evident that what appeared to be the 'spherical macromolecules' in earlier preparations are really subunits of a more complicated layer.

Although no well-defined microfibrillar structure of the type so clearly shown in the cell wall of *Neurospora crassa* (Fig. 23) has been found in walls of bacteria, there is some suggestion of a fibrous appearance in the wall of *Bacillus megaterium* (Fig. 28). There is little doubt that the outer wall surface of this organism has a rough texture as shown in Fig. 22. A different kind of microstructure was observed in the wall of an unidentified organism studied by LABAW AND MOSLEY (1954). Macromolecules (100 Å periodicity) were arranged in a rectangular fashion in the wall. Yet another kind of fine structure is that observed for one of the layers of the complex surface structure of the organism *Lampropedia hyalina*. As suggested in Chapter 1, this fine structured layer may not be part of the cell wall proper, but may represent an integument or wall surrounding a 'colony' of cells. The appearance of the 'wall' fraction isolated from *Lampropedia hyalina* is shown in Fig. 29 and illustrates the complexity of the surface envelope of this organism. The structure of the outer layer of the envelope of *Lampropedia hyalina* is complex and Fig. 30 shows the appearance of negatively stained (PTA) fragments of the layer and the interpretation of the fine structure is illustrated diagrammatically in Fig. 31 (from CHAPMAN, MURRAY AND SALTON, 1963).

An extremely interesting type of structure covering the wall or cell envelope of a *Mycobac-*

FIGURE 25. Macromolecular components in the wall or surface envelope of intact cells of *Halobacterium halobium* (a) × 44,000 prepared by Drs. HOUWINK, MOHR AND SPIT and *Selenomonas palpitans* (b) × 35,000 examined by VAN ITERSON, 1954.

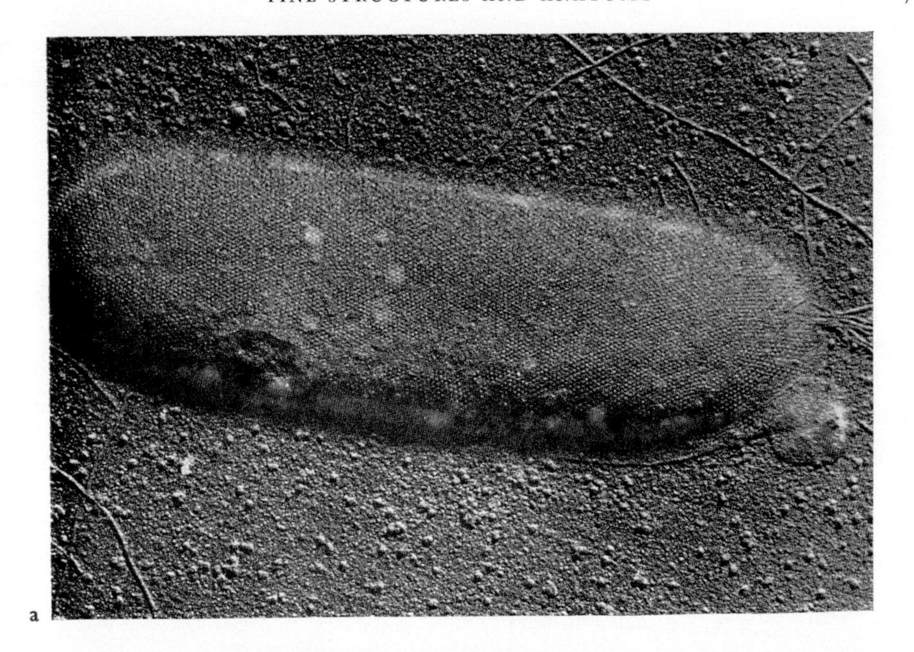

a

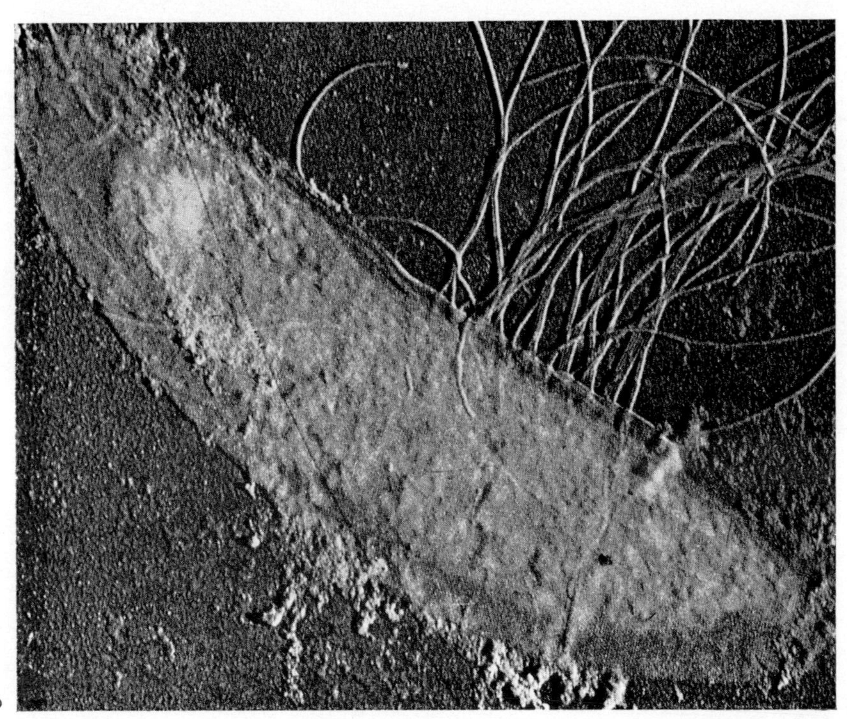

b

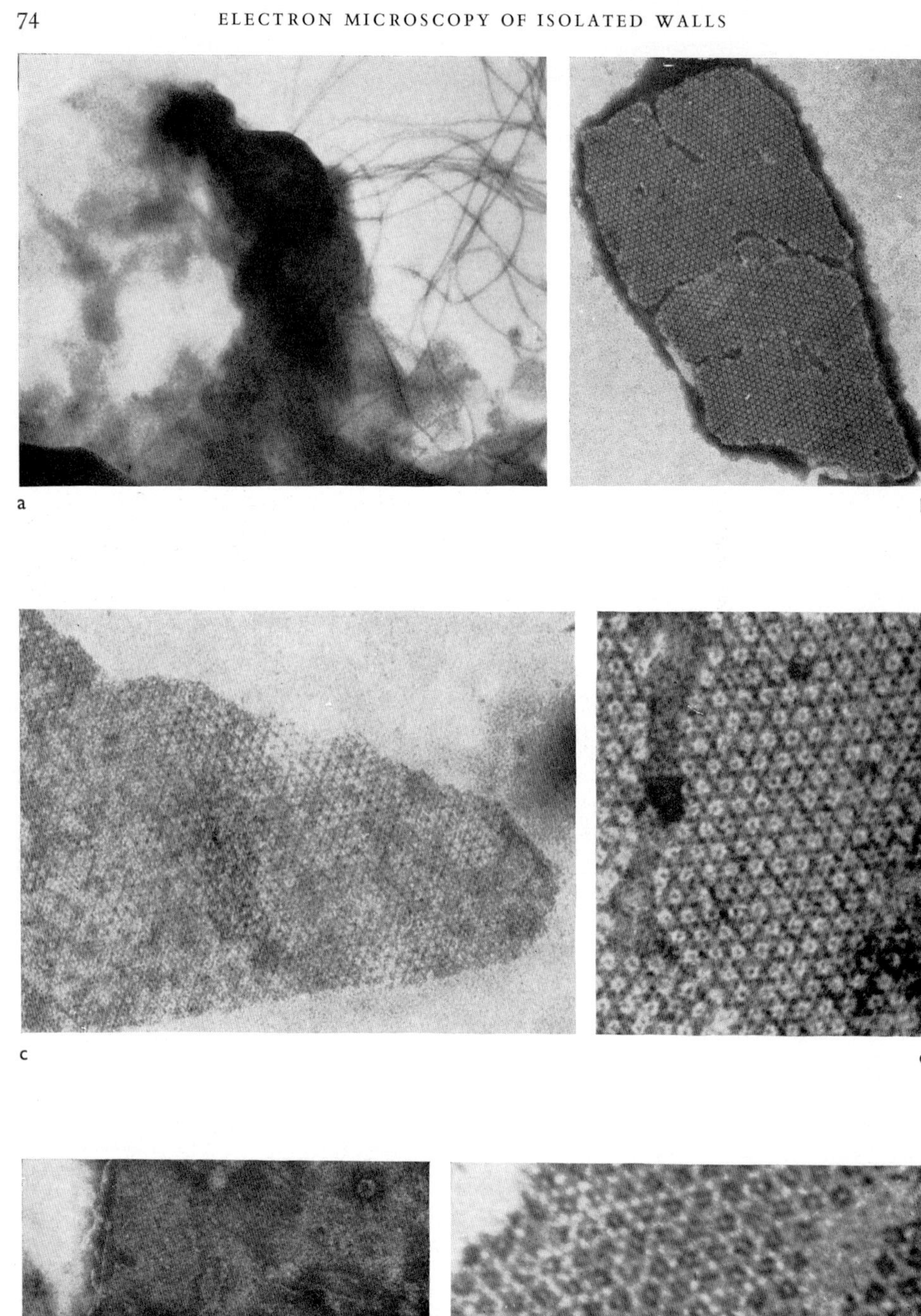
a
b
c
d
e
f

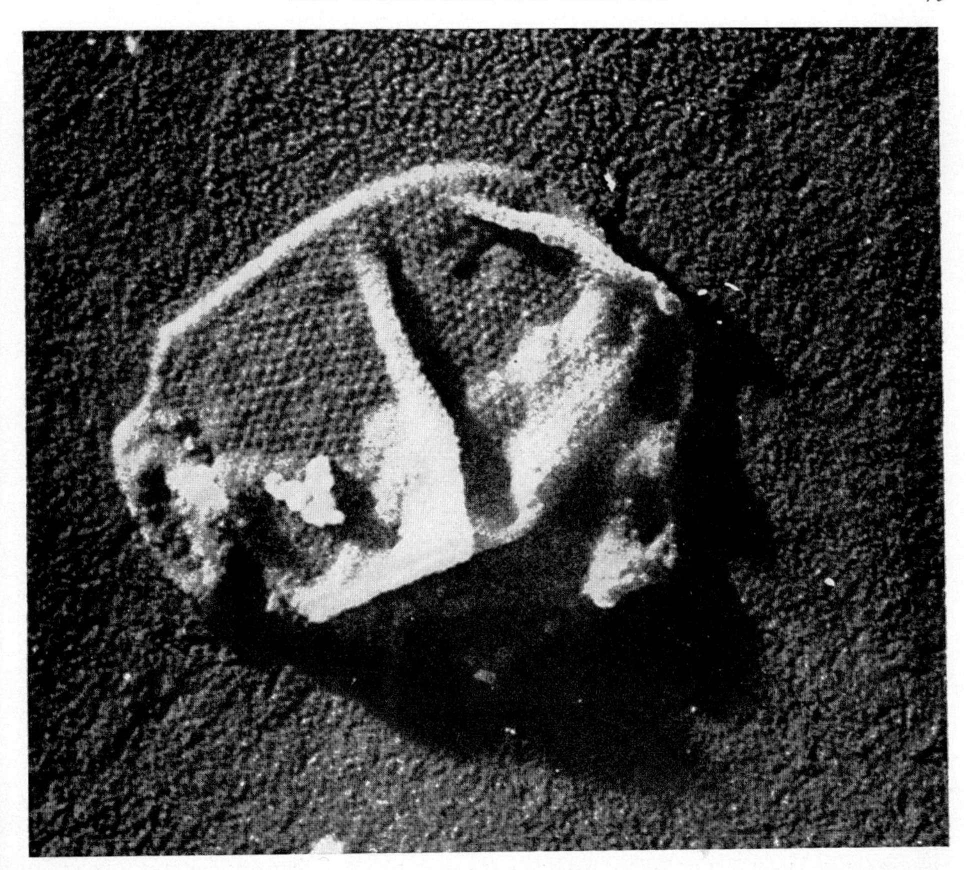

FIGURE 27. Isolated cell wall of *Rhodospirillum rubrum* showing the presence of macromolecular components, ×100,000. (SALTON AND WILLIAMS, 1954).

← FIGURE 26. The structure of *Spirillum serpens* cell wall.

(a) Envelope preparation from autolysed cells showing fine structure, unpublished observation, SALTON AND HORNE. ×75,000.

(b) Cell-wall fragment from cells disrupted by ultra-sound; phosphotungstic acid (PTA) preparation showing regular packing of the polygonal structure and underlying component, ×55,000.

(c) Cell wall fragment extracted with lauryl sulphate in the presence of PTA revealing the linked meshwork, ×102,000.

(d) Micrograph showing 'Y-linkage' of units.

(e) Cell fragments embedded in PTA after exposure to lauryl sulphate, ×55,000.

f) High magnification of cell-wall fragment with contrast reversed by an intermediate negative – the wall structure resolved in this case is shown as dark portions.

(Preparations b–f, MURRAY, 1963).

FIGURE 28. Isolated cell wall of *Bacillus megaterium;* no well-defined fine structure detectable but wall has a somewhat 'fibrous' appearance, ×32,000. (SALTON AND WILLIAMS, 1954).

terium species was observed by TAKEYA, MORI, TOKUNAGA, KOIKE AND HISATSUNA (1961). The outer wall surface of *Mycobacterium* Jucho strain is covered with paired fibrous structures as shown in Fig. 32. These structures could be removed from the wall by extraction with KOH – alcohol, a treatment which CUMMINS AND HARRIS (1958) also found necessary for the isolation of clean walls of mycobacteria. The fibrous surface structures observed by TAKEYA *et al.* (1961) were not removed by trypsin digestion. The appearance of the wall of *Mycobacterium* Jucho strain is contrasted in Fig. 33 with the walls of *Mycobacterium tuberculosis* (BCG strain) isolated after disintegration in the pressure cell apparatus described by RIBI, PERRINE, LIST, BROWN AND GOODE (1959). Striated bands running along the length of the cells of a *Spirillum* sp. were reported by BISSET (1955) although the appearance differs somewhat from those observed on the surface of the *Mycobacterium* sp. It would be of interest to know to what extent they are part of the cell wall.

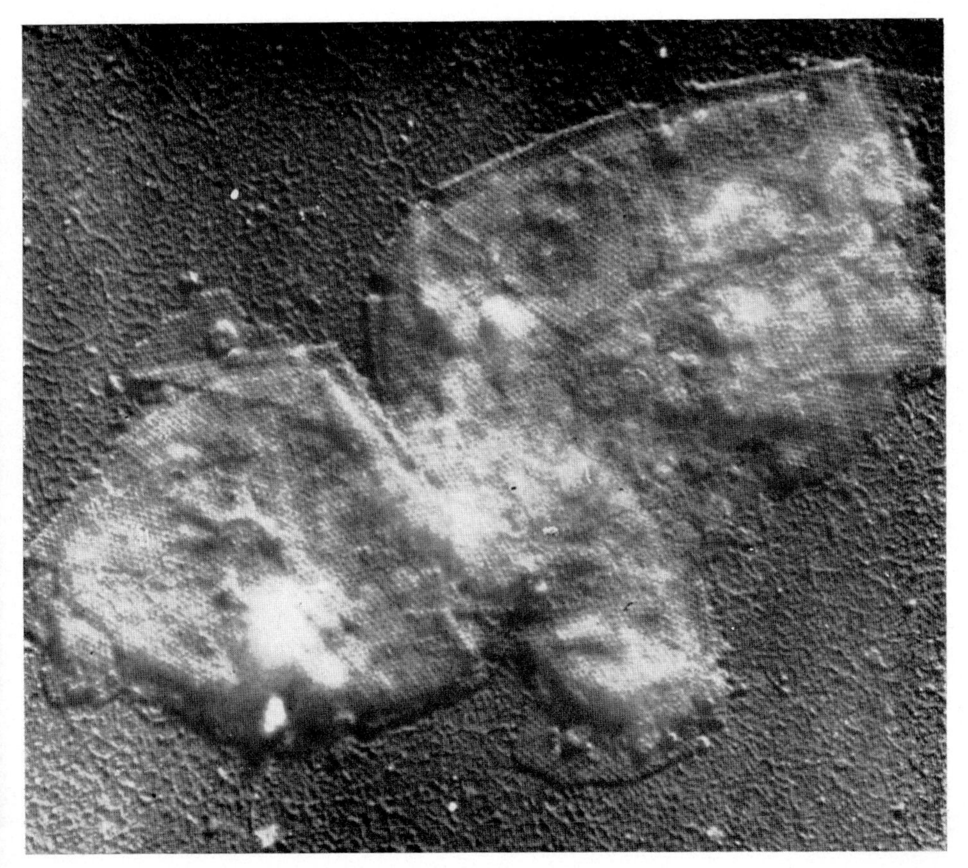

FIGURE 29. 'Wall fraction' prepared from *Lampropedia hyalina* cells disintegrated in the Micke shaker shows the complex multilayered and structured envelope and residual electron-dense material, $\times$ 62,800.

It has been of considerable interest to see whether the fine structures found in shadowed preparations of isolated walls would be apparent in thin sections of the organisms. The fine structure in the surface envelope of *Lampropedia hyalina* has been detected also in thin sections of this organism (MURRAY, 1963b; CHAPMAN, MURRAY AND SALTON, 1963). The microfibrillar components in the wall of *Neurospora crassa* were also seen in thin sections published by SHATKIN AND TATUM (1959). However, thin sections of organisms possessing the spherical macromolecular components in the wall or envelope have not yet shown the presence of these subunits. Thus the outer components of the envelopes of *Spirillum serpens*, *Rhodospirillum rubrum* and *Halobacterium halobium* examined in thin sections by MURRAY (1962), HICKMAN AND FRENKEL (1959) and BROWN AND SHOREY (1962) respectively, have had the typical profile of 'unit' or 'compound membranes' with no resolution into layers of spherical macromolecules. The profiles of the outer walls or membranes of these organisms have been indistin-

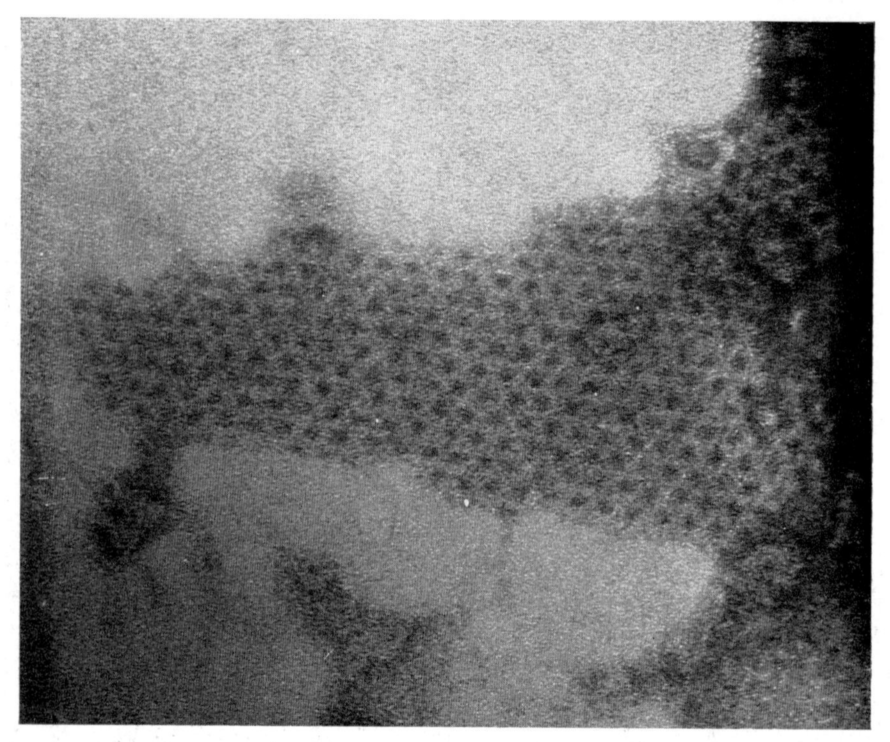

FIGURE 30. Isolated fragment of the surface envelope of *Lampropedia hyalina* examined by the phosphotungstate negative staining technique. This layer has the appearance of a 'honeycomb network' with a nearest neighbour repeat distance of 130–140Å, the holes being 75–80Å in size; × 323,300. (CHAPMAN, MURRAY AND SALTON, 1963).

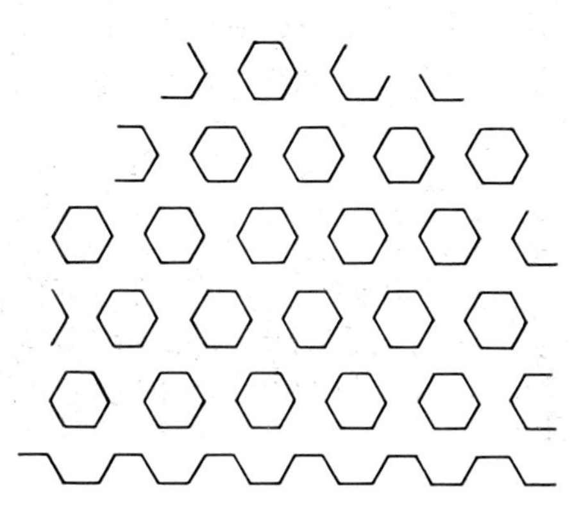

FIGURE 31. A diagram of the interpretation of the fine-structure of the outer envelope layer of *Lampropedia hyalina* based on the appearance of fragments negatively stained with phosphotungstate as shown in Fig. 30 (CHAPMAN, MURRAY AND SALTON, 1963).

FIGURE 32. Ghost cells of *Mycobacterium* Jucho strain showing the presence of paired, fibrous structures on the cell envelope surface, × 52,000. (TAKEYA *et al.*, 1961).

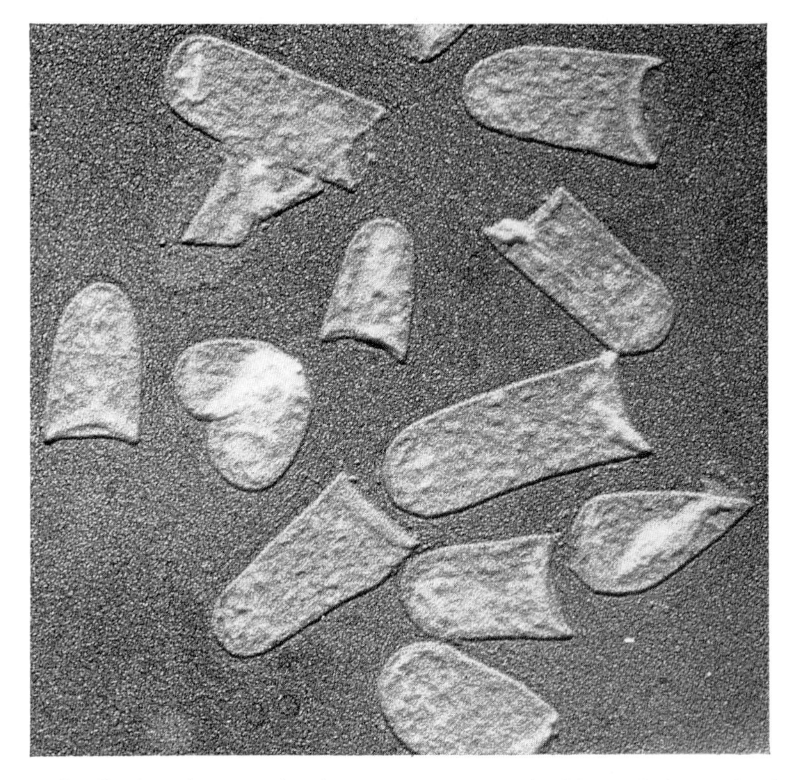

FIGURE 33. Cell walls of *Mycobacterium tuberculosis* (BCG), × 32,500, isolated from cells disrupted in the pressure cell used by RIBI *et al.*, 1959.

guishable from that of *Escherichia coli* (KELLENBERGER AND RYTER, 1958) which is devoid of a hexagonally packed layer of 'spheres'. However, MURRAY (1963*a*) has now obtained sections of cell-wall fragments of *Spirillum serpens* which have shown some resolution of the subunits in the outer structured layer of the wall.

The cell-wall profiles seen in thin sections of Gram-positive bacteria are not usually multi-layered and their rather uniform appearance agrees well with the amorphous, homogeneous cell-wall structures observed in shadowed preparations examined in the electron miroscope. Although there is a tendency for electron-dense material to accumulate at the interfaces of the walls of fixed, stained and embedded cells, it should not be assumed that this effect is due to chemically different layers. Where Gram-positive bacteria have shown differences in electron-density of the sectioned walls, the layering effect is rarely as clear-cut or sharp as that seen in the walls or membranes of Gram-negative bacteria. The difficulty encountered in correlating the chemical components of the wall with appearance in thin sections or shadowed preparations was further emphasized when ARCHIBALD, ARMSTRONG, BADDILEY AND HAY (1961) failed to detect any differences after extraction of the teichoic acid from walls of *Lactobacillus arabinosus*. This result was particularly surprising in view of the large contribution of the teichoic acid and soluble polysaccharide fractions to the weight of the wall (about 50% for *Lactobacillus arabinosus*).

However, it is probable that walls of some Gram-positive bacteria may have their different polymeric constituents (e.g. glycosaminopeptides, polysaccharides, teichoic acids) arranged in layers and sections of *Mycobacterium* Jucho strain illustrated in Fig. 34 do show a convincing multilayered cell-wall structure well differentiated from the underlying plasma membrane. The walls of mycobacteria are probably more complex chemically than those of other Gram-positive species and the presence of lipopolysaccharide-peptide compounds may contribute to the multilayered appearance in thin sections. As already pointed out, methods used for obtaining 'clean' cell walls of mycobacteria (CUMMINS AND HARRIS, 1958) probably involve the removal of the hydrophobic compounds from the wall surface.

Attempts to correlate the various chemical entities of the wall of *Escherichia coli* with the macromolecular structure have been made with considerable success by WEIDEL and his colleagues. WEIDEL, FRANK AND MARTIN (1960) isolated a rigid layer from the wall of *Escherichia coli*, strain B, which on treatment with egg-white lysozyme lost its structural rigidity and broke up into small macromolecules. They suggested that this rigid mucopeptide component represented the inner layer of the complex wall or envelope structure of *Escherichia coli* and that the outermost layer is composed of the more plastic protein-lipid-polysaccharide portion of the wall. The existence of a layer conferring structural rigidity upon the walls of a Gram-negative bacterium has thus been well established for *Escherichia coli* as illustrated by the elec-

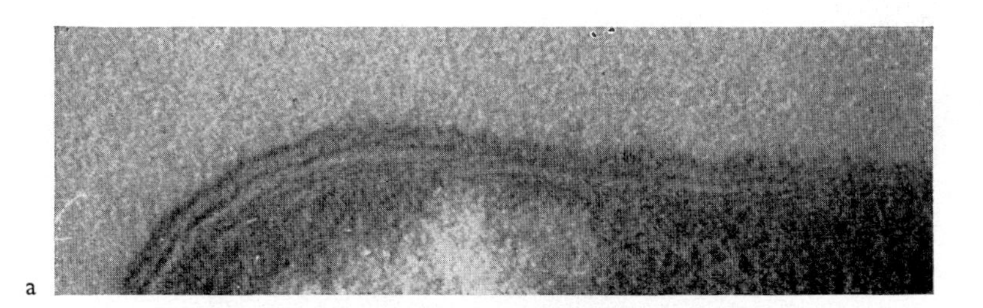

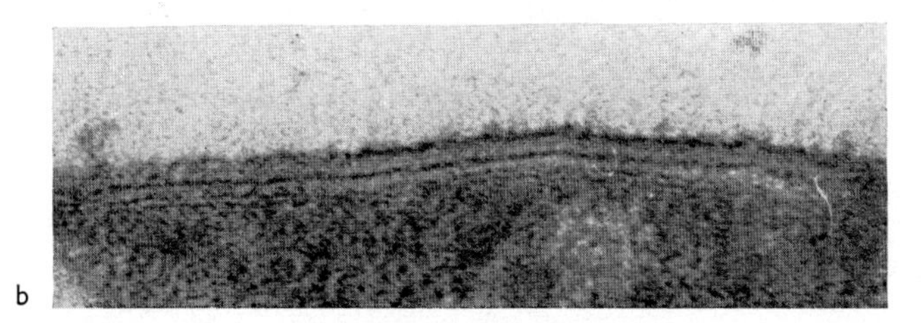

FIGURE 34. Thin sections of *Mycobacterium* strain Jucho showing the multilayered appearance of the wall together with the underlying membrane. (a) ×145,000; (b) ×120,000. Preparations by K. TAKEYA AND M. KOIKE.

tron micrographs in Fig. 35 taken from the studies of WEIDEL, FRANK AND MARTIN (1960). From these investigations, studies of the action of lysozyme and EDTA (ethylenediaminetetraacetic acid) and the induction of spheroplast formation by penicillin, MARTIN (1963) has suggested a model for the anatomy of the cell wall or compound membrane of *Escherichia coli* and this is illustrated in Fig. 36.

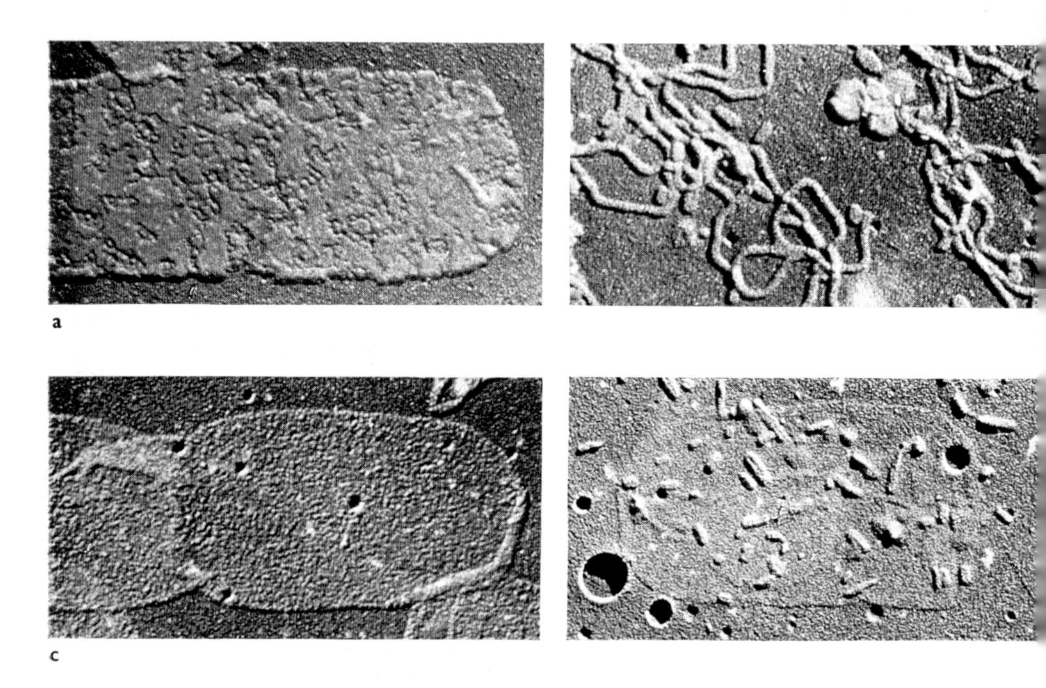

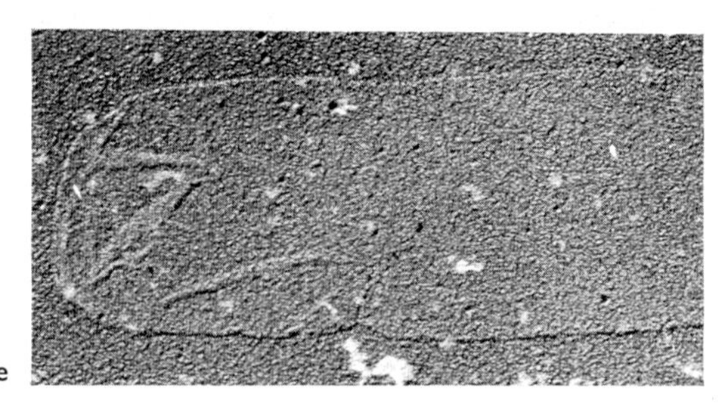

FIGURE 35. The structure of *Escherichia coli* B cell wall.
(a) Walls after treatment in the Mickle disintegrator with dodecylsulphate, × 37,300.
(b) Aggregates of wall lipopolysaccharide resulting from treatment of walls with hot phenol/water, × 36,600.
(c) Wall after phenol treatment, × 36,600.
(d) Wall on digestion with trypsin, × 36,600.
(e) The protein-free rigid layer sub-structure of the wall, × 36,600. (MARTIN AND FRANK, 1962).

MARTIN AND FRANK (1962) have extended their studies to the physico-chemical structure of walls of other Gram-negative bacteria and the separation of the macromolecular spherical subunits from the rigid layer of the *Spirillum* sp. originally studied by HOUWINK (1953) has been elegantly achieved as demonstrated by the electron micrographs in Fig. 37.

From these observations it seems likely that the multilayered appearance in the electron microscope of the walls of Gram-negative bacteria is due to the presence of chemically discreet layers, one of which confers structural rigidity upon the cell and can be isolated as a rigid mucopeptide component. To what extent all Gram-negative bacteria possess a well defined mucopeptide layer cannot be said at the moment. At least this component is not present in the surface membranes of *Mycoplasma* spp. and *Halobacterium* spp. as indicated in Chapter 1.

On the basis of the disaggregation of isolated cell walls by the anionic detergent, sodium

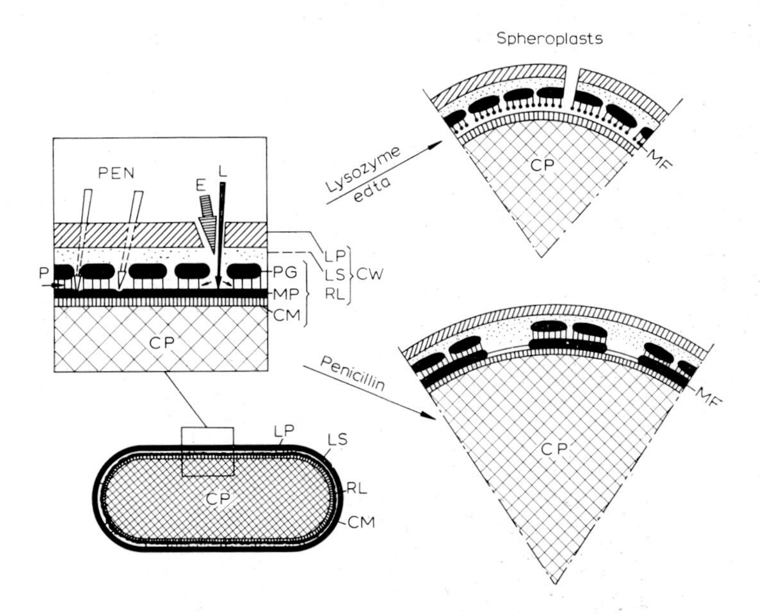

FIGURE 36. A diagrammatic representation of a Gram-negative bacterium with a complex, triple-layered cel wall. Separation of the layers can be achieved by solvent extraction and by treatment with proteolytic enzyme (P). Both lysozyme (L), aided by EDTA (E), and penicillin induce depolymerization of the rigid mucopolymer (MP), although probably to a different degree and in a different way. Only small mucopolymer fragments (MF), which are covalently linked to other cell wall components, remain in the wall after lysozyme treatment. Penicillin spheroplasts and the related L-forms may retain larger mucopolymer fragments or a modified, non-rigid mucopolymer in their stretched cell walls which surround protoplasts of greatly increased dimensions. Abbreviations used: LP = lipoprotein layer, LS = lipopolysaccharide layer, RL = rigid layer; PG = protein granula, MP = mucopolymer, CM = cytoplasmic membrane, CP = cytoplasm, MF = mucopolymer fragments, PEN = penicillin, L = lysozyme, E = ethylenediamine-tetraacetic acid, P = proteolytic enzyme. (Schematic model devised by MARTIN, 1963).

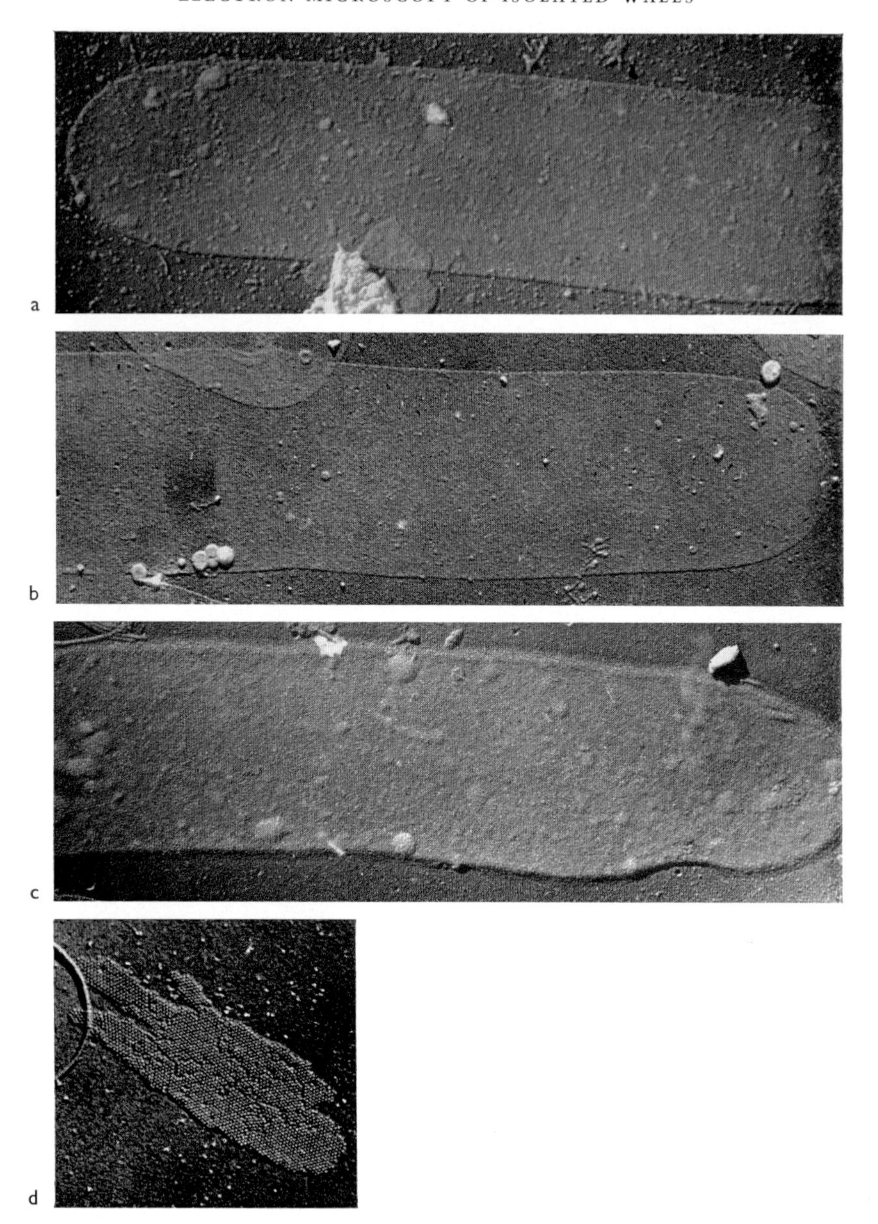

FIGURE 37. Structure of the wall of *Spirillum* sp.

(a) Wall isolated by disintegration of log-phase cells in the Mickle apparatus with dodecylsulphate.

(b) Rigid layer preparation from wall of log-phase cells.

(c) Wall isolated from stationary-phase cells.

(d) Macromolecular components from stationary-phase cells. All preparations ×21,500. From MARTIN AND FRANK.

dodecyl sulphate (SDS), SHAFA AND SALTON (1960) suggested that the mucopeptide component of the walls of some Gram-negative bacteria may be present as a reinforcing network, rather than a separate rigid layer. This suggestion seemed to be in accord with the change in shape of isolated walls of *Rhodospirillum rubrum* after treatment with egg-white lysozyme as shown in Fig. 38. No change in the appearance of the surface of the walls could be seen. However, if the rigid layer had been thin and on the inside rather than a network of mucopeptide then little alteration would have been expected to be visible from the outside surface. With the isolation of rigid layers from walls of several Gram-negative bacteria (*E. coli*, *Salmonella gallinarum* and *Spirillum* sp.) the existence of the network type of structure now seems unlikely. The failure of SHAFA AND SALTON (1960) to detect a rigid structure on disaggregation of walls with SDS puzzled these authors at the time and may have been due to the autolytic enzymes being active on the mucopeptide layer during the isolation of walls. Recent results obtained by WEIDEL, FRANK AND LEUTGER (1963) suggest that the activity of autolytic enzymes can prevent the successful isolation of the rigid layer of *Salmonella gallinarum* walls, unless special attempts are made to inhibit these enzymes. It will be recalled that the presence of such enzymes in Gram-negative bacteria was suggested by SALTON AND SHAFA (1958) from the structural and chemical changes occurring in the organism *Vibrio metchnikovi* (see also Chapter 1).

In contrast to the results of the investigations of WEIDEL, FRANK AND MARTIN (1960) and MARTIN AND FRANK (1962), BROWN, DRUMMOND AND NORTH (1962) have suggested that both external and internal multilayered 'wall' or membrane components of a marine pseudomonad contained the typical compounds of the mucopeptide. Until the purified mucopeptide fractions from a wide variety of Gram-negative bacteria have been examined in the electron microscope and their structural features determined, it will not be possible to reach a conclusion as to whether or not the majority of Gram-negative groups of bacteria possess a rigid layer. It is however, of considerable interest to note that MANDELSTAM (1962) isolated mucopeptide fractions from six different groups of Gram-negative bacteria and found that the relative amounts of the chemical constituents were the same.

SCHOCHER, BAYLEY AND WATSON (1962) purified the mucopeptide from the walls of *Aerobacter cloacae* and examined these fractions for their chemical and physical properties. Cell walls were prepared by disintegrating the bacteria with Ballotini glass beads and separating them from the rest of the cell constituents by differential centrifugation and enzyme digestion. The isolated walls were exhaustively treated with saturated aqueous sodium lauryl sulphate (sodium dodecyl sulphate) and after repeated washings on the centrifuge a highly purified mucopeptide was obtained. The mucopeptide fraction so isolated was examined in the electron microscope by SCHOCHER (unpublished observations) and the appearance of the purified material is shown in Fig. 39.

After the isolated mucopeptide from the walls of *Aerobacter cloacae* was subjected to a 10–15 minute period of shaking in the Mickle apparatus (without glass beads) the suspension was centrifuged and the supernatant fraction was examined in the electron microscope. As

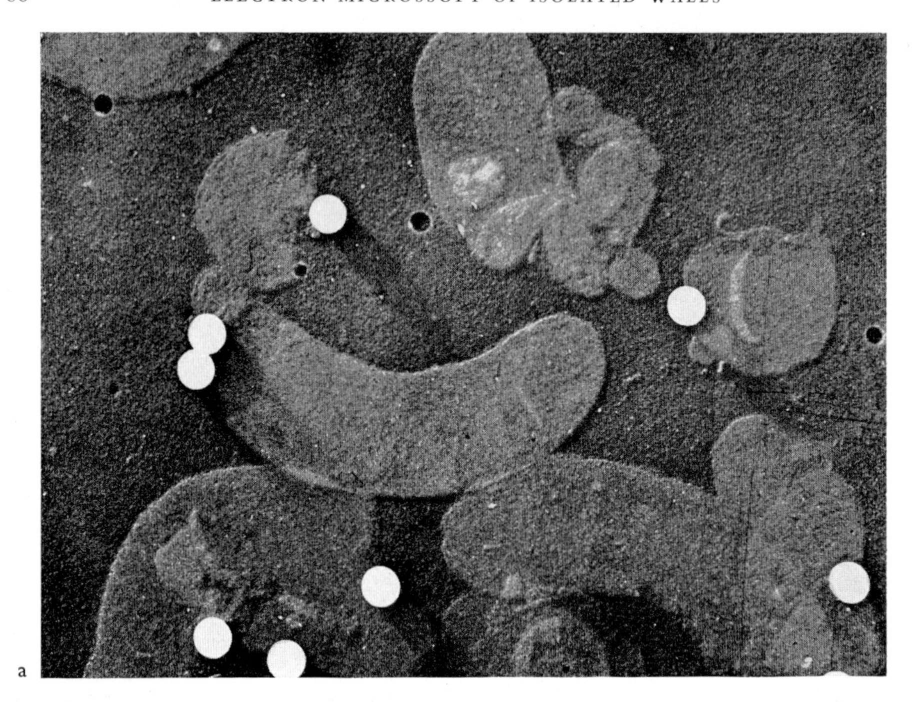

a

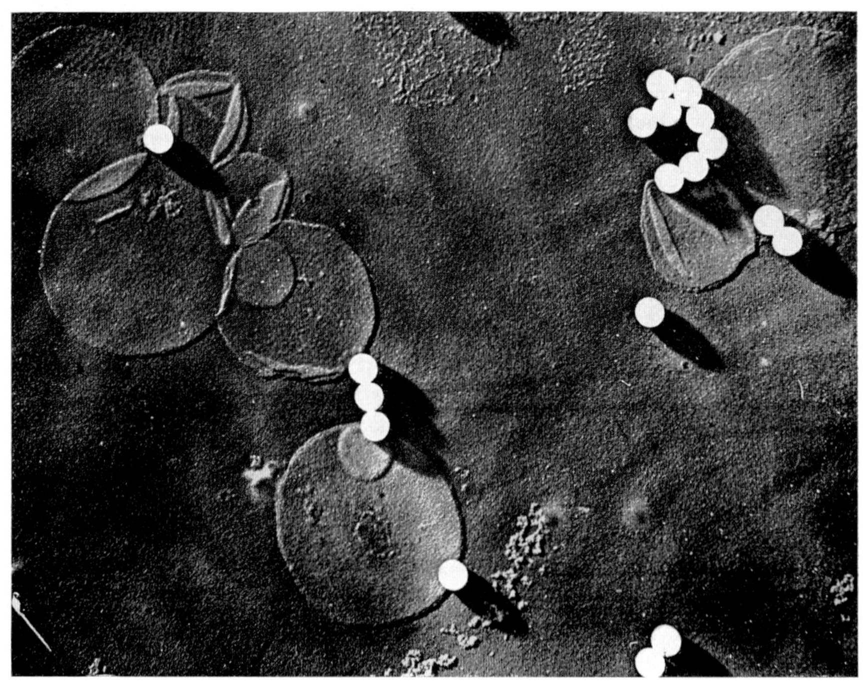

b

FIGURE 39. The purified glycosaminopeptide fraction from the cell walls of *Aerobacter cloacae* as isolated by SCHOCHER *et al.*, 1962.

shown in Fig. 40 this light mucopeptide fraction is in the form of 'discs' (SCHOCHER, unpublished observations) of about 500 Å in diameter, requiring a centrifugal force of 80,000 × g for more than 2 hours to sediment the bulk of the material. It seems likely that these 'discs' may represent the macromolecular subunits of the rigid mucopeptide layer. Just how these discs are joined together to form a continuous, rigid sheet of the type isolated from several Gram-negative bacteria (*Escherichia coli* and *Spirillum* sp.) is not apparent at the moment. Further studies of this type by SCHOCHER and those of WEIDEL, FRANK AND MARTIN (1960) and MARTIN AND FRANK (1962) should yield a great deal of information about the microanatomy of the walls of Gram-negative bacteria.

←

FIGURE 38. Isolated cell walls of *Rhodospirillum rubrum* before (a), × 21,500, and after incubation with lysozyme (b) × 16,000, showing the direct conversion of 'spiral' wall fragments into spherical structures.

The only other investigation relevant to the present discussion of the anatomy of isolated bacterial cell walls is that performed by FRANK, LEFORT AND MARTIN (1962) on the walls of a species of the closely related blue-green algae. SALTON (1960) reported the presence of mucopeptides (glycosaminopeptides) in blue-green algae, a discovery which further strengthens the association of this group with the bacteria. Because of the presence of mucopeptide in the wall these structures in blue-green algae would be potential substrates for lysozyme and may also be sensitive to the biosynthetic block brought about by penicillin treatment. FRANK, LEFORT AND MARTIN (1962) found that both direct treatment of isolated walls with lysozyme and growth in the presence of penicillin could be used effectively in studying the anatomy of the cell walls of the blue-green alga, *Phormidium uncinatum*.

Prolonged shaking of the isolated walls of *Phormidium uncinatum* in the Mickle apparatus apparently removed plastic surface material of an unknown nature revealing the 'zipper-like' appearance of the cell-wall septum. These rows of pores along the line of septa formation are shown in Fig. 41 together with the untreated, isolated walls. The fine structure revealed at the zone of septation in *Phormidium uncinatum* has not been detected so far in bacteria but examination of the walls of a trichome-forming organism such as *Caryophanon latum* may be a more likely source of this type of structure. The changes resulting from treatment of the walls of *Phormidium uncinatum* with lysozyme and growing the organism in the presence of penicillin are illustrated in Fig. 42 and have been taken from the study of FRANK, LEFORT AND MARTIN (1962).

The variety of fine structures so far found in bacterial cell walls is contrasted with other microorganisms in Table 12 which lists in addition details of the substructures of certain cell walls.

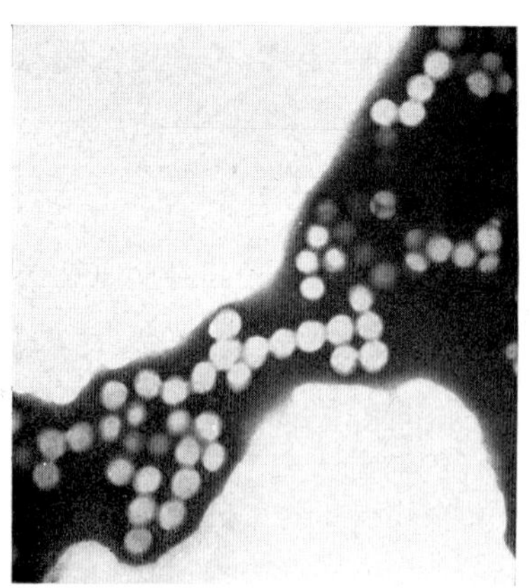

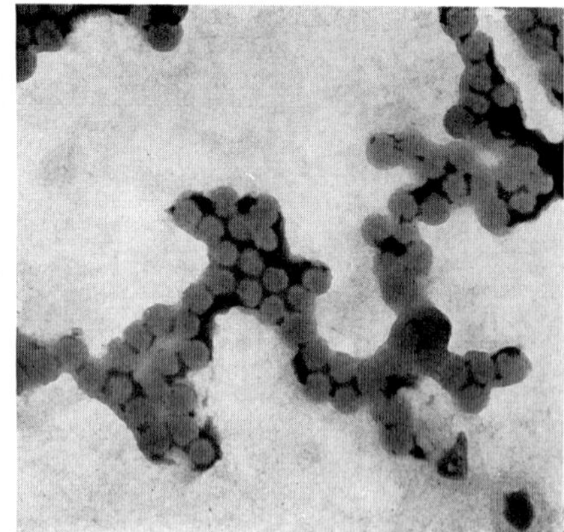

FIGURE 40. A negatively stained preparation of discs obtained by disintegrating the purified glycosaminopeptide fraction (see Fig. 39) from *Aerobacter cloacae* in the Mickle apparatus. The discs are about 500Å in diameter (SCHOCHER AND WATSON, unpublished observations).

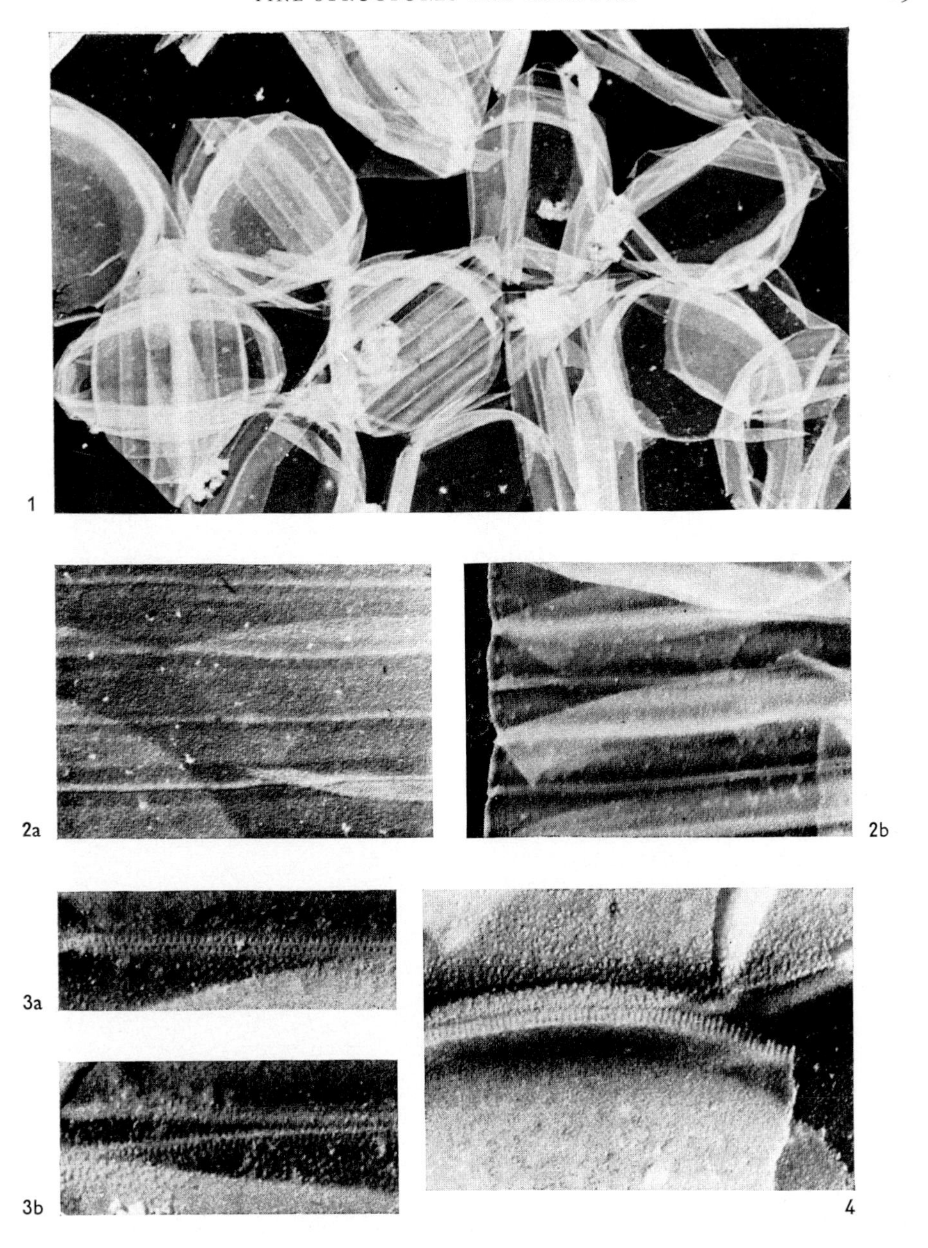

FIGURE 41. Structure of the cell wall of the blue-green alga, *Phormidium uncinatum*. (1) Isolated cell walls, × 5,350; (2a, b) change in the appearance of the outer layer of the wall during prolonged shaking in the Mickle apparatus, × 16,000; (3a, b and 4) structure of the pores and cell-wall septa, × 42,750. (FRANK, LEFORT AND MARTIN, 1962).

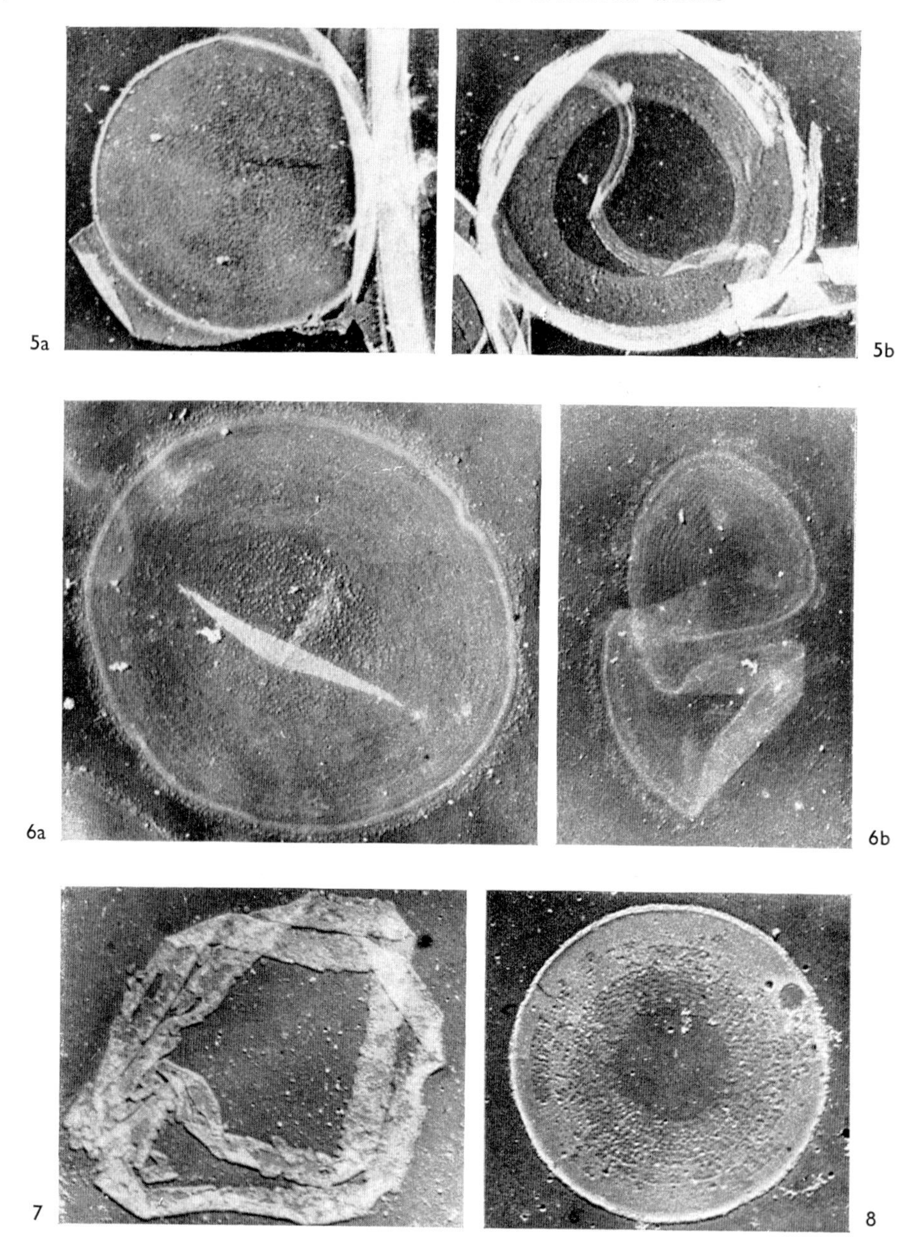

FIGURE 42. Structure of the cell wall of the blue-green alga, *Phormidium uncinatum*. (5a, b) Normal cell-wall septa, (a) fully grown, (b) partially grown, × 107,000; (6a, b) Distended and flexible middle lamella of septa after exposure to lysozyme, × 107,000; (7 and 8) effects on the cell wall of growing *Phormidium* in the presence of penicillin; outer walls disintegrate (7) and localized defects seen in developing septum (8). × 107,000. (FRANK, LEFORT AND MARTIN, 1962).

REFERENCES

ARCHIBALD, A. R., J. J. ARMSTRONG, J. BADDILEY and J. B. HAY, *Nature*, 191 (1961) 570.

BISSET, K. A., *The cytology and Life-History of Bacteria*, E. & S. Livingstone, Ltd., Edinburgh and London, 1955, p. 38.

BROWN, A. D., D. G. DRUMMOND and R. J. NORTH, *Biochim. Biophys. Acta*, 58 (1962) 514.

BROWN, A. D. and C. D. SHOREY, *Biochim. Biophys. Acta*, 59 (1962) 258.

CHAPMAN, G. B. and J. HILLIER, *J. Bacteriol.*, 66 (1953) 362.

CHAPMAN, J. A., R. G. E. MURRAY and M. R. J. SALTON, *Proc. Roy. Soc. (London)* B, (1963). In the press.

CUMMINS, C. S. and H. HARRIS, *J. Gen. Microbiol.*, 18 (1958) 173.

DAWSON, I. M., in *Soc. Gen. Microbiol. Symposium* No. 1, 1949, p. 119.

FRANK, H., M. LEFORT and H. H. MARTIN, *Z. Naturforsch.* 17b (1962) 262.

GLAUERT, A. M., *Brit. Med. Bull.*, 18 (1962) 245.

HICKMAN, D. D. and A. W. FRENKEL, *J. Biophys. Biochem. Cytol.*, 6 (1959) 277.

HOUWINK, A. L., *Biochim. Biophys. Acta.*, 10 (1953) 360.

HOUWINK, A. L., *J. Gen. Microbiol.*, 15, (1956) 146.

KELLENBERGER, E. and A. RYTER, *J. Biophys. Biochem. Cytol.*, 4 (1958) 323.

KELLENBERGER, E. and A. RYTER, and J. SECHAUD, *J. Biophys. Biochem. Cytol.*, 4 (1958) 671.

KNAYSI, G., J. HILLIER and C. FABRICANT, *J. Bacteriol.*, 60 (1950) 423.

MANDELSTAM, J., *Biochem. J.*, 84 (1962) 294.

MARTIN, H. H., *J. Theoret, Biol.*, 5 (1963) 1.

MARTIN, H. H. and H. FRANK, *Zentr. Bakt. Parasitenk.* (Orig.), 184 (1962) 306.

MUDD, S., F. HEINMETS and T. F. ANDERSON, *J. Bacteriol.*, 46 (1943) 205.

MUDD, S. and D. B. LACKMAN, *J. Bacteriol.*, 41 (1941) 415.

MUDD, S., K. POLEVITSKY, T. F. ANDERSON and L. A. CHAMBERS, *J. Bacteriol.*, 42 (1941) 251.

MURRAY, R. G. E., in *Soc. Gen. Microbiol. Symposium* No. 12, 1962, p. 119.

MURRAY, R. G. E., *Proc. Roy. Soc. (London)* B. Manuscript submitted.

RIBI, E., T. PERRINE, R. LIST, W. BROWN and G. GOODE, *Proc. Soc. Exptl. Biol. Med.*, 100 (1959) 647.

ROMANO, A. H. and W. J. NICKERSON, *J. Bacteriol.*, 72 (1956) 478.

ROMANO, A. H. and A. SOHLER, *J. Bacteriol.*, 72 (1956) 865.

SALTON, M. R. J., *J. Gen. Microbiol.*, 9 (1953) 512.

SALTON, M. R. J., in *Soc. Gen. Microbiol. Symposium* No. 6, 1956, p. 81.

SALTON, M. R. J., *Microbial Cell Walls*, John Wiley and Sons, New York, 1960

SALTON, M. R. J. and R. W. HORNE, *Biochim. Biophys. Acta*, 7 (1951) 177.

SALTON, M. R. J. and F. SHAFA, *Nature*, 181 (1958) 1321.

SALTON, M. R. J. and R. C. WILLIAMS, *Biochim. Biophys. Acta*, 14 (1954) 455.

SCHOCHER, A. J., S. T. BAYLEY and R. W. WATSON, *Canad. J. Microbiol.*, 8 (1962) 89.

SHAFA, F. and M. R. J. SALTON, *J. Gen. Microbiol.*, 22 (1960) 137.

SHATKIN, A. J. and E. L. TATUM, *J. Biophys. Biochem. Cytol.*, 6 (1959) 423.

TAKEYA, K., R. MORI, T. TOKUNAGA, M. KOIKE and K. HISATSUNE, *J. Biophys. Biochem. Cytol.*, 9 (1961) 496.

VAN ITERSON, W., *Proc. Intern. Congr. Elect. Mic.* (London), 1954, p. 602.

VINCENZI, L., *Hoppe-Seyl. Z.*, 11 (1887) 181.

WAMOSCHER, L., *Z. Hyg. Infekt.*, 111 (1930) 422.

WEIDEL, W., H. FRANK and W. LEUTGEB, *J. Gen. Microbiol.*, 30 (1963) 127.

WEIDEL, W., H. FRANK and H. H. MARTIN, *J. Gen. Microbiol.*, 22 (1960) 158.

WILLIAMS, R. C., *Exptl. Cell Res.*, 4 (1953) 188.

CHAPTER 4

Physico-chemical properties and chemical composition of walls

GENERAL PHYSICO-CHEMICAL PROPERTIES OF WALLS

The successful adaptation of bacteria to a wide variety of physico-chemical environments during the course of their evolution can be attributed in part to the development of suitable surface structures. As pointed out by SALTON (1956) a robust mechanically rigid cell wall has considerable survival value for an organism. A bacterial protoplast requires a rather specialized environment for survival whereas an intact cell can withstand a variety of habitats that would be lethal for the protoplasts. The surface membrane structure possessed by the obligate halophilic organisms *Halobacterium halobium* and *Halobacterium salinarium* has rather precise cation requirements for its stability (BROWN, 1963) and this factor together with the absence of the rigid glycosaminopeptide component (BROWN AND SHOREY, 1962) may readily explain the inability of this organism to survive low-salt environments.

The mechanical strength of the cell wall or envelope membrane structures and the ability to withstand the physico-chemical effects of various natural and synthetic lytic agents are factors of prime importance to the continued survival of the bacterial cell. MITCHELL AND MOYLE (1956) estimated that the solute concentration in several Gram-positive cocci would subject the cell envelope to an osmotic pressure equivalent to 20 atmospheres. The cell walls of these organisms have a high enough tensile strength to protect the cells against osmotic explosion when the external environment becomes very dilute. It is now known that the chemical component responsible for the rigidity of the walls of both Gram-positive and Gram-negative bacteria is the glycosaminopeptide (mucopeptide), a substance which contributes a great deal to the stability of the cell and confers a number of general properties on the bacterial walls.

Solubility properties

Isolated cell walls of Gram-positive bacteria appear to be largely, if not completely, insoluble in a variety of solvents used for dissolving various proteins and polysaccharides. The isolated walls of *Streptococcus faecalis* (SALTON, 1952) and *Corynebacterium diphtheriae* (HOLDSWORTH, 1952) were insoluble in 90% w/w phenol, a solvent which does however, solubilize about 80% of the isolated wall or membrane of *Escherichia coli* as shown in the studies of WEIDEL (1951). Walls of many Gram-positive bacteria are insoluble in many organic solvents including alcohols, ethers, aceton and pyridine (SALTON AND HORNE, 1951). They are also insoluble in protein solvents such as urea, saturated solutions of guanidine, formamide, diethylformamide and dimethylformamide in the cold. Unlike cellulose, the bacterial cell walls were insoluble in Schweizer's reagent (SALTON AND HORNE, 1951).

The tables are printed together at the end of the book.

The detection and identification of different kinds of polymers in cell walls has led to the discovery that some of the components can be selectively extracted with various reagents. Thus HOLDSWORTH (1952) showed that an oligosaccharide could be extracted from the walls of *Corynebacterium diphtheriae*, with a boiling saturated solution of picric acid, leaving behind an insoluble residue containing the wall amino acids and amino sugars. BADDILEY, BUCHANAN AND CARSS (1958) were able to extract the teichoic acids (see Chapter 6) from the walls of Gram-positive bacteria by treatment with 5% trichloroacetic acid (TCA) in the cold. As well as extracting the teichoic acid this treatment has recently been found to extract cell-wall polysaccharide from *Lactobacillus arabinosus* (ARCHIBALD, BADDILEY AND BUCHANAN, 1961). Hot formamide has proved to be a useful selective solvent and has been recently applied to cell-wall fractionation by KRAUSE AND MCCARTY (1961). They found that treatment of isolated group A streptococcal cell walls with formamide at 170° for 20 minutes extracted the polysaccharide material leaving an insoluble mucopeptide residue which still possessed the morphological characteristics of the original wall.

The immunologically active surface O antigens on the cell walls of Gram-negative bacteria can be extracted with 45% phenol solutions at 60°, a method used for the isolation of the lipopolysaccharide complexes studied so comprehensively by WESTPHAL AND LÜDERITZ (1960). These antigenic complexes may also be partially extracted from isolated walls by treatment

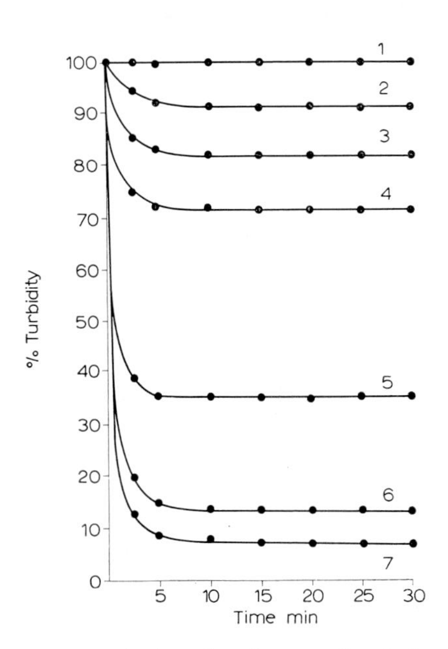

FIGURE 43. The course of disaggregation of *Salmonella gallinarum* cell walls determined by measuring ehanges in turbidity on the addition of sodium dodecyl sulphate to give final concentrations of: 0.006% (curve 2); 0.012% (curve 3); 0.025% (curve 4); 0.05% (curve 5); 0.1% (curve 6) and 0.2% (curve 7); untreated walls, curve 1. (SHAFA AND SALTON, 1960).

with diethylene glycol (SALTON, 1960) a method originally used in their separation from whole bacterial cells by MORGAN (1936).

The cell walls of a number of Gram-negative bacteria can be disaggregated by treatment with solutions of sodium dodecyl sulphate, SDS (SALTON, 1957a; SHAFA AND SALTON, 1960) Table 13 summarizes the data on the extent of disaggregation determined by measuring the decrease in turbidity of cell-wall suspensions treated with SDS. As illustrated in Fig. 43 the dissolution is extremely rapid and of a non-enzymic nature (SHAFA AND SALTON, 1960). The relationship between alkyl chain length and ability to disaggregate the cell walls of the Gram-negative organism, *Salmonella gallinarum* is shown in Fig. 44 and correlates well with the action of these materials on lipo-protein membranes (SCHULMAN, PETHICA, FEW AND SALTON, 1955). Treatment with SDS under similar conditions was without effect on the isolated cell walls of several Gram-positive bacteria (SALTON, 1957a). The dissolution or disaggregation of the walls of Gram-negative bacteria was attributed to the action of the anionic compounds on weak molecular forces involved in lipid-protein associations. SHAFA AND SALTON (1960) failed to detect any major cell-wall substructure after disaggregation with SDS and interpreted the action of the anionic detergent as involving a fairly complete dispersion of the wall as a whole. BOLLE AND KELLENBERGER (1958) did, however, present evidence for the existence of a continuous layer richer in mucopeptide constituents than the original wall, remaining after treatment with sodium lauryl sulphate (SDS). The removal of non-mucopeptide components from walls of Gram-negative bacteria by extraction with SDS solutions has

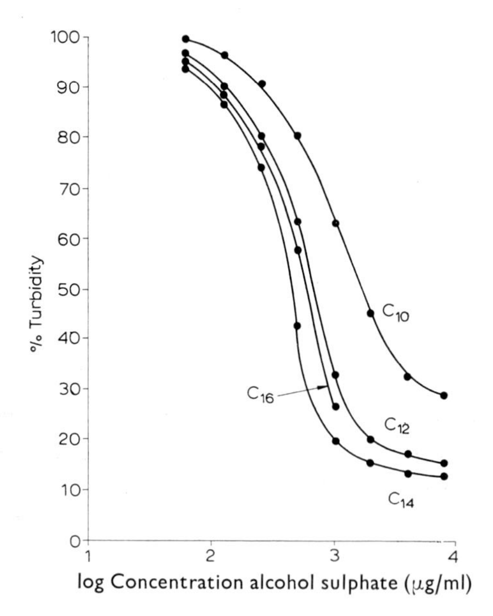

FIGURE 44. The influence of chain length of alcohol sulphates on the dissolution of *Salmonella gallinarum* cell-walls at pH 7, 37° (SHAFA AND SALTON, 1960).

provided one of the steps in isolating the rigid, mucopeptide layer (WEIDEL, FRANK AND MARTIN, 1960). Indeed highly purified mucopeptide was isolated from cell-wall preparations of *Aerobacter cloacae* by successive treatments with aqueous saturated solutions of SDS (SCHOCHER, BAYLEY AND WATSON, 1962).

The unexpected complete disaggregation of the isolated walls of several Gram-negative bacteria by SDS in the experiments performed by SHAFA AND SALTON (1960) and the failure to detect the rigid layer has now been explained satisfactorily in recent experiments by WEIDEL, FRANK AND LEUTGEB (1963). Using the same strain of *Salmonella gallinarum* WEIDEL, FRANK AND LEUTGEB (1963) found that walls prepared by disintegration of freshly harvested cells in the Mickle apparatus were completely disaggregated as shown by SHAFA AND SALTON (1960). However, if the harvested cells were treated with 4% w/v SDS and allowed to stand overnight before washing and disintegration then the rigid mucopolymer layer could be isolated.

MANDELSTAM (1962) has outlined a procedure for isolating mucopeptide material from the cell walls of Gram-negative bacteria. To achieve complete extraction of protein components of the walls 5 steps including disintegration in the Mickle shaker and extraction of the lipids, treatment with pepsin, extraction with 90% v/v formic acid, 90% w/w phenol and 'copper ethylenediamine' (6% w/v $Cu(OH)_2$ and 8% w/v ethylene diamine) were required.

Complete dissolution of isolated walls of Gram-positive bacteria can be achieved with more drastic reagents such as treatment with 10% sodium hypoclorite in the cold, concentrated hydrogen peroxide and 0.5*N* NaOH in an atmosphere of N_2 at room temperature for 2 weeks (the latter being conditions used for solubilization of the lysozyme substrate by MEYER, PALMER, THOMPSON AND KHORAZO, 1936). These reagents probably act by breaking some covalent bonds holding the wall complexes together yielding water soluble material of sufficiently high molecular weights to give viscous solutions. To what extent hydrogen bonding is involved in the rigidity of the wall components cannot be said with certainty, although BROWN (1958) has suggested that solubilization of some wall material by repeated extraction with trichloroacetic acid and by alternate drying and rehydration may be the result of disruption of H-bonds.

The solubility properties of cell walls and cell-wall polymers in various reagents are summarized in Table 14.

Ultra-violet, visible and infra-red spectroscopy of walls

Examination of absorption spectra of cell-wall preparations in ultra-violet, visible and infra-red regions can be used for the detection of certain classes of compounds and absorption bands of specific chemical groupings. Spectroscopy of such complex structures as cell walls is of rather limited value and is in most instances of diagnostic value only. However, the marked difference between the U-V absorption spectra of cell wall and 'cytoplasmic' fractions of *Streptococcus faecalis* illustrated in Fig. 20 (Chapter 2) is of some practical value in following the fractionation of disrupted cells. Contamination of the wall fractions with 'cytoplasmic' nucleic acid can be readily detected by this means. Moreover, the relative freedom of the isolated walls from 260 mμ-absorbing material established that nucleic acids were not major components of the rigid cell-wall structure (SALTON AND HORNE, 1951). BARKULIS AND JONES (1957) have shown that the high scattering of the cell walls can indeed mask the presence of 260 mμ-absorbing components for they were able to extract such materials from the wall preparations by treatment with hot water. It is, nonetheless, generally agreed that nucleic acid is not a structural component of the cell wall and that the bulk of the RNA is in the form of ribosomes or soluble-RNA and that the DNA is localized in the chromatin body of the bacterial cell (MCQUILLEN, 1961; KELLENBERGER, 1960).

The isolated cell walls of pigmented Gram-positive bacteria are usually devoid of the pigments. The carotenoids of various micrococci are localized in intracellular particle and membrane-mesosome fractions but not in the walls. However, walls isolated from certain photosynthetic bacteria were pigmented as mentioned by SALTON (1956a) and the absorption spectrum of *Rhodospirillum rubrum* walls in Fig. 45 indicates the presence of both bacteriochlorophyll and carotenoids. The amounts of pigment are much lower in the wall fractions than in the chromatophores. The presence of pigments normally found in the chromatophores indicated that some material still adhered to the cell wall or alternatively as suggested in Chapter 1 the chromatophores originating from the plasma membrane may also form one of the layers of the multilayered outer component of the envelope. Walls isolated from several blue-green algae also contained carotenoids but in contrast to the photosynthetic bacteria, they were completely devoid of the chlorophyll pigments (SALTON, 1960).

MASON AND POWELSON (1958) were the first to demonstrate the presence of carotenoid in isolated bacterial cell walls when they examined the spectrum of walls of the organism *Myxococcus xanthus*. The Gram-negative organism, *Serratia marcescens* also has the pigment,

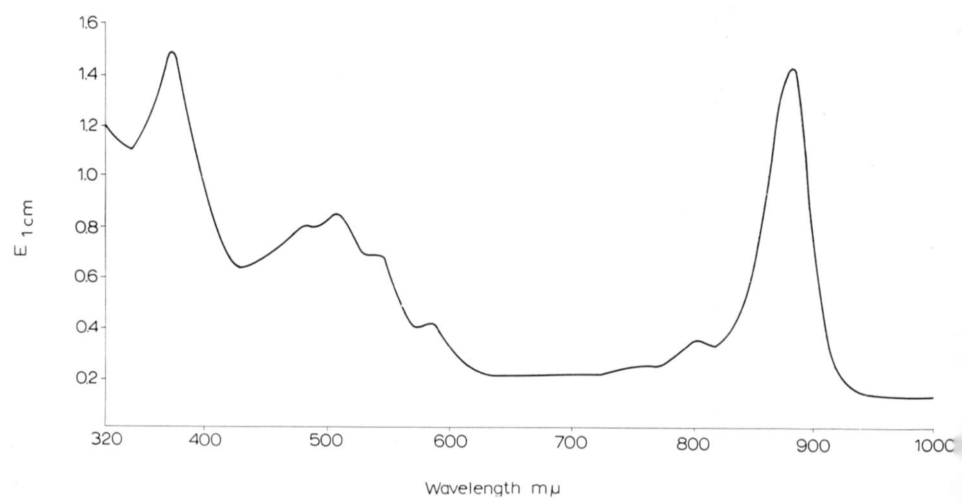

FIGURE 45. Absorption spectrum of the isolated wall preparation from *Rhodospirillum rubrum* showing the presence of peaks corresponding to the carotenoids and bacteriochlorophyll.

prodigiosin, associated with the cell envelope fraction as shown by PURKAYASTHA AND WILLIAMS (1960).

Infra-red spectroscopy has been applied to isolated bacterial cell walls (SALTON, unpublished observations) and the presence of certain characteristic groups from the substances already known to be in the walls from chemical analysis was confirmed.

X-ray diffraction study of walls

The application of X-ray diffraction analysis to a study of the submicroscopic structure of bacterial cell walls was investigated by HURST (1952) but interpretation of the results in terms of cell-wall structure was not possible. HURST (1952) used whole cells of *Escherichia coli* and yeast extracted in various ways and assumed that the walls were responsible for the spacings indicating the presence of oriented lipid. It is certain that the materials studied by HURST (1952) were structurally heterogeneous. GROSSBARD AND PRESTON (1957) used the technique in examining intact cells and cell walls of *Escherichia coli* and their results suggested that lipids are oriented in the wall surface and that the hydrocarbon chains lie normal to the cell surface. In further studies by GROSSBARD AND PRESTON (1958), the interpretation of X-ray diffraction analysis of *Escherichia coli* cell walls indicated that the length of the hydrocarbon chains may be 46.6 Å, thus consisting of 36 carbon atoms. It will be of interest to see to what extent these measurements can be correlated with chemical studies of cell-wall lipids. BROWN AND FRASER (personal communication) have also observed a 46 Å spacing in an X-ray diffraction analysis of the membrane of the marine pseudomonad, NCMB 845.

CHEMICAL COMPOSITION OF WALLS

1. General chemical properties

Cell walls from a variety of both Gram-positive and Gram-negative bacteria as well as isolated spore coats or walls have been analysed for their nitrogen and phosphorus contents, ash and mineral constituents, reducing substances liberated on acid hydrolysis and lipid contents. Analysis of walls of Gram-positive bacteria revealed the presence of both nitrogen and phosphorus and in the walls of some species (e.g. *Bacillus subtilis* and *Staphylococcus aureus*) the phosphorus contents were quite high, a feature now known to be due to the presence of the teichoic acids (see Chapter 6). Typical results for N and P contents together with reducing values determined after acid hydrolysis are given for a variety of Gram-positive bacteria in Table 15. Similar analytical data for various Gram-negative bacteria have been collected together and are presented in Table 16.

Spore coats or walls from a number of *Bacillus* spp. have now been analysed and general analytical data from investigations by STRANGE AND DARK (1956), YOSHIDA, IZUMI, TANI, TANAKA, TARAISHI, HASHIMOTO AND FUKUI (1957), SALTON AND MARSHALL (1959), MURRELL (1960) are presented in Table 17. A comparison of the chemical analysis of the spore and vegetative walls of *Bacillus subtilis* is given in Table 18 and the marked difference in N and P contents is in accord with the greater complexity of the amino acid composition of the spore wall and the presence of a major teichoic acid component in the vegetative cell walls (SALTON AND MARSHALL, 1959).

Little is known about the ash contents of bacterial cell walls. SALTON (1952) found an ash content of 2.7% for *Streptococcus faecalis* cell walls. Rather higher values (10%) were obtained for the % ash in the walls isolated from *Corynebacterium diphtheriae* by HOLDSWORTH (1952). Identification of the ash constituents had not been investigated until recently. The studies of KOZLOFF AND LUTE (1957) suggested that in *Escherichia coli* cell walls, Zn may be present in the form of a metallo-protein complex and they calculated that there were probably 3,000 atoms of Zn per cell wall. HUMPHREY AND VINCENT (1962) have carried out direct analyses of the Ca^{2+} and Mg^{2+} contents of cell walls of *Rhizobium trifolii* grown under 'normal' and 'calcium-deprived' conditions. VINCENT AND COLBURN (1961) had previously shown that walls of calcium-deprived cells of *Rhizobium trifolii* developed an abnormal morphology suggestive of a weakened double-layered wall structure. The contents of Ca^{2+} and Mg^{2+} in whole cells and isolated walls of *Rhizobium trifolii* from the investigations of HUMPHREY AND VINCENT (1962) are given in Table 19.

Bacterial cell walls have been analysed for their contents of reducing substances in order to obtain some idea of the amounts of polysaccharide or carbohydrate components. Results such as those shown in Tables 15 and 16 can only be regarded as fairly crude estimates, for the application of a variety of methods for estimating carbohydrates in cell walls of several Gram-positive species has shown very little agreement between the procedures. SALTON AND PAVLIK (1960) determined the reducing values, anthrone values and glucose by glucose oxidase in cell walls containing glucose and the two amino sugars glucosamine and muramic acid as the sole monosaccharide constituents. The poor agreement between these analytical methods is illustrated in Table 20. The conditions used for hydrolysis of the polysaccharide components of the walls undoubtedly lead to the liberation of free amino acids which have been shown to interfere with reductimetric methods by STRANGE, DARK AND NESS (1955). The accurate measurement of polysaccharide contents by determination of the reducing values (usually given as glucose equivalents) therefore presents a problem which presumably can only be overcome by specifically estimating each individual sugar and amino sugar component of the various polymers in the cell walls.

Major classes of chemical substances in bacterial cell walls

From quantitative and qualitative studies of the composition of bacterial cell walls which have been reported during the past ten years, it is now possible to summarize the principal classes of chemical compounds occurring in these structures. Thus the classification of the major types of polymeric substances in bacterial walls has only become possible as a result of the identification of the various amino acid, amino sugar, sugar and polyol building blocks liberated by hydrolysis of isolated cell walls and components extracted from the cell walls. Detailed discussions of the various monomeric constituents will be presented in succeeding sections of this chapter and the proposed structures of some of the cell-wall polymers will be dealt with in subsequent chapters.

One of the interesting features emerging from the comparative chemical and biochemical studies of microbial cell walls is the complexity of these structures. Although the walls of some microorganisms (e.g. algae, fungi and yeasts) may be predominantly polysaccharide in nature, it is now evident that some species contain in addition significant protein and lipid constituents. Isolation of microbial cell walls by extraction of whole organisms with alkali has led to the impression that cellulose or chitin, or both in some instances (FULLER AND BARSHAD, 1960) are the principal components. While there is no doubt that both cellulose, chitin and other polysaccharides are important structural polymers, other constituents have been overlooked as a result of the isolation methods used. The investigations of NICKERSON and his colleagues have shown that yeast wall glucans and mannans occur as protein complexes and that they are not present as simple polysaccharides (KESSLER AND NICKERSON, 1959).

In contrast to algae, fungi and yeasts, the walls of bacteria are very different in chemical structure. A new class of structural heteropolymers, called the mucocomplexes by SALTON (1956a) and the mucopeptides by MANDELSTAM AND ROGERS (1959) has been found in isolated cell walls of all of the Gram-positive and most of the Gram-negative bacteria so far examined. With the discovery that cell walls of certain Gram-positive bacteria were composed of a small variety of amino acid, amino sugar and sugar constituents (SALTON, 1952, 1953) the similarity of the wall complex to other mucopolysaccharides and their distinction from known mucoproteins was emphasized by SALTON (1952, 1956a). Although the bacterial cell walls and their rigid components are not 'mucoid' in physico-chemical properties, their affinities with other natural 'muco' substances was immediately obvious (SALTON, 1952). As there is still considerable uncertainty about the variety of polymers in even the simpler walls of the Gram-positive bacteria, the classification of the major chemical constituents cannot be considered final at this stage, though some tentative outlines can be made.

There is now good evidence to suggest that there is a major chemical component which is responsible for the rigidity of the bacterial cell wall and that this material is a complex of amino acids and amino sugars. The term mucopeptide (MANDELSTAM AND ROGERS, 1959) has been widely adopted for the description of these components previously referred to more accurately as amino sugar-peptide complexes (SALTON, 1952). However, the need for a short, descriptive word for this class of polymer has led to the use of the prefix 'muco' to indicate chemical relationships with mucopolysaccharides, mucoproteins, etc. 'Mucopeptide' and 'mucocomplex' are both trivial terms from the viewpoint of chemical nomenclature. Since there is no general agreement as to what chemical implications are associated with the use of the term 'muco' (MEYER, 1938; STACEY, 1943; ROSEMAN, 1959; JEANLOZ, 1960) it is the author's opinion that terms such as mucopeptide and mucocomplex are chemically vague and physico-chemically misleading as the wall fraction responsible for its rigidity is far from 'mucoid'. SHARON (1963) has adopted the more chemically precise name, 'peptido-polysaccharide', for this class of constituent. The use of terms such as 'peptido-polysaccharide',

(SHARON, 1963), 'glycopeptide' (STROMINGER, 1962) instead of mucopeptide or mucocomplex is to be encouraged. Following the nomenclature of JEANLOZ (1960) for polysaccharides containing amino sugars, the present author has used the term 'glycosaminopeptide' as an alternative to 'mucopeptide'. Indeed, any of the terms 'peptido-polysaccharide', 'glycopeptide' or 'glycosaminopeptide' are preferable, the last of the three having the added advantage of indicating the amino sugar nature of the 'polysaccharide' part of the complex. All of the available evidence suggests that the rigid structure of the bacterial wall contains only amino sugars and amino acids and that other sugars found in the isolated walls may occur as oligo- or polysaccharides more readily solubilized than the glycosaminopeptide. The extraction of the oligosaccharide from *Corynebacterium diphtheriae* walls (HOLDSWORTH, 1952) and complete removal of the polysaccharide portion of the group A streptococcal wall leaving an insoluble mucopeptide (KRAUSE AND MCCARTY, 1961) indicate that where polysaccharides are present they may be less firmly attached to the backbone structure of the glycosaminopeptide (mucopeptide) than are the peptides of the latter compounds.

In addition to glycosaminopeptide and polysaccharide or mucopolysaccharide components, BADDILEY, BUCHANAN AND CARSS (1958) discovered a new type of cell-wall polymer, the teichoic acids (see Chapter 6). The solubility of the teichoic acids and other cell-wall polysaccharides in trichloroacetic acid has already been mentioned and again the glycosaminopeptide residue retains the structural properties of the original wall (ARCHIBALD, BADDILEY AND BUCHANAN, 1961). Another class of wall component discovered in a Gram-positive organism is the teichuronic acid isolated and characterized by JANCZURA, PERKINS AND ROGERS (1961). This latter component has so far only been described in walls, of *Bacillus subtilis*. A polymer rich in glucose and an amino uronic acid (probably 2-amino-2-deoxymannuronic acid) has been isolated recently from *Micrococcus lysodeikticus* walls by PERKINS (1963).

The more complex walls or compound membranes of Gram-negative bacteria also possess the glycosaminopeptide or mucopeptide components which may constitute almost the entire wall of some Gram-positive organisms. Although the mucopeptide layer of the wall of Gram-negative bacteria is undoubtedly responsible for the rigidity of the wall, it may represent as little as 10–20% of the weight of the isolated wall structure (WEIDEL AND PRIMOSIGH, 1958; SALTON, 1958a). The teichoic acids have been conspicuous in the walls of certain Gram-positive bacteria and usually believed to be absent from the envelope structures of Gram-negative organisms. However, ARMSTRONG, BADDILEY, BUCHANAN, DAVISON, KELEMEN AND NEUHAUS (1959) did record traces of the glycerol type of polymer in *Escherichia coli* B cell walls and more recently LILLY (1962) presented evidence for the presence of a ribitol teichoic acid in strain 26–26 of *Escherichia coli*. There is little doubt that the bulk of the wall in Gram-negative bacteria is accounted for by the protein, lipid and polysaccharide constituents. Macromolecular complexes of all 3 classes of material have been isolated from the surface and cell walls of Gram-negative organisms and it is evident from the work of WEIDEL AND WESTPHAL and their colleagues that some of the lipid and polysaccharide constituents can be isolated as lipopolysaccharide complexes.

Apart from bacteria, the only other microbial groups possessing some of the characteristic cell-wall constituents (e.g. muramic acid, DAP) are the blue-green algae (SALTON, 1960; FRANK, LEFORT AND MARTIN, 1962), Rickettsiae (ALLISON AND PERKINS, 1960) and the agent of meningopneumonitis of the psittacosis group of viruses (JENKIN, 1960). Although SCHAECTER, TOUSIMIS, COHN, ROSEN, CAMPBELL AND HAHN (1957) had isolated membranes from *Rickettsia mooseri* and reported the presence of amino acids, hexosamine and sugars, it was not until ALLISON AND PERKINS (1960) performed careful analyses that the presence of the 'key' compound muramic acid was established. The latter workers also found muramic acid in mouse pneumonitis agent but failed to find any traces of it in highly concentrated suspensions of vaccinia virus. JENKIN (1960) also reported the presence of small but identifiable amounts of muramic acid in isolated walls of meningopneumonitis agent and from the overall composition of the structures he concluded that they were chemically similar to walls of Gram-negative bacteria.

The principal classes of chemical constituents so far found in bacterial cell walls and related groups of organisms are summarized in Table 21; the pigments which have been discussed above are not included.

The remainder of this chapter will deal with qualitative and quantitative aspects of the amino acid, amino sugar, sugar and lipid constituents of the bacterial cell wall. It should be realized that a great deal of work has been done on the identification and estimation of these types of compounds in whole bacterial cell walls rather than in homogeneous polymers isolated from these structures and unless otherwise stated it can be assumed that analytical results refer to the former kind of material. The proposed chemical structures for some of the cell-wall polymers will be dealt with in separate chapters.

2. *Amino acid composition of cell walls and spore walls*

One of the most striking features to emerge from the comparative studies of the composition of the walls of a number of Gram-positive and Gram-negative bacteria was the difference in amino acid composition. Walls of certain Gram-positive bacteria contained as few as three or four major amino acid constituents, whereas those isolated from Gram-negative bacteria gave a variety of amino acids similar to that normally encountered in most proteins (SALTON, 1952). This marked difference was apparent without treating the wall preparations with proteolytic enzymes (SALTON, 1953). Thus the walls of *Micrococcus lysodeikticus* isolated by mechanical disintegration and washed only with *M* NaCl and water, contained the amino acids alanine, glutamic acid, lysine and glycine (SALTON, 1953). Aromatic and sulphur-containing amino acids were present in walls of Gram-negative bacteria but were completely absent from walls of the Gram-positive species.

The retention of surface proteins (including those possessing antigenic properties) during the cell-wall isolation procedures was clearly demonstrated when walls of group A streptococci *(Streptococcus pyogenes)* were prepared (SALTON, 1953). The M-protein antigen was detectable in the isolated cell-wall preparations and the variety of amino acids found on acid hydrolysis was wider than that detected in walls of the other Gram-positive organisms. These results showed that surface protein components may contribute to the amino acids detectable in wall preparations. However, the investigations of LANCEFIELD (1943) showed that on incubation of intact cells of group A streptococci with trypsin the M-protein could be removed from the surface without affecting the integrity and viability of the cells. In a similar fashion it was shown that treatment of the isolated walls possessing adherent M-protein with trypsin gave trypsin-resistant residues of simpler amino acid composition (SALTON, 1953). The careful investigations of BARKULIS AND JONES (1957) indicated that M-protein was not the only trypsin labile protein retained on isolation of the walls of group A streptococci. Subsequent studies by TEPPER, HAYASHI AND BARKULIS (1960) with virulent and avirulent strains showed that the avirulent strain Q 496 possessed a cell-wall protein of similar physical properties to the M-protein, but obviously lacking in the characteristic serological activity of the latter. The quantitative amino acid composition of the virulent and avirulent strains before and after treatment with trypsin is presented in Table 22.

The results of TEPPER, HAYASHI AND BARKULIS (1960) and the earlier study of SALTON (1953) are in general agreement and have shown that amino acid composition of the trypsin-treated group A streptococcal cell walls is simpler than that of the original wall fractions. All of the valine, arginine, methionine and proline had been removed by enzyme action (TEPPER, HAYASHI AND BARKULIS, 1960). It is evident then that in some instances 'non-wall' amino acids (presumably present as proteins) can be removed from cell-wall preparations by treatment with proteolytic enzymes. MCCARTY (1952), CUMMINS AND HARRIS (1956) were the first to adopt proteolytic digestion as a routine step in the preparation of cell walls. Such treatments would remove any proteins normally found on the wall surface as well as contaminating 'cytoplasmic' proteins. Pre-treatment of the wall fractions with proteolytic enzymes used in the extensive survey of the qualitative composition of the walls of a great variety of Gram-positive bacteria enabled CUMMINS AND HARRIS (1956a and b; 1958) to emphasize the relationships between the patterns of cell-wall amino acid composition and the taxonomic groupings.

Despite the use of trypsin-digestion in the routine preparation of walls of *Bacillus subtilis* by JANCZURA, PERKINS AND ROGERS (1961) they found 7–10% of protein material irrespective of the method of cell-wall preparation. They concluded that an insoluble protein is one of the four types of polymers present in the wall of this organism. There were apparently no features of the amino acid composition of this material (lysozyme-insoluble and resistant to prolonged incubation with trypsin and pepsin) which would distinguish it from other proteins. The possibility that insoluble proteins in wall preparations of spore-forming bacilli could originate from spore coat structures should be considered before any final conclusion is made about these components.

With cell walls of Gram-positive bacteria other than the spore formers, minor amounts of amino acids such as aspartic acid, valine, leucine and isoleucine, serine and threonine may still be present after digestion with proteolytic enzymes and as a consequence it is sometimes difficult to decide whether such amino acids are part of a cell-wall structure or whether they are derived from contaminating particulate material. Aspartic acid in certain lactic acid bacteria (IKAWA AND SNELL, 1960) and serine in *Staphylococcus* spp. (SALTON AND PAVLIK, 1960) are so consistently present that we must conclude that they are more than just contaminants. Their mode of attachment in the cell wall is however, not known. This problem is just one facet of the general difficulty of defining the homogeneity of cellular structures such as bacterial walls. An added complication has been pointed out by BRITT AND GERHARDT (1958) who reported that bacterial cell walls can adsorb some amino acids (e.g. lysine) quite strongly. Such adsorption effects would also interfere with both qualitative and quantitative amino acid analysis of the isolated cell walls.

Digestion of the cell walls of Gram-negative bacteria with trypsin does not modify the overall qualitative amino acid composition; the outstanding difference betweeen walls of the Gram-positive and Gram-negative groups persists after the action of proteolytic enzymes. This should not be taken to mean that proteolytic enzymes are without action on the isolated walls of Gram-negative bacteria. So far as the author is aware there are no quantitative data on the amino acid composition of walls of Gram-negative bacteria before and after digestion with proteolytic enzymes. As would be expected digestion with trypsin does not bring about a release of soluble compounds containing muramic acid or DAP from the walls of Gram-negative organisms (SALTON, unpublished observations). On the other hand there is often a substantial loss of material when crude wall fractions of Gram-negative bacteria are subjected to tryptic digestion. Whether products released from walls under these conditions include any cell-wall components cannot be said at present.

The initial investigations of the amino acid composition of bacterial cell walls were largely qualitative, the relative proportions of amino acids being judged from the areas of 'spots' of amino acids separated on two-dimensional paper chromatograms (HOLDSWORTH, 1952; SALTON, 1952, 1953; CUMMINS AND HARRIS, 1956a, b, 1958). MITCHELL AND MOYLE (1951) performed semi-quantitative analysis of the amino acids in the walls from *Staphylococcus aureus*. Several interesting features emerged from these early qualitative or semi-quantitative analyses. Firstly it was obvious that the walls of certain Gram-positive bacteria may contain 3 or 4 principal amino acids with smaller amounts or even only faint traces of additional ones. The second important result of these studies was the identification of the amino acid, α,ε-diaminopimelic acid (DAP), which was discovered, isolated and characterized by WORK (1949, 1951) and later shown to be a typical component of bacterial cell walls. HOLDSWORTH (1952) established the presence of DAP in the walls of *Corynebacterium diphtheriae* and SALTON (1953) found this amino acid in walls of *Bacillus subtilis* and several Gram-negative bacteria. α,ε-DAP is of course now widely known as an amino acid constituent of the glycosaminopeptide or mucopeptide and is not found in cellular proteins. STRANGE AND POWELL (1953) also found DAP in the 'spore peptides'. The isolation and characterization of these peptides secreted during spore germination led to the discovery and elucidation of the chemical structure of another very important wall compound, namely muramic acid (STRANGE AND DARK, 1956; STRANGE AND KENT, 1959).

A great deal of qualitative information on the amino acid composition of the cell walls of Gram-positive bacteria is now available and much of the information has come from the extensive surveys of CUMMINS AND HARRIS (1956a, b, 1958), CUMMINS, GLENDENNING AND HARRIS (1957) and CUMMINS (1962). The major taxonomic groups of the Eubacteria and Actinomycetales have been studied and the principal combinations of major amino acid constituents found in the walls of Gram-positive bacteria are presented in Table 23. It should be emphasized again that these results refer only to the major amino acid components and do not exclude the possibility that the walls of some Gram-positive bacteria possess components containing additional amino acids in minor amounts. The evidence available suggests that these major amino acid constituents are principally derived from the peptide portion of the cell-wall mucopeptide, although alanine is also present in the teichoic acid polymer. The amino acids, aspartic acid, serine and threonine all occur in walls in more than trace amounts, but is is not known at present whether or not they are in special compounds or covalently linked to the glycosaminopeptides. The relatively simple amino acid composition of the walls of Gram-positive bacteria led to the concept of a cell-wall basal structure (WORK, 1957) which has now been identified with the mucopeptide or glycosaminopeptide fraction of the walls of both Gram-positive and Gram-negative bacteria. Another interesting feature of these studies has been the exclusive occurrence of either DAP or lysine in the walls. In none of the walls so far examined does DAP or lysine occur together as major amino acid constituents. With Gram-negative bacteria the situation is not entirely clear. Products have been obtained by enzymic digestion of *Escherichia coli* walls and these have contained both DAP and lysine although there was uncertainty that these fragments were single compounds (WORK AND LECADET, 1960; WORK, 1961). The analysis of the mucopeptide fragments released from *Escherichia coli* walls with lysozyme are in agreement with the general picture shown in Table 23 as only DAP was present (PRIMOSIGH, PELZER, MAASS AND WEIDEL, 1961).

All of the cell walls of Gram-positive bacteria thus contain alanine, glutamic acid and lysine or DAP. There is insufficient information at present to indicate whether these amino acids occur in one or more peptide species, but some aspects of this problem and the possible location of other amino acids in the wall components will be discussed in relation to glycosaminopeptide structure (Chapter 5). Qualitative identification of cell-wall amino acids has been established for a number of Gram-negative species but unlike amino acids of certain taxonomic groups of Gram-positive bacteria, there are no distinctive patterns in the Gram-negative groups. With the complete separation of wall or membrane protein from wall glycosaminopeptide, quantitative analysis would be the best way of detecting any relationships between the various species.

Before passing on to discuss the quantitative aspects of the amino acid composition of walls it is probably worth emphasizing that all of the major amino acids have been satisfactorily identified. An unknown substance found in hydrolysates of lactobacilli was identified by CUMMINS AND HARRIS (1956) as α-aminosuccinoyllysine which originated from an aspartic acid-lysine peptide. Other unknown ninhydrin-positive substances reported in cell-wall hydrolysates (SALTON, 1960) have not been characterized so it is not known at present whether they are wall amino acids.

Quantitative amino acid analysis of cell walls

Although a great deal is known about the qualitative amino acid constitution of bacterial cell walls, precise analytical data are available for only several organisms. Furthermore the analyses available for Gram-negative bacteria have usually been for the whole cell-wall structure and have not permitted direct distinction between the contribution due to the protein and glycosaminopeptide fractions of the wall.

Cell-wall preparations of both Gram-positive and Gram-negative bacteria have generally been hydrolysed under very similar conditons, e.g. 6*N* HCl for 12–24 hours at temperatures ranging from 100°–105°. It has been assumed, without any experimental evidence, that recovery of amino acids from mucopeptide and from wall protein would be about the same for a standard period of hydrolysis in 6*N* HCl. However, with differing amounts of amino sugars and sugars in the walls it would not be surprising if marked variations in the recoveries of amino acids were encountered. SALTON AND PAVLIK (1960) found that the liberation of lysine and DAP from walls of several Gram-positive bacteria was maximal in 8–16 hours hydrolysis at 105° with either 4*N* or 6*N* HCl. There is not enough quantitative data available at present to suggest whether any of the analytical differences observed are due to variations in amino acid destruction during hydrolysis of walls from various species.

Quantitative amino acid analysis by column chromatography of hydrolysates of cell walls of Gram-positive bacteria have been performed by IKAWA AND SNELL (1956) and their results for *Lactobacillus casei* and *Streptococcus faecalis* are compared with the trypsin-resistant residues of group A streptococcal walls in Table 24. These results further emphasize the general conclusion, based on paper chromatographic identification, that 3 or 4 of the wall amino acids occur in much larger amounts than all of the others.

Several detailed amino acid analyses of cell walls of Gram-negative bacteria have been performed by KOCH AND WEIDEL (1956) and by SALTON (unpublished data quoted in 1960a) on strains of *Escherichia coli* and by COLLINS (1963) with *Pseudomonas aeruginosa* and *Salmonella bethesda*. In addition SCHOCHER, BAYLEY AND WATSON (1962) have analysed whole cell walls of *Aerobacter cloacae* and the purified mucopeptide from these walls. These data on walls of Gram-negative bacteria are summarized in Table 25.

Spore coats of *Bacillus coagulans* have been analysed for their amino acid composition by HUNNELL AND ORDALL (1961) using column separation of hydrolysed material. Their quantitative results are compared in Table 26 with the molar ratios estimated by colorimetric determination of the amino acids of *Bacillus subtilis* spore walls (SALTON AND MARSHALL, 1959). Data on integuments of *Bacillus cereus* and *Bacillus stearothermophilis* (WARTH, OHYE AND MURRELL, 1963) separated on paper chromatograms show that the spore walls or integuments are obviously more complex than the walls of the parent vegetative cells.

In addition to analysis by separation on chromatography columns, many quantitative results have been obtained by estimating the mole ratios of amino acids in cell walls after subjecting hydrolysates to paper chromatographic separations. Because of the small variety of amino acids in walls of Gram-positive bacteria and the greater ease of separation, the measurement of mole ratios has been largely confined to this group.

The mole ratios of the principal amino acids in cell walls of a variety of Gram-positive bacteria have been determined independently in a number of laboratories, and there is fairly good agreement for results of a given species, suggesting that the various modifications used in isolating walls have given a similar end product. Perhaps the greatest source of variation in determining the mole ratios of amino acids in walls is due to differences in techniques of paper chromatographic separation and development of the ninhydrin colour of the separated amino acids. Hydrolysis conditions varying from one laboratory to another would also add to the differences of results. The data for several strains of *Staphylococcus aureus* walls from investigations of HANCOCK AND PARK (1958), MANDELSTAM AND ROGERS (1958), ROGERS AND PERKINS (1959), SALTON AND PAVLIK (1960) are presented in Table 27 and as suggested by ROGERS AND PERKINS (1959) some strain differences are to be expected. Results from strains of group A streptococci from studies of TEPPER, HAYASHI AND BARKULIS (1960) and those of KRAUSE AND MCCARTY (1961) also showed good agreement; in the former investigations the virulent and avirulent strains both contained mole ratios of 1 : 0.7 : 2.2 for glutamic acid, lysine and alanine (relative to glutamic acid) and the values for walls and insoluble mucopeptide in the latter work were approximately 1 : 1 : 3 for glutamic acid, lysine and alanine.

SALTON AND PAVLIK (1960) investigated the mole ratios of the major amino acids in walls of Gram-positive bacteria including various micrococci and *Bacillus* spp. and the results are presented in Tables 28 and 29. It is of considerable interest to note that in many cell walls the mole ratios of glutamic acid, lysine and alanine are approximately 1 : 1 : 3 and that walls from related organisms contain amino acids in very similar proportions. Thus all of the staphylococci (with the exception of one strain studied by ROGERS AND PERKINS, 1959) possessed substantial amounts of glycine in sharp contrast to the smaller proportions in the wall of the yellow-pigmented micrococci. Other significant departures from the 1 : 1 : 2–3 ratios for glutamic acid, lysine and alanine were apparent for walls of the two closely related species *Micrococcus citreus* and *Micrococcus varians* both of which have high mole ratios of glutamic acid. The mole proportions of glutamic acid in walls of some of the Bacillus species also tended to be high.

The results for the mole ratios of amino acids in walls isolated by disintegration and differential centrifugation (Tables 27–29) differ quantitatively from those obtained by KANDLER AND HUND (1959) for wall residues isolated after extraction of cells with 1*N* NaOH. A selection of the results of KANDLER AND HUND (1959) is summarized in Table 30 and they are presented for comparative purposes.

The amino acid composition as indicated by the mole ratios in the whole cell walls cannot, at the moment, be interpreted in terms of the peptide structures present in the various species. Although it is tempting to suggest that related organisms possess very similar peptide structures, at present it is not possible to claim that the results represent anything more than the gross amino acid composition of the walls. So far as the amino acid alanine is concerned, the determination of mole ratios in the walls gives no information about its distribution between components such as the glycosaminopeptides and the teichoic acids. However, in those cell walls devoid of teichoic acids or other special structures it is probable that the mole ratios of amino acids are of real structural significance. Despite the fact that the main features of the amino acid composition of the walls of Gram-positive bacteria have now been well established, it is obvious that they have not yet been interpreted satisfactorily in relation to peptide structure and this aspect of cell-wall chemistry is awaiting further detailed investigation.

MANDELSTAM (1962) has isolated and purified the glycosaminopeptide fractions from the cell walls of several different groups of Gram-negative bacteria. In contrast to the individuality displayed by the proportions of amino acids in the cell walls of Gram-positive bacteria, the amino acid composition of the mucopeptide from the Gram-negative organisms is remarkably constant as illustrated by MANDELSTAM'S (1962) results shown in Table 31. The results given in Table 31 include those obtained by SCHOCHER, BAYLEY AND WATSON (1962) for the purified mucopeptide from *Aerobacter cloacae* and those for the rigid 'mucopolymer' isolated from *Spirillum* sp. by MARTIN AND FRANK (1962).

The occurrence of mucopeptide in walls of blue-green algae was reported by SALTON (1960) and quantitative results for the molar ratios of amino acids in *Phormidium uncinatum* walls (FRANK, LEFORT AND MARTIN, 1962) are given in Table 32 together with unpublished results (SALTON) for walls of *Microcoleus vaginatus*.

D- and L- isomers of amino acids in walls

The occurrence of substantial amounts of certain D-amino acids in hydrolysates of bacterial cells had been known for some time (STEVENS, HALPERN AND GIGGER, 1951) before it became apparent that the cell walls were the principal sources of these compounds. SNELL and his colleagues (SNELL, RADIN AND IKAWA, 1955; IKAWA AND SNELL, 1956) discovered a large quantity of D-alanine in a TCA-insoluble fraction of *Streptococcus faecalis* and the localization of this material was traced to the cell wall by direct isolation and examination of hydrolysates. Although some racemization occurred when an amino acid such as L-alanine was treated with hydrochloric acid in the presence of carbohydrate (conditions which may simulate those existing during hydrolysis of bacterial walls), the amounts of the D-isomer in the isolated walls were higher than would be expected from chemical conversion of L to D isomer (IKAWA AND SNELL, 1956). The existence of a high proportion of cell-wall alanine and glutamic acid as the D-isomers was thus established by IKAWA AND SNELL (1956) and the widespread occurrence of D-alanine in the walls of bacterial groups other than the lactic acid organisms was reported by SALTON (1957b). The presence of D-alanine and D-glutamic acid therefore became one of the distinctive features of the chemistry of bacterial walls. The discovery of D-alanine in the bacterial cell wall also helped to focus attention on the probable functions of the nucleotides containing D-alanine. These nucleotides isolated by PARK (1952) and considered further in relation to the mode of action of penicillin (PARK AND STROMINGER, 1957) frequently had the same proportions of L and D-alanine as found in the cell walls of staphylococci (STROMINGER AND THRENN, 1959). Since this finding that cell walls and nucleotides from *Staphylococcus aureus* have approximately the same proportions of the D and L-isomers of alanine, ARMSTRONG, BADDILEY AND BUCHANAN (1960) have found that the ester-linked alanine of the teichoic acid is the D-isomer. Thus D-alanine may occur in both glycosaminopeptide and teichoic acid polymers of the bacterial walls.

Until the investigations of IKAWA AND SNELL (1956) demonstrated that much of the glutamic acid in the walls of *Lactobacillus casei* and *Streptococcus faecalis* was the D-isomer, D-glutamic acid had only been found in nature as the D-glutamyl capsular polypeptide of *Bacillus anthracis*. SALTON (1957b) confirmed the presence of D-glutamic acid in cell walls and reported that virtually all of the glutamic acid in *Micrococcus lysodeikticus* wall was the D-isomer. It became apparent then, that organisms devoid of capsular polypeptides contained substantial quantities of D-glutamic acid localized in the rigid cell-wall structures. PARK (1958) reported that D-glutamic acid and muramic acid occurred in walls in 1:1 ratios.

The list of D-amino acids in bacterial cell walls was extended to aspartic acid when TOENNIES, BAKAY AND SHOCKMAN (1959) found that this amino acid occurred partly as the D-isomer in the cell wall of *Streptococcus faecalis*. D-aspartic acid had only previously been found in nature in the antibiotic bacitracin (ABRAHAM, 1957) and in hydrolysates of certain bacteria, yeasts and fungi (STEVENS, HALPERN AND GIGGER, 1951).

IKAWA AND SNELL (1960) investigated the amino acid composition of various lactic acid bacteria in some detail, estimating both total amounts of amino acids and the L-isomers of alanine, glutamic acid, aspartic acid and lysine. Some typical analyses from their studies are presented in Table 33 and the percentages of glutamic acid, aspartic acid and alanine in the D-configuration in the walls of the various organisms they used are summarized in Table 34. From the studies of IKAWA AND SNELL (1960) it can be concluded that in the lactic acid bacteria about half of the alanine occurs as the D-isomer, D-aspartic acid residues constitute roughly 75% of the total aspartic acid contents and usually all of the glutamic acid is in the D-form. Apart from determinations on *Staphylococcus aureus* walls (STROMINGER AND THRENN, 1959) there is very little quantitative data for other bacterial groups. However, the available evidence indicates that the proportions of D- and L-isomers in other walls are very similar to those so clearly shown for the lactic acid bacteria.

The distribution of the various isomers of α, ε-diaminopimelic acid in both Gram-positive and Gram-negative bacteria has been studied extensively by HOARE AND WORK (1957). DAP can occur in bacteria as the LL-, meso (DL)-, or DD- isomers and occasionally the LL- and meso isomers together. The detection of DAP in hydrolysates of whole bacteria (DEWEY AND WORK, 1953) closely agrees with the occurrence of this amino acid in isolated cell walls and only one exceptional divergence has been noted where no DAP was detected in vegetative cells of *Bacillus sphaericus* but was found in the bacterial spores (POWELL AND STRANGE, 1957). Similarly for the distribution of the isomers of DAP there has been excellent agreement between the forms found in whole cell hydrolysates with those detected in the isolated bacterial cell walls. The meso-isomer is most widely distributed in bacteria and walls and so far as the author is aware, it is the only form detected in the walls of Gram-negative bacteria. The distribution of the different isomers of DAP in walls isolated from various groups of bacteria is presented in Table 35.

Evidence so far available suggests that only L-lysine is present in walls (see quantitative results of IKAWA AND SNELL (1960) in Table 33). Apart from the principal amino acid constituents some cell walls contain smaller, but significant amounts of serine. Although D-serine is known to occur in Nature, preliminary experiments with hydrolysates of staphylococcal walls suggest that most of the cell-wall serine behaves as the L-isomer in microbiological assays (SALTON, unpublished results).

One of the unusual features of the amino acid composition of bacterial cell walls has, therefore, been the widespread occurrence of alanine, glutamic acid and aspartic acid as the D or DL-isomers.

Apart from bacterial cell walls and their conspicuous D-isomers of amino acids, the only other group of substances which has attracted attention recently for similar reasons is the peptidoglycolipid class of compounds isolated from mycobacteria (LEDERER, 1961). These peptidoglycolipids appear to be closely associated with the 'native' cell wall of mycobacteria but little can be said about their attachment to the rigid component. The methods used in 'cleaning up' walls of mycobacteria could well remove these constituents and leave only the glycosaminopeptide and firmly bound polysaccharides. Mycoside C possesses a heptapeptide composed of D-phenylalanine, D-*allo*-threonine and D-alanine in molar ratios of 1:3:3.

The peptide portion of the complex peptidoglycolipid wax D isolated from *Mycobacterium tuberculosis* contained 2 moles of *meso*-DAP, 2 of L-alanine, 1 of D-alanine and 2 of D-glutamic acid (ASSELINEAU, BUC, JOLLES AND LEDERER, 1958). With the presence of these typical 'wall' amino acid components and the monosaccharides arabinose, mannose and galactose and amino sugars in wax D, a very close relationship of this peptidoglycolipid with the cell-wall structure of mycobacteria seems likely. The list of D-amino acids in these bacterial peptidolipids has now been extended to D-*allo*-isoleucine found in the compound from *Nocardia asteroides* and D-leucine in the mycoside C fraction of *Mycobacterium avium* (IKAWA AND SNELL, 1962).

Identification of free amino groups and C-terminal amino acids

The first attemps to identify cell-wall amino acids possessing free amino groups were made by INGRAM AND SALTON (1957) using the method introduced by SANGER (1945) for determining the N-terminal residues in proteins as the dinitrophenyl-amino acids. By reacting the cell walls with 1-fluoro-2,4-dinitrobenzene (FDNB), INGRAM AND SALTON (1957) were able to identify the free amino groups in walls from *Micrococcus lysodeikticus*, *Sarcina lutea* and *Bacillus megaterium* as belonging to alanine (detected as DNP-alanine), the ε-NH_2 group of lysine and the ε-NH_2 group of diaminopimelic acid in the case of walls from the latter organism. The walls of *Micrococcus lysodeikticus* and *Sarcina lutea* (now known to be devoid of the teichoic acids) yielded a small number of 'N-terminal' alanine residues (approximately 1 per 50,000) but they did contain appreciable amounts of lysine with its ε-NH_2 group free for reaction with FDNB.

Not all of the residues of lysine and α, ε-diaminopimelic acid of bacterial walls have their ε-NH_2 groups free (INGRAM AND SALTON, 1957; SALTON, 1961). About 1/3rd of the DAP residues in the walls of *Bacillus megaterium* were free for reaction, thus suggesting that the remaining 2/3 rds were either 'cross-linked' or substituted in some way (INGRAM AND SALTON, 1957). The possibility that amino acids such as lysine and DAP may form cross-linked peptides was also contemplated by WORK (1957). In the wall preparations of *Micrococcus lysodeikticus* used by INGRAM AND SALTON (1957) most of the lysine had its ε-NH_2 group unsubstituted.

The free amino groups of cell walls from a variety of Gram-positive and Gram-negative bacteria have been identified by BROWN (1958), SALTON (1961) and a selection of data is given in Table 36. With the variety of 'N-terminal' groups found in walls of certain species it is evident that no simple interpretation of these results in terms of peptide structure is possible at this stage. The investigations of the structure of bacitracin (ABRAHAM, 1957) illustrated some of the unusual features to be encountered in bacterial peptides and it would be surprising if novel structures were not present in the bacterial walls. The possibility that single amino acid substituents on the ε-NH_2 of lysine (or DAP) in walls of some species could add to the complexity of the variety of free amino groups detected in cell walls has already been contemplated (SALTON, 1961). The data presented in Table 36 do make it clear that a great deal of structural work will have to be performed before it is possible to interpret such results.

Although the walls or compound membranes of Gram-negative bacteria contain more protein than glycosaminopeptide, the contribution of N-terminal amino acids other than those found in the rigid components is not very great. At present there is no information about the size or mode of linkage of the 'protein' components with other constituents of the wall.

In bacterial walls containing polymers of teichoic acids, the O-ester-linked D-alanine (ARMSTRONG, BADDILEY, BUCHANAN, CARSS AND GREENBERG, 1958) will have its amino group free for reaction with FDNB. This complicates attempts to obtain an estimate of subunit size based on quantitative estimation of 'N-terminal' amino acids. The effects of removing the teichoic acids from the walls of *Staphylococcus aureus* and *Lactobacillus arabinosus* are shown in Table 37 and indicate the decrease in the contribution of alanine to the free amino groups. An increase in free amino groups belonging to other amino acids accompanied the loss of alanine due to removal of the teichoic acid; such observations indicate that under the conditions of extraction used in these studies some hydrolysis or exposure of new amino groups occurred. Wall preparations of *Staphylococcus aureus* were extracted for about 3 weeks in the cold with TCA to remove the teichoic acids and the residues and original walls supplied by Dr. J. L. STROMINGER showed a decrease in alanine with free amino groups by 127 μM/g but a corresponding increase of 138 μM glycine/g and 12 μM ε-NH_2-lysine/g was observed.

The development of specific enzymes for removing teichoic acids without degrading the residual wall structure would be of great value in further structural analysis of cell-wall peptides.

The two techniques commonly used for determining C-terminal groups in proteins are the carboxypeptidase and hydrazinolysis methods. Both SALTON (1958b) and PERKINS AND ROGERS (1959) found that carboxypeptidase had no action on cell walls of Gram-positive bacteria, nor was there any action on the products of digestion with lysozyme. For structures resistant to carboxypeptidase, the chemical method of determining the C-terminal amino acids by reaction with anhydrous hydrazine is available. This technique was introduced by AKABORI, OHNO AND NARITA (1952) and although the recoveries of C-terminal residues are poor, modifications by NIU AND FRAENKEL-CONRAT (1955) and BRADBURY (1958) can give consistent and satisfactory results.

Proteins subjected to hydrazinolysis usually give the maximum yields of C-terminal groups on heating for 8–12 hours at temperatures of up to 100°. BRADBURY (1958) found that better recoveries were obtained at lower temperatures (60°) when hydrazinolysis was carried out in the presence of hydrazine sulphate. The C-terminal groups are liberated as the free amino acids and after removal from the amino acid hydrazides they are recovered and determined as the DNP derivatives. SALTON (1961) followed the reaction of cell walls with hydrazine, determining the C-terminal amino acids released as the DNP compounds using the detailed technique described by NIU AND FRAENKEL-CONRAT (1955). Very clean reaction products were obtained on hydrazinolysis of the cell walls of *Micrococcus lysodeikticus* and the time course for the release of products (the values uncorrected for losses) is shown in Fig. 46. By determining the recoveries of known amino acids and C-terminal groups of synthetic peptides

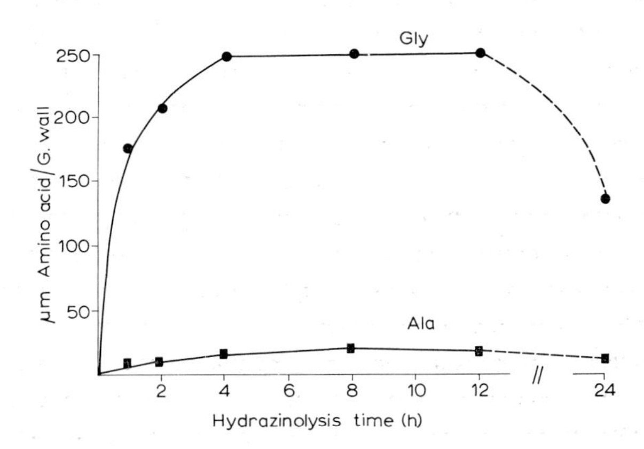

FIGURE 46. The course of liberation of 'C-terminal' amino acids from cell walls of *Micrococcus lysodeikticus* heated with anhydrous hydrazine at 100°. Curve A-glycine; Curve B-alanine. All values uncorrected for losses on heating with hydrazine and conversion to DNP amino acids (SALTON, 1961).

in the presence and absence of cell walls during heating with anhydrous hydrazine, sufficiently consistent values were obtained to conclude that the method could be applied to studies with walls (SALTON, 1961).

Identification and quantitative determinations of the C-terminal groups in a variety of cell walls (both Gram-positive and Gram-negative species) were performed by the author (SALTON 1961) and the results are given in Table 38. The presence of unknown compounds in the reaction mixtures indicated that degradation had occurred in certain walls reacted with anhydrous hydrazine. In some cell walls as many as four different 'C-terminal' amino acids were found, although in quite a few instances the numbers of such groups were small. As with the detection of free amino groups, the identification of C-terminal residues in whole cell walls has not provided data easy to interpret in terms of what is known of cell-wall peptide structure at present.

The only walls releasing substantial amounts of amino acids as 'C-terminal' groups were *Micrococcus lysodeikticus*, *Sarcina lutea* (both glycine and the latter glutamic acid as well). An appreciable amount of 'C-terminal' diaminopimelic acid was recovered on hydrazinolysis of *Bacillus megaterium* cell walls (see Table 38). Removal of the teichoic acid from two strains of *Staphylococcus aureus* also resulted in increases of the 'C-terminal' groups released on hydrazinolysis as shown in Table 39. Relatively small changes accompanied the removal of the teichoic acid from the walls of *Lactobacillus arabinosus* (SALTON, 1961) as seen from the data in Table 39.

SCHOCHER, JUSIC AND WATSON (1961) have also identified alanine and DAP as the C-terminal groups of the purified glycosaminopeptide of *Aerobacter cloacae* by hydrazinolysis.

The possibility that bacterial walls may possess special structures such as single amino acid substituents on the carboxyl groups of muramic acid or the γ-carboxyl group of glutamic acid was suggested by SALTON (1960). Such amino acids would presumably behave as 'C-terminal' groups and would not of course be related to the peptide structures. The detection, by PERKINS AND ROGERS (1959), of a compound in the diffusable portion of partial acid hydrolysates of *Micrococcus lysodeikticus* walls added some support to this suggestion. Although obtained only in small yield the compound contained muramic acid, glucosamine and glycine in equimolar proportions. In addition to this compound, HEYMANN, ZELEZNICK AND MANNIELLO (1961) have released by lysozyme action on *Streptococcus pyogenes* Group A cell walls a product yielding on hydrolysis, a mixture of muramic acid, glucosamine and alanine in molar ratios of 0.77:0.94:1. This compound moreover, gave a substance containing muramic acid and alanine after N-acetylglucosamine had been liberated by β-N-acetylglucosaminidase. If the amino acids of these two compounds, isolated from *Micrococcus lysodeikticus* and group A streptococcal walls, are linked to the carboxyl group of muramic acid then on hydrazinolysis they would yield 'C-terminal' residues of glycine and alanine.

The results obtained from the structural analysis of cell walls using the conventional methods for protein studies by reaction with FDNB and by hydrazinolysis, emphasize the individuality of the cell walls, but at this stage it is not possible to use the data, without reservation, for subunit size determinations. Part of the difficulty arises from our lack of information of the number of polymers in the walls of even the simpler Gram-positive cell walls. Whether the glycosaminopeptide or mucopeptide is a mixture of polymers cannot be said at present. At least with the Gram-negative bacteria the recent work of MANDELSTAM (1962) seems to indicate that the mucopeptides of walls from these organisms have more constant ratios of principal amino acids and muramic acid than those observed for walls of Gram-positive bacteria.

3. Amino sugar composition

Amino sugars have long been known to be important constituents of structural polysaccharides and mucopolysaccharides (KENT AND WHITEHOUSE, 1955). The N-acetylglucosamine polymer, chitin is widely distributed in Nature and on the basis of microchemical tests it was thought to be present in bacteria (VIEHOEVER, 1912). Indeed, early investigations of the composition of isolated cell walls revealed the presence of amino sugars. By means of the Elson and Morgan reaction the occurrence of substantial amounts of amino sugar in walls of the organism *Streptococcus faecalis* (SALTON, 1952) was established. The Elson and Morgan reaction of hydrolysed walls has been used to give a quantitative measure of the amino sugar contents of walls of Gram-positive and Gram-negative bacteria and some typical results are presented in Table 40. It is of interest to note that walls of Gram-positive bacteria are generally richer in amino sugar than those of Gram-negative species and we now know that this property is associated with the amount of glycosaminopeptide (mucopeptide) component.

The estimation of amino sugars as 'glucosamine equivalents' by the Elson and Morgan reaction after acid hydrolysis of isolated walls has given an approximation of the 'total amino sugar contents since most of the determinations performed in the past have not been corrected for the individual amino sugars known to be present in cell walls. Moreover, 'total' amino sugar contents will represent a balance between the release of free amino sugar by acid hydrolysis from a wall polymer and the destruction of the amino sugar by hydrolytic procedures. The extent to which these factors may contribute to the values is illustrated by the data for hydrolysis of *Micrococcus lysodeikticus* walls (SALTON AND PAVLIK, 1960) presented in Table 41.

The evidence so far available suggests that most, if not all of the amino sugar is present in the cell walls as the N-acyl (probably N-acetyl) derivative. INGRAM AND SALTON (1957) were unable to detect any free amino groups belonging to amino sugars in bacterial walls reacted with 1-fluoro-2,4-dinitrobenzene (FDNB). It seems unlikely that the hydrolytic conditions used after treatment with FDNB were sufficiently drastic to destroy all of the dinitrophenyl-amino sugars. Small amounts of a compound corresponding to DNP – muramic acid were detected in lysozyme digests reacted with FDNB and hydrolysed in the usual way (INGRAM AND SALTON, 1957). N-acetyl amino sugar compounds have been detected and isolated from partial acid hydrolysates and enzymic digests of bacterial cell walls (SALTON, 1956; PERKINS AND ROGERS, 1959; SALTON, 1959; PERKINS, 1960a; SALTON AND GHUYSEN, 1960) and there is good evidence that in some bacterial species the cell-wall amino sugars may also be O-acetylated (BRUMFITT, WARDLAW AND PARK, 1958). Only N-acetyl amino sugar compounds are found in enzymic digests of walls, although false positive reactions for free amino sugars have been reported but have now been shown to be due to removal of N-acetyl groups on heating under alkaline conditions (PERKINS, 1960b).

Unlike chitin which is a homopolymer of N-acetylglucosamine, the bacterial cell walls yield several different amino sugars in hydrolysates. Both glucosamine and galactosamine previously known to occur in natural products have been identified as cell-wall constituents. In addition one of the key substances in bacterial wall glycosaminopeptides is the amino sugar muramic acid, discovered and isolated by STRANGE and his colleagues (see below). Another amino sugar recently reported in a wall compound is an amino uronic acid found in a polymer extracted from the cell wall of *Micrococcus lysodeikticus* by PERKINS (1962). This amino uronic acid and glucose appear to be the two sugar components of the polymer and evidence has been presented which suggests that the amino uronic acid is probably 2-amino-mannuronic acid (PERKINS, 1963).

In addition to the detection of glucosamine, galactosamine, muramic acid and an amino uronic acid in isolated cell walls several other amino sugars have been reported in bacterial polysaccharides. The occurrence and distribution of amino sugars in bacterial products have been admirably reviewed recently by SHARON (1963) and it is of interest to note that several new amino sugars have been isolated from surface and capsular polysaccharides of bacteria. CRUMPTON AND DAVIES (1958) discovered that the lipopolysaccharide from *Chromobacterium violaceum* contained the N-acetyl derivative of D-fucosamine. An unidentified amino sugar was found in a different strain of this organism by CORPE (1958). SHARON AND JEANLOZ (1959, 1960) discovered an extremely interesting amino sugar, a diaminohexose, in a polysaccharide isolated from *Bacillus subtilis*. More recently the C-substances of pneumococci have been investigated by SHABAROVA, BUCHANAN AND BADDILEY (1962) and mannosamine detected in certain compounds. There is now a growing body of information about the occurrence of amino sugars in walls, bacterial polysaccharides and polymeric substances associated with isolated cell walls and some of the data available on their distribution is presented in Table 42.

Muramic acid

STRANGE AND POWELL (1953) detected an unknown amino sugar in the hydrolysis products obtained from the peptides secreted by germinating bacterial spores. The peptides contained alanine, glutamic acid, α,ε-diaminopimelic acid, glucosamine and the 'unknown' amino sugar. The similarity between the composition of these spore peptides and bacterial cell walls was striking and CUMMINS AND HARRIS (1954, 1956) reported the detection of an amino sugar of apparently identical behaviour on paper chromatograms of hydrolysed wall preparations. STRANGE AND DARK (1956) isolated the 'unknown' amino sugar in pure form as a crystalline substance from spore peptides and bacterial walls and the structure was elucidated by STRANGE (1956) as 3-O-carboxyethyl glucosamine. SALTON (1956) had also confirmed the presence of a carboxyl group in the 'unknown' amino sugar which was present in the N-acetylated disaccharide released from cell walls by digestion with lysozyme. The structure proposed by STRANGE (1956) received further support from the study of the mode of action of penicillin by PARK AND STROMINGER (1957). The relationship of 'muramic acid' (the name proposed by STRANGE for the cell-wall amino sugar and used by WORK, 1957) to glucosamine and galactosamine is illustrated in Fig. 47.

The structure of muramic acid as the 3-O-lactyl ether of D-glucosamine (3-O-carboxyethyl-

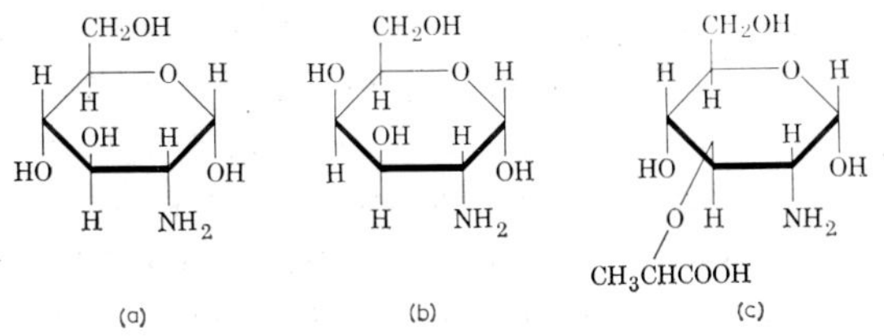

FIGURE 47. A comparison of the structures of the 2-amino sugars found in cell-wall polymers. (a) D-glucosamine; (b) D-galactosamine; (c) muramic acid, 3-O-carboxyethyl-D-glucosamine (3-O-D-lactyl-D-glucosamine).

D-glucosamine) was later confirmed in the chemical synthesis of this compound by STRANGE AND KENT (1959). The synthetic route involved the preparation of 4:6-O-benzylidine-α-methyl-D-glucosaminide (MOGGRIDGE AND NEUBERGER, 1938), its reaction with sodium hydride and the introduction of the carboxyethyl side chain by treating with ethyl bromopropionate. The series of reactions used by STRANGE AND KENT (1959) are illustrated in Fig. 48. ZILLIKEN (1959) and LAMBERT AND ZILLIKEN (1960) confirmed the synthesis of muramic acid using slight modifications in the procedure as indicated in Fig. 49.

Although the overall yields of muramic acid from D-glucosamine were not reported by STRANGE AND KENT (1959) nor by LAMBERT AND ZILLIKEN (1960) it is generally recognized

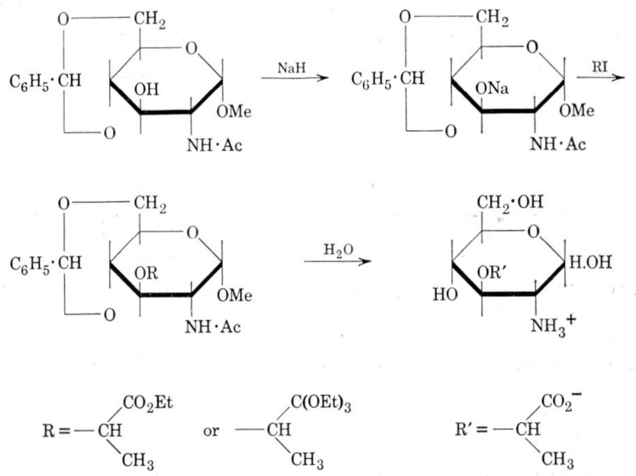

FIGURE 48. The sequence of reactions used in the chemical synthesis of muramic acid (STRANGE AND KENT, 1959).

that they were comparatively low. Improvements in the chemical synthesis of muramic acid have been introduced by MATSUSHIMA AND PARK (1962) with better yields of the final product. The method used 4,6-benzylidene-α-methyl N-acetyl-D-glucosaminide as an intermediate which is then converted to the muramide by reaction with L-α-chloropropionic acid.

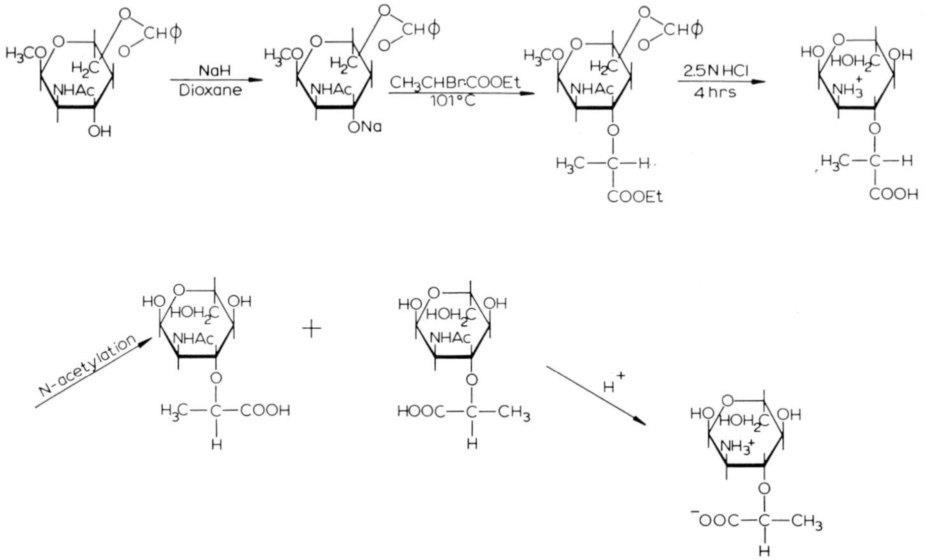

NaH
Dioxane
$CH_3CHBrCOOEt$
101°C
2.5N HCl
4 hrs
N-acetylation
+
H^+

3-O-L(I)-lactyl-

3-O-D(-)-lactyl

3-O-D(-)-lactyl-D-glucosamine

FIGURE 49. The chemical synthesis of 3-O-lactyl ethers of N-acetyl-D-glucosamine used by LAMBERT AND ZILLIKEN, 1960.

A new method of synthesis of muramic acid and related 3-O substituted glucosamine compounds has been developed by GIGG AND CARROL (1961). Their synthesis involved conversion of a Zervas intermediate (prepared by converting N-benzoyl-D-glucosamine into 2-phenyl-4,5-[5,6 isopropylidine-D-glucofurano] Δ^2-oxazolium) into a 3-O-substituted compound after reaction with NaH and introduction of the substituent groups. The structure of the Zervas intermediate used in the synthesis is illustrated in Fig. 50. By this method several 3-O-substituted glucosamine derivatives were synthesized in quite high yields.

The properties of synthetic and naturally occurring muramic acid were investigated by STRANGE AND KENT (1959) and a comparison of some of their characteristics is given in Table 43.

In 1957 PARK AND STROMINGER also confirmed the identity of the acidic amino sugar originally reported in the nucleotides isolated from penicillin-treated *Staphylococcus aureus* cells by PARK (1952) as muramic acid. With the identification of muramic acid as an important amino sugar constituent of bacterial cell walls and the occurrence of the N-acetyl compound of muramic acid in the Park nucleotides, the probable biochemical functions of the latter and the mechanism of penicillin action became much clearer (Park and Strominger, 1957).

Muramic acid has generally been found only in microbial cell-wall glycosaminopeptides and the nucleotide wall 'precursors'. An unknown amino sugar was reported by Guex-Holzer and Tomcsik (1956) in the capsular polysaccharide which was serologically related to the cell-wall 'polysaccharide'. Whether this substance was indeed muramic acid has not yet been clarified. Thus at present the natural occurrence of muramic acid seems to be confined to the mucopeptide structures and biosynthetic precursors.

Of all the constituents of glycosaminopeptides, muramic acid is probably the most distinctive and the most widely found compound, and its detection in microbial cells or wall or envelope structures can be used as a fairly reliable indicator of the presence of the wall glycosaminopeptides. DAP is the next best indicator but it is not as universally found in bacterial walls. Muramic acid has been detected in cell-wall preparations of most of the Gram-positive

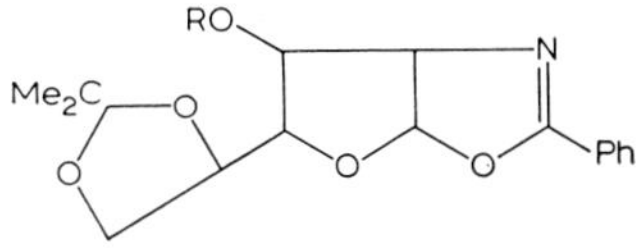

Figure 50. The structure of the Zervas intermediate (2-Phenyl-4,5-(5,6 isopropylidine-D-glucofurano) $\triangle^2$-oxazolium) used by Gigg and Carroll (1961) for the synthesis of 3-o-substituted glucosamine derivatives.

and Gram-negative bacteria examined as well as in the walls of blue-green algae (Salton, 1960; Frank, Lefort and Martin, 1962) and in Rickettsias (Allison and Perkins, 1960). Neither muramic acid nor diaminopimelic acid was detectable in the pleuropneumonia-like organism, *Mycoplasma mycoides* (Plackett, 1959) or in *Halobacterium halobium* (Brown and Shorey, 1962).

Because of its possession of a carboxyl group and a reducing moiety, muramic acid is capable of performing the dual function of covalently linking peptides (through amide bonds) and amino sugars (through glycosidic linkages) in the glycosaminopeptide heteropolymers. Muramic acid may occur in the cell-wall mucopeptide as an unsubstituted N-acetyl amino sugar glycosidically linked to N-acetylglucosamine as well as N-acetyl amino sugar carrying peptide or amino acid substituents. A more detailed discussion of the key role muramic acid plays in glycosaminopeptide structure is given in Chapter 5.

Glucosamine and galactosamine

Glucosamine is the only other amino sugar which is as widely distributed in bacterial cell walls as muramic acid. Unlike muramic acid it may be present in cell-wall polymers other than the mucopeptide. Thus it is present in the glycosaminopeptide and the teichoic acid of *Staphylococcus aureus* (Baddiley, Buchanan, Rajbhandary and Sanderson, 1962). In all of these cell-wall compounds glucosamine appears to be present as the N-acetyl compound.

From the studies of CUMMINS AND HARRIS (1956a, b; 1958) it is apparent that galactosamine is much less widely distributed as a cell-wall constituent than glucosamine and muramic acid. Galactosamine has been detected in wall preparations from various taxonomic groups of Gram-positive bacteria including *Lactobacillus* spp., *Streptococci*, *Arthrobacter* spp., *Corynebacterium* spp. (CUMMINS AND HARRIS, 1956a, b; 1958; 1959; CUMMINS, 1962). It is not known at present if all of the galactosamine occurs in bacterial cell walls as the N-acetyl compound. The galactosamine has not been found in purified glycosaminopeptide fractions and it seems likely that it may be present in cell-wall components other than the mucopeptide.

A mucopolysaccharide extracted from cell-wall preparations of *Bacillus subtilis* (strain 6346) with hot 5% trichloroacetic acid was isolated by JANCZURA, PERKINS AND ROGERS (1961). This wall polymer consisted of equivalent amounts of N-acetylgalactosamine and glucuronic acid and has accordingly been called 'teichuronic acid' by its discoverers. The relationship of teichuronic acid to chondroitin, a mucopolysaccharide of identical composition, is of some interest. The *Bacillus subtilis* polymer differs from chondroitin in its ease of hydrolysis, direction of optical rotation and resistance to testicular hyaluronidase (JANCZURA, PERKINS AND ROGERS, 1961). From a study of the properties of the teichuronic acid, JANCZURA, PERKINS AND ROGERS (1961) suggested that the glycosidic bonds were α in configuration and that the glycosidic linkage between at least part of the glucuronic acid and the N-acetylgalactosamine was likely to be on C-3 of the amino sugar. Teichuronic acid has so far only been reported in strain 6346 of *Bacillus subtilis* and there is at present no indication of its occurrence in walls of other bacterial species.

N-acetylgalactosamine has been found recently in the glycerol teichoic acid of the wall of *Staphylococcus albus* and in the ribitol teichoic acid of *Streptococcus faecalis* cell walls (BADDILEY, 1962). It will be of great interest to see whether galactosamine occurs only in certain types of teichoic acid or whether it will be present also in oligo-and polysaccharides of bacterial cell walls.

4. Monosaccharides of bacterial walls and lipopolysaccharides

Bacteria have long been known for the variety of polysaccharides they can synthesize. Capsular polysaccharides, bacterial gums and extracellular polysaccharides have commanded a great deal of attention in the past. The pneumococcal polysaccharides have been studied in considerable detail and much is now known about their structure and specificity of immunological reactions (HEIDELBERGER, 1956; HEIDELBERGER AND PLESCIA, 1960; STACEY AND BARKER, 1960). Interest in bacterial 'somatic' polysaccharides was stimulated by the early work of MORGAN (1936) when he began his investigations on the isolation and chemical characterization of the specific polysaccharides of the dysentery group of Gram-negative bacteria. The chemistry of the somatic O-antigens of Gram-negative bacteria continued to receive attention from MORGAN (1949) and his colleagues and more recent studies on these polysacharides and lipo-polysaccharides have come from the laboratories of DAVIES, GOEBEL AND WESTPHAL.

Although quite a lot was known about the chemical composition of the polysaccharides and their complexes from Gram-negative bacteria, little was known about their precise anatomical status in the bacterial cell, apart from the fact that they were obviously 'surface antigens'. The 'somatic' antigens of Gram-positive bacteria had only been touched upon by work on the C-carbohydrates of pneumococci and streptococci. As for cell-wall polysaccharides these were ill-defined until methods were perfected for isolating the walls as homogeneous structures. Since then the location of bacterial polysaccharides in the wall or in association with wall or membrane components has been determined by direct experimentation.

The early investigations on isolated walls by HOLDSWORTH (1952) and by SALTON (1952) established the presence of polysaccharide or oligosaccharide components giving mixtures of sugars upon hydrolysis. Studies with cell walls of Gram-negative bacteria (SALTON, 1953) also revealed the presence of the monosaccharide components of the serologically specific polysaccharides. Following these first studies with isolated cell walls, extensive investigations have been carried out by CUMMINS AND HARRIS (1956a, b; 1958) and CUMMINS (1962) on the identification of the sugars in relation to the various taxonomic groups of Gram-positive bacteria. In this work, isolated cell walls have been used exclusively. On the other hand most of the studies on the chemistry of the polysaccharides of Gram-negative bacteria have involved the isolation of the lipopolysaccharides from whole cells and the confirmation that the typical monosaccharides of these substances are also present in the walls followed later (SALTON, 1960). With the elegant and comprehensive studies on the isolation and characterization of the dideoxy sugars by WESTPHAL, STAUB AND DAVIES and their colleagues, and the detection and isolation of heptoses (JESAITIS AND GOEBEL, 1952; DAVIES, 1957 and WEIDEL, 1955) this field has become a very extensive one and only selected aspects can be reviewed in this book.

The discussion of the monosaccharide components of bacterial walls will of course be confined to sugars other than the amino sugars, unless the amino sugars have been shown to be part of the homogenous isolated soluble oligo- or polysaccharides or the lipopolysaccharides. The monosaccharide constituents of the walls will be dealt with separately under Gram-positive and Gram-negative groups. As there are only several studies of the structures of polysaccharides from walls of Gram-positive organisms they will be referred to in this section, for at present there are insufficient data to warrant a separate chapter. By contrast a great deal has been done on the chemistry of the sugar constituents and the structure of the lipopolysaccharides of Gram-negative bacteria and their relationships to immunochemical phenomena and various facets of this subject have been reviewed admirably by WESTPHAL (1956), DAVIES (1960) AND STAUB (1959) and no attempt will be made to cover much of this material here.

Monosaccharides of walls of Gram-positive bacteria

Following the systematic studies of CUMMINS AND HARRIS (1956a, b; 1958; 1959) and CUMMINS (1962), it has become apparent that the monosaccharides identified on hydrolysis of bacterial walls are of considerable taxonomic importance. Thus certain sugars may be typical for most species of a genus e.g. rhamnose is the characteristic monosaccharide identified in streptococcal walls. Most of the major groups of Gram-positive bacteria have been covered in the survey of CUMMINS AND HARRIS and in investigations of other workers (SALTON AND PAVLIK, 1960; SALTON, 1960a). The principal combinations of monosaccharides detected in various groups of Gram-positive bacteria are given in Table 44.

Very little has been done on the quantitative estimation of cell-wall polysaccharides and their constituent sugars. HOLDSWORTH (1952) was the first investigator to isolate an oligosaccharide from a bacterial wall. The oligosaccharide had a molecular weight of about 1,000 and it contained D-arabinose, D-mannose and D-galactose in molar ratios of 3:1:2.

MCCARTY (1952) determined the amount of rhamnose in isolated walls of group A streptococci and in the purified 'group-specific' polysaccharide prepared by degrading the walls with the enzymes from *Streptomyces* sp. Polysaccharides obtained by digestion with the Streptomyces enzymes were incompletely freed of the amino acid constituents of the wall glycosaminopeptides. In later investigations, KRAUSE AND MCCARTY (1961) discovered that the wall polysaccharides could be solubilized by extraction with hot formamide. The sugar and amino sugar composition of the formamide extracted polysaccharide from group A and group A- variant streptococcal walls is compared in Table 45 with that of the original wall and formamide-insoluble residue.

IKAWA (1961) extracted the cell walls of *Lactobacillus plantarum (L. arabinosus)*, *Lactobacillus casei* and *Streptococcus faecalis* with various solvents including 0.1*N* NaOH, and trichloroacetic acid (TCA) in the hot and cold. The products so obtained were complex mixtures but the teichoic acids and fractions containing cell-wall rhamnose and amino sugars constituted the major products. From these investigations IKAWA (1961) concluded that the cell walls of certain organisms contained 'non-teichoic acid-type polysaccharides'. In the products derived from *Lactobacillus plantarum* and *Streptococcus faecalis* walls rhamnose was found in fractions which also contained the compounds characteristic of the teichoic acid polymers e.g. ribitol, phosphate, or glycerol phosphate and D-alanine. It would be nice to know whether these fractions behave as homogeneous compounds on electrophoresis or by examination with immunochemical agar-gel diffusion methods. There was good evidence for the presence of two distinct polysaccharides in the wall of *Lactobacillus casei;* one contained rhamnose and glucose and the other yielded rhamnose, glucose and amino sugar on hydrolysis. ARCHIBALD, BADDILEY AND BUCHANAN (1962) have also found that a polysaccharide containing rhamnose, galactose and glucosamine is extracted together with the teichoic acid from the walls of *Lactobacillus arabinosus.*

KNOX AND BRANDSEN (1962) found that soluble compounds containing rhamnose were released on autolysis of isolated cell walls of *Lactobacillus casei* strains. The fractionation procedures employed in these investigations did not separate the mucopolysaccharide and mucopeptide, and KNOX AND BRANDSEN (1962) concluded that these components are joined covalently Two different polysaccharides are believed to be present in the wall of *Lactobacillus casei*, the principal differences being due to the preponderance of galactose and glucose in one fraction and rhamnose and hexosamine in the other.

Although uronic acids are fairly common constituents of bacterial capsular polysaccharides and extracellular gums and slimes they have not been encountered very often in cell wall preparations. The first well-defined product yielding uronic acid on hydrolysis was the 'teichuronic acid' polymer extracted from the cell walls of *Bacillus subtilis* with TCA by JANCZURA, PERKINS AND ROGERS (1961). This mucopolysaccharide was composed of equimolar amounts of acetylgalactosamine and glucuronic acid which constituted 80–90% of the dried material. This mucopolysaccharide thus contained the same components as chondroitin. The properties of the teichuronic acid strongly suggested that the glycosidic bonds were α in configuration and that the linkage between at least part of the glucuronic acid and the N-acetylgalactosamine is likely to be on C-3 of the amino sugar.

Another substance given the general name of teichuronic acid has been isolated by PERKINS (1962) from the walls of *Micrococcus lysodeikticus*. This polysaccharide could be extracted from the walls with dilute TCA or with formamide. It contained glucose and what appeared to be small amounts of hexosamine. Further investigations of the small amount of 'hexosamine' showed that it was a residue of an acidic compound of marked lability to acid hydrolysis and that it now appears to be an aminohexuronic acid. This polymer therefore seems to be composed of equimolar amounts of glucose and probably aminomannuronic acid (PERKINS, quoted by ROGERS, 1962; PERKINS, 1963) and thus differs from the *Bacillus subtilis* teichuronic acid and the use of this term for these components should not be taken at this stage to infer any general structural relationships between the two polysaccharides. PERKINS (1963) does not refer to the aminouronic acid-glucose polymer as a 'teichuronic acid'.

The anatomical location of the polysaccharide isolated from *Bacillus subtilis* cells by SHARON (1957) and SHARON AND JEANLOZ (1960) under similar extraction conditions to those yielding the cell-wall teichuronic acid, has not yet been established. However, the monosaccharides of this polymer, identified as galactose, glucosamine and a new amino sugar, 4-acetamido-2-amino-2,4,6-trideoxyhexose (SHARON, 1957; SHARON AND JEANLOZ, 1960) differ from the constituent sugars of the teichuronic acid.

Finally, mention should be made of the complex glycolipids isolated from mycobacteria. Although the chemically-defined glycolipids have not been isolated directly from the cell walls of mycobacteria, their relationship to these structures must be close since both peptide and sugar constituents detected in isolated walls are also found in these substances. Much of the chemical work on these glycolipids has come from LEDERER's laboratory and has been reviewed recently by LEDERER (1961). Wax D has a molecular weight of about 54,000, of which about half can be accounted for by ester linked, mycolic acid and the remaining portion a peptidopolysaccharide. The heptapeptide of α, ε-diaminopimelic acid, alanine and glutamic acid is believed to be linked through amino sugar to a polysaccharide containing D-arabinose, D-mannose, D-galactose, glucosamine and galactosamine, the whole polysaccharide having a molecular weight of 26,000.

The mycosides of mycobacteria are of considerable chemical interest but at present the location of these substances in the bacterial cell wall has only been recently established. In addition to containing D-amino acids (LEDERER, 1961), mycoside C also yields the following three different 6-deoxyhexoses: 6-deoxytalose, 3-O-methyl-6-deoxytalose and 3,4-di-O-methylrhamnose (MACLELLAN, 1962). The detection of 6-deoxy-L-talose in the cell-wall carbohydrate of a strain of *Actinomyces bovis* in an earlier study by MACLELLAN (1961) suggests that the mycosides could well be derived from the walls of mycobacteria.

Monosaccharides of walls and lipopolysaccharides of Gram-negative bacteria

An examination of the monosaccharides of isolated cell walls of *Salmonella pullorum* revealed the presence of glucose, galactose, mannose and rhamnose (SALTON, 1953). Such sugars had been identified in the immunologically specific polysaccharides isolated as lipopolysaccharide complexes from Gram-negative bacteria by the phenol extraction method of LÜDERITZ AND WESTPHAL (1952a, b). These early studies indicated that the surface antigens were still attached or adsorbed to the wall and this together with the detection of protein, lipid and α, ε-diaminopimelic acid pointed to the chemical heterogeneity of these structures from Gram-negative bacteria (SALTON, 1953).

The suggestion that these antigenic polysaccharides should be classified anatomically as 'microcapsules' has already been discussed in Chapter 1. It is now fairly evident that these polysaccharides constitute part of the protein-polysaccharide-lipid complex of the outer layers of the multilayered envelope, wall or compound membrane of the cells of Gram-negative bacteria (WEIDEL AND PRIMOSIGH, 1958; WEIDEL, FRANK AND MARTIN, 1960; SALTON, 1960). So far as the author is aware there is no quantitative information on the amount of these polysaccharides in bacterial cells and it is not known whether any losses of these components occur on cell-wall isolation or upon treatment of crude wall fractions with trypsin. Where wall or membrane fractions are isolated by sonic disintegration of Gram-negative bacteria the fragmentation of the envelope accompanying cell rupture is bound to result in loss of the lipopolysaccharide-protein complexes.

Very little systematic work has been carried out on the identification of the monosaccharides of cell-wall preparations from Gram-negative bacteria. Studies have been confined to the elucidation of the detailed chemical architecture of the walls of *Escherichia coli* B by WEIDEL and his colleagues and to an examination of the sugars in isolated walls of a limited variety of Gram-negative bacteria by SALTON (1960a). The monosaccharides detected in 'wall' or 'membrane' fractions from these investigations together with those of BROWN (1960), NIKAIDO (1961), BROWN AND SHOREY (1962) and HUMPHREY AND VINCENT (1962) are summarized in Table 46.

The extent to which the composition of the cell-wall polysaccharides may be governed by the growth medium can now be judged from the investigations of NIKAIDO (1961; 1962). The consequences of growing galactose-negative mutant strains of *Salmonella enteritidis* on media devoid of galactose clearly established the deletion of the sugars galactose, mannose, rhamnose and tyvelose in the cell-wall polysaccharides (NIKAIDO, 1961; 1962). Thus the composition of the wall polysaccharides of 'M mutants' of *Salmonella enteritidis* and *Salmonella typhimurium* is governed by a single gene controlling the uridine diphosphogalactose-4-epimerase (epimerase). The addition of galactose to the growth medium on which the mutant strains were grown resulted in a restoration of the deleted sugars to the polysaccharide and in 20 minutes exposure to 0.1% galactose medium, the sugar composition of the wall carbohydrate was qualitatively indistinguishable from that of the wild types (NIKAIDO, 1962).

In addition to the differences observed in the composition of wall lipopolysaccharides of the *Salmonella* spp., it should be recalled that WEIDEL, FRANK AND MARTIN (1959) reported that growth of *Escherichia coli* B on media containing glucose instead of lactate, resulted in the production of a wall polysaccharide free from the heptose component. Whether the possession of a simpler variety of monosaccharide constituents in the walls of certain organisms shown in Table 46 is due to genetic losses of certain enzymes or phenotypic variations, cannot be said at present.

These studies showing variation in the monosaccharide composition of the lipopolysaccharides which are probably localized in the outer component of the wall or compound external membrane of Gram-negative bacteria, suggest that the possession of these components is not directly concerned with the stability of the structures. However, the lysis of the galactose-negative mutants of Salmonella by the addition of galactose could be relevant to this problem, although it would seem more likely that the accumulation of uridine diphosphogalactose could limit the availability of uridine triphosphate needed for the synthesis of uridine nucleotides concerned with the biosynthesis of the glycosaminopeptide portion of the cell wall (FUKASAWA AND NIKAIDO, 1961).

In contrast to the limited amount of work done on the direct examination of the polysaccharides of the isolated walls of Gram-negative bacteria, a great deal has been done on the chemistry and biological properties of the complexes separated from whole cells. Indeed, within the space of ten years, the 3,6-dideoxy hexoses of bacterial lipopolysaccharides have been characterized (WESTPHAL AND LÜDERITZ, 1960) and the main outlines of the immunochemistry of these substances established (STAUB, TINELLI, LÜDERITZ AND WESTPHAL, 1959; WESTPHAL, KAUFMANN, LÜDERITZ AND STIERLIN, 1960). As well as dideoxysugars, the aldoheptoses have been recognized as important monosaccharide components of these bacterial compounds and a number of these sugars have been detected and chemically characterized (DAVIES, 1960). It is now generally agreed that although these lipopolysaccharides have been isolated from whole cells, they occur in or on the cell-wall. KRÖGER, LÜDERITZ AND WESTPHAL (1959) have shown that the sugars of these complexes derived from smooth and rough strains of *Escherichia coli* O18 and *Salmonella paratyphi* B can also be detected in hydrolysates of cell walls isolated from these organisms.

Altogether five naturally occurring dideoxysugars have been isolated and chemically characterized. These sugars have been given names indicating their origins e.g. abequose was isolated from *Salmonella abortus equi*, tyvelose from *Salmonella typhi*, ascarylose from *Paras-*

caris equorium, paratose from *Salmonella paratyphi* A and colitose from *Escherichia coli* O111 (WESTPHAL AND LÜDERITZ, 1960). All of these dideoxysugars have been found in bacterial lipopolysaccharides, including ascarylose, which was first isolated from the glycolipid of the egg membrane of *Parascaris equorium* by FOUQUEY, POLONSKY AND LEDERER (1957). The distribution of the dideoxy sugars in various bacterial groups and the specific serological groupings determined by the presence of these sugars are shown in Table 47 (LÜDERITZ, 1960).

The configuration and the nomenclature of the dideoxysugars have been summarized by LÜDERITZ (1960) and their trivial and systematic names and structures are given in Fig. 51.

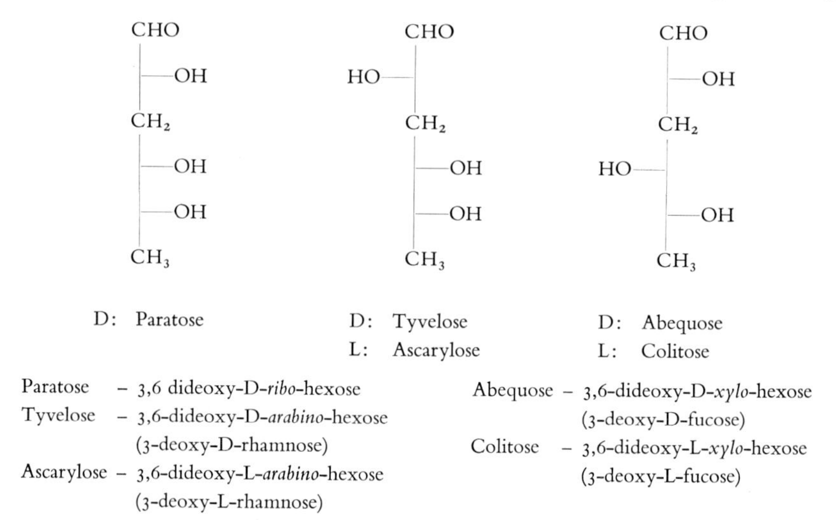

Paratose – 3,6 dideoxy-D-*ribo*-hexose
Tyvelose – 3,6-dideoxy-D-*arabino*-hexose (3-deoxy-D-rhamnose)
Ascarylose – 3,6-dideoxy-L-*arabino*-hexose (3-deoxy-L-rhamnose)
Abequose – 3,6-dideoxy-D-*xylo*-hexose (3-deoxy-D-fucose)
Colitose – 3,6-dideoxy-L-*xylo*-hexose (3-deoxy-L-fucose)

FIGURE 51. The configuration and nomenclature of 3,6-dideoxyhexoses.

The other 'novel' group of sugars found in bacterial lipopolysaccharides are the aldoheptoses. Although ketoheptoses have been known in Nature for some time (LA FORGE AND HUDSON, 1917; ROBISON, MACFARLANE AND TAZELAAR, 1938; BENSON, BASSHAM AND CALVIN, 1951; HORECKER AND SMYRNIOTIS, 1952) their occurrence in bacterial polysaccharides had not been reported until JESAITIS AND GOEBEL (1952) detected in lipocarbohydrate from *Shigella sonnei* Phase II, a sugar which behaved as a heptose when subjected to the DISCHE (1953) reaction.

SLEIN AND SCHNELL (1953) found an aldoheptose in the specific polysaccharide isolated from *Shigella flexneri* type 3. The aldoheptose occurred as the phosphorylated compound, the free heptose being liberated by treatment with seminal phosphatase. It was concluded that the heptose was L-*glycero*-D-*manno*-heptose (SLEIN and SCHNELL, 1953) which was later isolated also from the lipocarbohydrate fraction of *Escherichia coli* B walls by WEIDEL (1955).

The occurrence and nature of aldoheptoses in bacterial polysaccharides has been investigated in some detail by DAVIES and his colleagues (DAVIES, 1955; 1957a, b; 1958; MACLELLAN AND DAVIES, 1957; CRUMPTON, DAVIES AND HUTCHISON, 1958; MACLELLAN, 1957). It is likely that these compounds occur in the somatic lipopolysaccharides although they have also been found in organisms known to possess capsular polysaccharides (e.g. *Azotobacter indicum*, QUINNELL, KNIGHT AND WILSON, 1957). The presence of aldoheptoses in bacterial lipopolysaccharides may be more widespread than originally believed, for they have also been detected in all of the *Salmonella* spp. examined by KAUFFMANN, LÜDERITZ, STIERLIN AND WESTPHAL (1960). DAVIES (1960) has reviewed the distribution of heptoses in various bacterial groups and some of the data on their occurrence are presented in Table 48. The readiness with which the heptoses can be detected in small amounts of isolated cell walls was demonstrated by SALTON (1960) and Fig. 52 shows the characteristic absorption spectrum of heptoses when the walls are directly subjected to the DISCHE (1953) reaction.

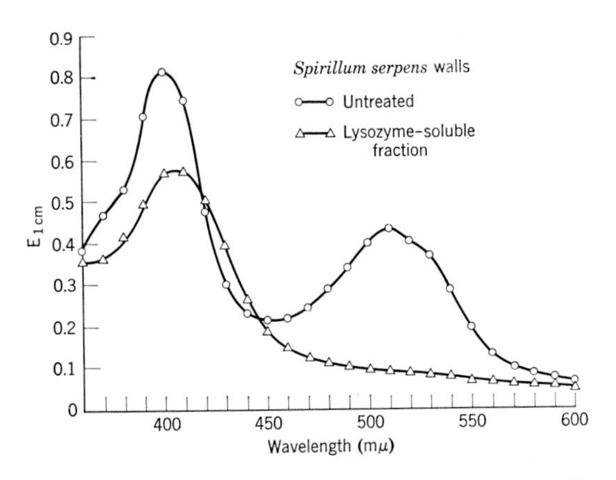

FIGURE 52. Absorption spectra of the Dische reaction for heptoses performed on cell walls of *Spirillum serpens* and on the lysozyme soluble fraction from the walls.

The monosaccharide constituents of over 300 bacterial polysaccharides isolated from Gram-negative bacteria have been summarized by DAVIES (1960) and his admirable review of this subject indicates the formidable amount of work which has gone into the investigations of these polymers. The important role that a single genetic lesion may have on the composition of these bacterial polysaccharides has been clearly established by the work of NIKAIDO (1961, 1962) and as a consequence the dramatic effect the medium may have on the qualitative constitution is amply illustrated. Thus if the medium included the sugar involved in the genetic lesion then qualitatively 'normal' polysaccharide is produced (NIKAIDO, 1962). From an inspection of the data collected by DAVIES (1960) it is evident that monosaccharide composition may vary as much within species as between species and genera of Gram-negative bacteria. DAVIES (1960) concluded that 'the distribution of kinds of sugars is somewhat characteristic of different genera, but is not likely to be of great taxonomic significance.' A selection of data on the monosaccharide units detected in different taxonomic groups of bacteria is given in Table 49.

Several points have emerged from the investigations of the lipopolysaccharides of Gram-negative bacteria. Glucose and to a slightly lesser extent galactose occur with much greater frequency than any of the other monosaccharides or groups of sugars e.g. mannose, 6 deoxy-hexoses, 3,6-dideoxyhexoses, pentoses, aldoheptoses and amino sugars. In Salmonellas a general picture of a heteropolysaccharide being built up from a basal structure of glucose and aldoheptose phosphate is beginning to emerge. Galactose thus seems to be essential for the introduction of the other monosaccharide components of the polysaccharide. To what extent this type of structure will be found in other groups of Gram-negative bacteria remains to be seen from future investigations in this field.

As pointed out by DAVIES (1960) very little direct work has been done on the chemical structure of these polysaccharides, especially those containing aldoheptoses. Immunochemical studies have on the other hand given a great deal of information and have enabled STAUB (1959) and WESTPHAL (1956) to deduce structures for some of the Salmonella polysaccharides.

5. *Cell-wall lipids*

One of the most conspicuous features emerging from the comparative studies of bacterial cell-wall composition, was the marked difference in lipid contents of walls isolated from Gram-positive and Gram-negative organisms (SALTON, 1953). Relatively small amounts of lipid could be extracted from the walls of many Gram-positive bacteria and even after acid hydrolysis the amount of ether-extractable material seldom exceeded about 1–2% (SALTON, 1952; 1953).

Walls of Gram-negative bacteria on the other hand usually contained substantial quantities of lipid and in several instances it accounted for as much as 20% of the weight of the wall. The lipid in the walls of certain Gram-negative organisms may be in two forms – one easily extracted by ether or methanol/ether and the other form more firmly bound and only released after drastic treatment of the wall (SALTON, 1953).

Although the lipid contents of the majority of Gram-positive bacteria are low, as may be expected there are certain organisms within this group possessing large amounts of lipid in the wall fractions e.g. mycobacteria. KOTANI, KITAURA AND TANAKA (1959) isolated walls from *Mycobacterium* BCG and obtained satisfactory preparations without resorting to the use of saponifying agents such as alcoholic-KOH (CUMMINS AND HARRIS, 1958). The cell-wall preparations contained a high proportion of the various wax fractions in addition to other lipids. The total amount of lipid accounted for about 60% of the weight of the cell wall and included wax fractions A, B, C and D, phosphatides and bound lipids. Walls of *Corynebacterium diphtheriae* also contained large quantities of lipid (30.5%) in the analyses performed by MORI, KATO, MATSUBARA AND KOTANI (1960). Major lipid components have not been mentioned by other investigators who have isolated walls of this organism, but it should be kept in mind that complex lipid or waxy materials have been found in coryneform organisms (LEDERER, 1961).

To estimate the 'total lipid' contents of walls, SALTON (1953) found that a preliminary hydrolysis with 6*N* HCl for 30 minutes–2 hours at 100° followed by other extraction gave the maximum values. Some typical results for a variety of walls of Gram-positive and Gram-negative bacteria are presented in Table 50. Determinations using such methods for the release of the firmly bound lipid would be far from absolute since some hydrolysis of the lipid and loss of certain groups are likely to have occurred under the drastic conditions required to liberate the material.

Very little work has been done on the chemistry of lipids derived from bacterial cell walls. FEW (1955) isolated lipid material from the cell envelope (outer and inner components, 'wall-membrane') of *Pseudomonas denitrificans* and from the wall and 'small particle fraction' (derived from disintegrated membranes) of *Staphylococcus aureus*. The lipid from *Pseudomonas denitrificans* envelope contained 1.1% N, 4.3% P and ethanolamine and serine were detected after hydrolysis. The results suggested that about half the lipid mixture consisted of phosphatidic acid and half nitrogen-containing phosphatides (FEW, 1955). Only 0.8% of the weight of wall of *Staphylococcus aureus* could be recovered as lipid and there was insufficient material for analysis (FEW, 1955).

Recent investigations of the lipids of several Gram-negative bacteria by KANESHIRO AND MARR (1962) have indicated that the principal lipid extractable from *Azotobacter agilis*, *Agrobacterium tumefaciens* and *Escherichia coli* with ethanol and methanol-chloroform was a phosphatidyl ethanolamine. Whole cells were used for these studies but the material for *Azotobacter agilis* was probably identical to that derived from the envelope preparations (COTA-ROBLES, MARR AND NILSON, 1958). The fatty acids of phospholipid from *Azotobacter agilis* were found to be myristic (7%), palmitic (35%), palmitoleic (41%) and octadecanoic (17%) acids. Phospholipid of *Escherichia coli* studied by KANESHIRO AND MARR (1961) contained a new major component of *cis*-9,10-methylene hexadecanoic acid, the C17 homologue of lactobacillic acid (*cis*-11,12-methylene octadecanoic acid) which was discovered earlier by HOFMANN AND LUCAS (1950). It will be of interest to learn whether this phospholipid is derived from the wall or inner membrane.

The structures of certain glycolipids isolated from the Gram-positive organism, *Mycobacterium tuberculosis*, have already been mentioned. Wax D contained about 50% of the complex glycolipid as mycolic acid (the approximate formula $C_{88}H_{176}O_4 \pm 5CH_2$). The structure for mycolic acid proposed by ASSELINEAU (1960) is in agreement with all the experimental findings and is illustrated below:

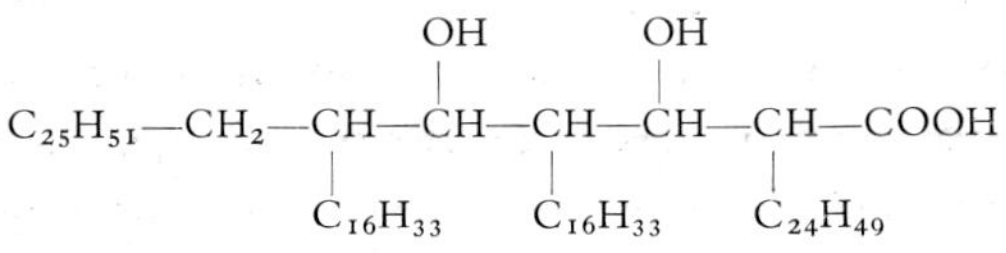

$$
\begin{array}{c}
\phantom{C_{25}H_{51}-CH_2-CH-}\mathrm{OH}\mathrm{OH}\\
C_{25}H_{51}-CH_2-\underset{\displaystyle C_{16}H_{33}}{\mathrm{CH}}-\overset{}{\mathrm{CH}}-\underset{\displaystyle C_{16}H_{33}}{\mathrm{CH}}-\mathrm{CH}-\underset{\displaystyle C_{24}H_{49}}{\mathrm{CH}}-\mathrm{COOH}
\end{array}
$$

Other glycolipids from mycobacteria include the 'mycosides', the lipid moieties of which differ e.g. Mycoside A contains a di- or tri-mycocerosate of an aromatic alcohol; mycoside B has a lipid moiety of two molecules of a branched-chain acid fraction with a methoxylated phenolic triol (LEDERER, 1961). Although some of these glycolipids have now been detected in isolated wall preparations (e.g. *Mycobacterium* BCG, prepared by KOTANI, KITAURA, HIRANO AND TANAKA, 1959), neither the chemically characterized waxes nor the mycosides have been isolated directly from cell walls. However with the detection of waxes in the isolated walls the chemical affinities of these substances with cell-wall substances is now quite convincing (LEDERER, 1961).

REFERENCES

ABRAHAM, E. P., *Biochemistry of Some Peptide and Steroid Antibiotics*, John Wiley, New York, 1957.

AKABORI, S., K. OHNO and K. NARITA, *Bull. Chem. Soc., Japan*, 25 (1952) 214.

ALLISON, A. C. and H. R. PERKINS, *Nature*, 188 (1960) 796.

ARCHIBALD, A. R., J. BADDILEY and J. G. BUCHANAN, *Biochem. J.*, 81 (1961) 124.

ARMSTRONG, J. J., J. BADDILEY and J. G. BUCHANAN, *Biochem. J.*, 76 (1960) 610.

ARMSTRONG, J. J., J. BADDILEY, J. G. BUCHANAN, B. CARSS and G. R. GREENBERG, *J. Chem. Soc.*, (1958) 4344.

ARMSTRONG, J. J., J. BADDILEY, J. G. BUCHANAN, A. L. DAVISON, M. V. KELEMEN and F. C. NEUHAUS, *Nature*, 184 (1959) 247.

ASSELINEAU, J., *Bull. Soc. Chim. France*, (1960) 135.

ASSELINEAU, J., H. BUC, P. JOLLES and E. LEDERER, *Bull. Soc. Chim. Biol.*, 40 (1958) 1953.

BADDILEY, J., *Biochem. J.*, 82 (1962) 36 P.

BADDILEY, J., J. G. BUCHANAN and B. CARSS, *Biochim. Biophys. Acta*, 27 (1958) 220.

BADDILEY, J., J. G. BUCHANAN, U. L. RAJBHANDARY and A. R. SANDERSON, *Biochem. J.*, 82 (1962) 439.

BARKULIS, S. S. and M. F. JONES, *J. Bacteriol.*, 74 (1957) 207.

BENSON, A. A., J. A. BASSHAM and M. CALVIN, *J. Am. Chem. Soc.*, 73 (1951) 2970.

BOLLE, A. and E. KELLENBERGER, *Schweiz. Z. Allgem. Pathol. Bakteriol.*, 21 (1958) 714.

BRADBURY, J. H., *Biochem. J.*, 68 (1958) 482.

BRITT, E. M. and P. GERHARDT, *J. Bacteriol.*, 76 (1958) 288.

BROWN, A. D., *Biochim. Biophys. Acta*, 28 (1958) 445.

BROWN, A. D., *Biochim. Biophys. Acta*, 44 (1960) 178.

BROWN, A. D., *Biochim. Biophys. Acta*, 75 (1963) 425.

BROWN, A. D. and C. D. SHOREY, *Biochim. Biophys. Acta*, 59 (1962) 258.

BRUMFITT, W., A. C. WARDLAW and J. T. PARK, *Nature*, 181 (1958) 1783.

COLLINS, F., Manuscript in preparation (1963).

CORPE, W. A., *Bacteriol. Proc.*, 126 (1958).

COTA-ROBLES, E. H., A. G. MARR and E. H. NILSON, *J. Bacteriol.*, 75 (1958) 243.

CRUMPTON, M. J. and D. A. L. DAVIES, *Biochem. J.*, 70 (1958) 729.

CRUMPTON, M. J., D. A. L. DAVIES and A. M. HUTCHISON, *J. Gen. Microbiol.*, 18 (1958) 129.

CUMMINS, C. S., *J. Gen. Microbiol.*, 28 (1962) 35.

CUMMINS, C. S., O. M. GLENDENNING and H. HARRIS, *Nature*, 180 (1957) 337.

CUMMINS, C. S. and H. HARRIS, *J. Gen. Microbiol.*, 13 (1955) iii.

CUMMINS, C. S. and H. HARRIS, *J. Gen. Microbiol.*, 14 (1956a) 583.

CUMMINS, C. S. and H. HARRIS, *Intern. Bull. Bacteriol. Nomen.*, 6 (1956b) 111.

CUMMINS, C. S. and H. HARRIS, *J. Gen. Microbiol.*, 18 (1958) 173.

CUMMINS, C. S. and H. HARRIS, *Nature*, 184 (1959) 831.

DAVIES, D. A. L., *Biochem. J.*, 59 (1955) 696.

DAVIES, D. A. L., *Biochem. J.*, 67 (1957a) 253.
DAVIES, D. A. L., *Biochim. Biophys. Acta*, 26 (1957b) 151.
DAVIES, D. A. L., *J. Gen. Microbiol.*, 18 (1958) 118.
DAVIES, D. A. L., *Advances in Carbohydrate Chemistry*, Vol. 15, p. 271, Academic Press, New York, 1960.
DAVIS, G. H. G. and J. H. FREER, *J. Gen. Microbiol.*, 23 (1960) 163.
DISCHE, Z., *J. Biol. Chem.*, 204 (1953) 983.
FEW, A. V., *Biochim. Biophys. Acta*, 16 (1955) 137.
FOUQUEY, C, J. POLONSKY and E. LEDERER, *Bull. Soc. Chim. Biol.*, 39 (1957) 101.
FRANK, H., M. LEFORT and H. H. MARTIN, *Biochem. Biophys. Res. Comm.*, 7 (1962) 322.
FUKASAWA, T. and H. NIKAIDO, *Biochim. Biophys. Acta*, 48 (1961) 470.
FULLER, M. S. and I. BARSHAD, *Am. J. Botany*, 47 (1960) 104.
GIGG, R. H. and P. M. CARROL, *Nature*, 191 (1961) 495.
GROSSBARD, E. and R. D. PRESTON, *Nature*, 179 (1957) 448.
GROSSBARD, E and R. D. PRESTON, *Abstr. 7th Intern. Congr. Microbiol. Stockholm*, 1958.
GUEX-HOLZER, S. and J. TOMCSIK, *J. Gen. Microbiol.*, 14 (1956) 14.
HANCOCK, R. and J. T. PARK, *Nature*, 181 (1958) 1050.
HEIDELBERGER, M., *Lectures in Immunochemistry*, Academic Press, New York, 1956.
HEIDELBERGER, M. and O. J. PLESCIA (Eds.) *Immunochemical Approaches to Problems in Microbiology*, Rutgers University Press, 1960.
HEYMANN, H., L. D. ZELEZNICK and J. A. MANNIELLO, *J. Am. Chem. Soc.*, 83 (1961) 4859.
HOARE, D. S. and E. WORK, *Biochem. J.*, 65 (1957) 441.
HOFMANN, K. and R. A. LUCAS, *J. Am. Chem. Soc.*, 72 (1950) 4328.
HOLDSWORTH, E. S., *Biochim. Biophys. Acta*, 9 (1952) 19.
HORECKER, B. L. and P. Z. SMYRNIOTIS, *J. Am. Chem. Soc.*, 74 (1952) 2123.
HUMPHREY, B. and J. M. VINCENT, *J. Gen. Microbiol.*, 29 (1962) 557.
HUNNELL, J. W. and Z. J. ORDAL, in *Spores II*, Ed. H. Orin Halvorson, p. 101, Burgess, Minneapolis, 1960.
HURST, H. J., *J. Exptl. Biol.*, 29 (1952) 30.
IKAWA, M., *J. Biol. Chem.*, 236 (1961) 1087.
IKAWA, M. and E. E. SNELL, *Biochim. Biophys. Acta*, 19 (1956) 576.
IKAWA, M. and E. E. SNELL, *J. Biol. Chem.*, 235 (1960) 1376.
IKAWA, M. and E. E. SNELL, *Biochim. Biophys. Acta*, 60 (1962) 186.
INGRAM, V. M. and M. R. J. SALTON, *Biochim. Biophys. Acta*, 24 (1957) 9.
JANCZURA, E., H. R. PERKINS and H. J. ROGERS, *Biochem. J.*, 80 (1961) 82.
JEANLOZ, R. W., *Arthritis and Rheumatism*, 3 (1960) 233.
JENKIN, H. M., *J. Bacteriol.*, 80 (1960) 639.
JESAITIS, M. A. and W. F. GOEBEL, *J. Exptl. Med.*, 96 (1952) 409.
KANDLER, O. and A. HUND, *Z. Bakt. Parasitenk.*, 113 (1959) 63.
KANESHIRO, T. and A. G. MARR, *J. Biol. Chem.*, 236 (1961) 2615.
KANESHIRO, T. and A. G. MARR, *J. Lipid Res.*, 3 (1962) 184.
KAUFFMANN, F., O. LÜDERITZ, H. STIERLIN and O. Westphal, *Z. Bakt. Parasitenk. (Orig)*, 178 (1960) 442.
KELLENBERGER, E., *Soc. Gen. Microbiol. Symposium* No. 10, (1960) 39.
KENT, P. W. and M. W. WHITEHOUSE, *Biochemistry of the Amino Sugars*, Butterworths, London, 1955.
KESSLER, G. and W. J. NICKERSON, *J. Biol. Chem.*, 234 (1959) 2281.
KNOX, K. W. and J. BRANDSEN, *Biochem. J.*, 85 (1962) 15.
KOCH, G. and W. WEIDEL, *Hoppe-Seyl. Z. physiol. Chem.*, 303 (1956) 213.
KOTANI, S., T. KITAURA, T. HIRANO and A. TANAKA, *Biken's Journal*, 2 (1959) 129.
KOZLOFF, L. M. and M. LUTE, *J. Biol. Chem.*, 228 (1957) 529.
KRAUSE, R. M. and M. MCCARTY, *J. Exptl. Med.*, 114 (1961) 127.

KROGER, E., O. LÜDERITZ and O. WESTPHAL, *Naturwissenschaften*, 46 (1959) 428.
LA FORGE, F. B. and C. S. HUDSON, *J. Biol. Chem.*, 30 (1917) 61.
LAMBERT, R. and F. ZILLIKEN, *Chem. Ber.*, 93 (1960) 187.
LANCEFIELD, R. C., *J. Exptl. Med.*, 78 (1943) 465.
LEDERER. E., *Pure and Applied Chemistry*, Vol. 2, Butterworths, London, 1961, p. 587.
LILLY, M. D., *J. Gen. Microbiol.*, 28 (1962) ii–iii.
LÜDERITZ, O., *Bull. Soc. Chim. Biol.*, 42 (1960).
LÜDERITZ, O. and O. WESTPHAL, *Z. Naturforsch.*, 7b (1952a) 136.
LÜDERITZ, O. and O. Westphal, *Z. Naturforsch.*, 7b (1952b) 548.
MACLELLAN, A. P., *Biochem. J.*, 67 (1957) 3P.
MACLELLAN, A. P., *Biochim. Biophys. Acta*, 48 (1961) 600.
MACLELLAN, A. P., *Biochem. J.*, 82 (1962) 394.
MACLELLAN, A. P. and D. A. L. DAVIES, *Biochem. J.*, 66 (1957) 562.
MANDELSTAM, J., *Biochem. J.*, 84 (1962) 294.
MANDELSTAM, J. and H. J. ROGERS, *Nature*, 181 (1958) 956.
MANDELSTAM, J. and H. J. ROGERS, *Biochem. J.*, 72 (1959) 654.
MARTIN, H. H. and H. FRANK, *Z. Naturforsch.* 17b (1962) 190.
MASON, D. J. and D. POWELSON, *Biochim. Biophys. Acta*, 29 (1958) 1.
MATSUSHIMA, Y. and J. T. PARK, *J. Org. Chem.*, 27 (1962) 3581.
MCCARTY, M., *J. Exptl. Med.*, 96 (1952) 569.
MCQUILLEN, K. in *Progress in Biophysics and Biophysical Chemistry*, Vol. 12, p. 67, Pergamon Press, London. 1961
MEYER, K., *Cold Spr. Harb. Symp. quant. Biol.*, 6 (1938) 99.
MEYER, K., J. W. PALMER, R. THOMPSON and D. KHORAZO, *J. Biol. Chem.*, 113 (1936) 479.
MITCHELL, P. and J. MOYLE, *J. Gen. Microbiol.*, 5 (1951) 981.
MITCHELL, P. and J. MOYLE., *Soc. Gen. Microbiol. Symposium* No. 6 (1956).
MOGGRIDGE, R. C. G. and A. NEUBERGER, *J. Chem. Soc.*, (1938) 745.
MORGAN, W. T. J., *Biochem. J.*, 30 (1936) 909.
MORGAN, W. T. J., *Soc. Gen. Microbiol. Symposium* No. 1 (1949) 9.
MORI, Y., K. KATO, T. MATSUBARA and S. KOTANI, *Biken's Journal*, 3 (1960) 139.
MURRELL, W. G., *Soc. Gen. Microbiol. Symposium* No. 11, 1961, p. 100.
NIKAIDO, H., *Proc. Natl. Acad. Sci.*, (U.S.A.), 48 (1962) 1337.
NIU, C. I. and H. FRAENKEL-CONRAT, *J. Am. Chem. Soc.*, 77 (1955) 5882.
PARK, J. T., *J. Biol. Chem.*, 194 (1952) 877, 885, 897.
PARK, J. T., *Soc. Gen. Microbiol. Symposium* No. 8 (1958) 49.
PARK, J. T. and J. L. STROMINGER, *Science*, 125 (1957) 99.
PERKINS, H. R., *Biochem. J.*, 74 (1960a) 182.
PERKINS, H. R., *Biochem. J.*, 74 (1960b) 186.
PERKINS, H. R., *Biochem. J.*, 83 (1962) 5P.
PERKINS, H. R., *Biochem. J.*, 86 (1963) 475.
PERKINS, H. R. and H. J. ROGERS, *Biochem. J.*, 72 (1959) 647.
PLACKETT, P., *Biochim. Biophys. Acta*, 35 (1959) 260.
POWELL, J. F. and R. E. STRANGE, *Biochem. J.*, 65 (1957) 700.
PRIMOSIGH, J., H. PELZER, D. MAASS and W. WEIDEL, *Biochim. Biophys. Acta*, 46 (1961) 68.
PURKAYASTHA, M. and R. P. WILLIAMS, *Nature*, 187 (1960) 349.
QUINNELL, C. M., S. G. KNIGHT and P. W. WILSON, *Can. J. Microbiol.*, 3 (1957) 277.
ROBISON, R., M. G. MACFARLANE and A. TAZELAAR, *Nature*, 142 (1938) 114.
ROGERS, H. J., *Biochem. Soc. Symposium* No. 22 (1962) 55.
ROGERS, H. J. and H. R. PERKINS, *Nature*, 184 (1959) 52c.

ROSEMAN, S., *Ann. Rev. Microbiol.*, 28 (1959) 545.
ROSS, J. W., *Appl. Microbiol.*, 11 (1963) 33.
SALTON, M. R. J., *Biochim. Biophys. Acta*, 9 (1952) 334.
SALTON, M. R. J., *Biochim. Biophys. Acta*, 10 (1953) 512.
SALTON, M. R. J., *Soc. Gen. Microbiol. Symposium* No. 6 (1956a) 81.
SALTON, M. R. J., *Biochim. Biophys. Acta*, 22 (1956b) 495.
SALTON, M. R. J., *Proc. Intern. Congr. Surface Activity (London)*, p. 245 Butterworths, London, 1957a.
SALTON, M. R. J., *Nature*, 180 (1957b) 338.
SALTON, M. R. J., *J, Gen. Microbiol.*, 18 (1958a) 481.
SALTON, M. R. J., *Abstr. 7th Intern. Congr. Microbiol., Stockholm*, 1958b.
SALTON, M. R. J., *Biochim. Biophys. Acta*, 34 (1959) 308.
SALTON, M. R. J., *Biochim. Biophys. Acta*, 45 (1960a) 364.
SALTON, M. R. J., *Microbiol Cell Walls*, John WILEY & Sons, New York, 1960b.
SALTON, M. R. J., *Biochim. Biophys. Acta*, 52 (1961) 329.
SALTON, M. R. J. and J. M. GHUYSEN, *Biochim. Biophys. Acta*, 45 (1960) 355.
SALTON, M. R. J. and R. W. HORNE, *Biochim. Biophys. Acta*, 7 (1951) 177.
SALTON, M. R. J. and B. MARSHALL, *J. Gen. Microbiol.*, 21 (1959) 415.
SALTON, M. R. J. and J. G. PAVLIK, *Biochim. Biophys. Acta*, 39 (1960) 398.
SANGER, F., *Biochem. J.*, 39 (1945) 507.
SCHAECHTER, M., A. J. TOUSIMIS, Z. A. COHN, H. ROSEN, J. CAMPBELL and E. E. HAHN, *J. Bacteriol.*, 74 (1957) 822.
SCHOCHER, A. J., S. T. BAYLEY and R. W. WATSON, *Can. J. Microbiol.*, 8 (1962) 89.
SCHOCHER, A. J., D. JUSIC and R. W. WATSON, *Biochem. Biophys. Res. Comm.*, 6 (1961) 16.
SCHULMAN, J. H., B. A. PETHICA, A. V. FEW and M. R. J. SALTON, in *Progress in Biophysics and Biophysical Chemistry*, Vol. 5, p. 41, London, Pergamon Press.
SHABAROVA, Z. A., J. G. BUCHANAN and J. BADDILEY, *Biochim. Biophys. Acta*, 57 (1962) 146.
SHAFA, F. and M. R. J. SALTON, *J. Gen. Microbiol.*, 22 (1960) 137.
SHARON, N., *Nature*, 179 (1957) 919.
SHARON, N. in '*The Chemistry and Biochemistry of Amino Sugars*', Ed. R. W. Jeanloz, Academic Press, New York, 1963.
SHARON, N. and R. W. JEANLOZ, *Biochim. Biophys. Acta*, 31 (1959) 277.
SHARON, N. and R. W. JEANLOZ, *J. Biol. Chem.*, 235 (1960) 1.
SLADE, H. D. and W. C. SLAMP, *J. Bacteriol.*, 84 (1962) 345.
SLEIN, M. W. and G. W. SCHNELL, *Proc. Soc. Exptl. Biol.*, N.Y. 82 (1953) 734.
SNELL, E. E., N. S. RADIN and M. IKAWA, *J. Biol. Chem.*, 217 (1955) 803.
STACEY, M., *Advanc. Carbohyd. Chem.*, 2 (1946) 161.
STACEY, M. and BARKER, S. A., *Polysaccharides of Microorganisms*, Clarendon Press, Oxford, 1960.
STAUB, A. M. in *Polysaccharides in Biology, 5th Conf. Josiah Macy, Jr. Foundation*, 1959.
STAUB, A. M., R. TINELLI, O. LÜDERITZ and O. WESTPHAL, *Ann. Instit. Pasteur*, 96 (1959) 303.
STEVENS, C. M., P. E. HALPERN and R. P. GIGGER, *J. Biol. Chem.*, 190 (1951) 705.
STRANGE, R. E., *Biochem. J.*, 64 (1956) 23P.
STRANGE, R. E., F. A. DARK and A. G. NESS, *Biochem. J.*, 59 (1955) 172.
STRANGE, R. E. and F. A. DARK, *Biochem. J.*, 62 (1956a) 459.
STRANGE, R. E. and F. A. DARK, *Nature*, 177 (1956b) 186.
STRANGE, R. E. and L. H. KENT, *Biochem. J.*, 71 (1959) 333.
STRANGE, R. E. and J. F. POWELL, *Biochem. J.*, 58 (1954) 80.
STROMINGER, J. L., *Fed. Proc.*, 21 (1962) 134.
STROMINGER, J. L. and R. H. THRENN, *Biochim. Biophys. Acta*, 33 (1959) 280.

TEPPER, B. S., J. A. HAYASHI and S. S. BARKULIS, *J. Bacteriol.*, 79 (1960) 33.
TOENNIES, G., B. BAKAY and G. D. SHOCKMAN, *J. Biol. Chem.*, 234 (1959) 1376.
TOMCSIK, J. in *Soc. Gen. Microbiol. Symposium* No. 6 (1956) 41.
VIEHOEVER, A., *Ber. deutsch. bot. Gesellsch.*, 30 (1912) 443.
VINCENT, J. M. and J. COLBURN, *Aust. J. Sci.*, 23 (1961) 269.
WARTH, A. D., D. F. OHYE and W. G. MURRELL, *J. Cell. Biol.*, 16 (1963) 579.
WEIDEL, W., *Z. Naturforsch.*, 6b (1951) 251.
WEIDEL, W., *Hoppe-Seyler's Z. physiol. Chem.*, 299 (1955) 253.
WEIDEL, W., H. FRANK and H. H. MARTIN, *J. Gen. Microbiol.*, 22 (1960) 158.
WEIDEL, W., H. FRANK and W. LEUTGEB, *J. Gen. Microbiol.*, 30 (1963) 127.
WEIDEL, W. and J. PRIMOSIGH, *J. Gen. Microbiol.*, 18 (1958) 513.
WESTPHAL, O. in *Polysaccharides in Biology, 2nd Conf. Josiah Macy Jr. Foundation*, 1956, p. 115.
WESTPHAL, O., *Bull. Soc. Chim. Biol.*, 42 (1960).
WESTPHAL, O., F. KAUFFMANN, O. LÜDERITZ and H. STIERLIN, *Z. Bakt. Parasitenk. (Orig.)*, 179 (1960) 336.
WESTPHAL, O. and O. LÜDERITZ, *Angew. Chemie*, 72 (1960) 881.
WORK, E., *Biochim. Biophys. Acta*, 3 (1949) 400.
WORK, E., *Biochem. J.*, 49 (1951) 17.
WORK, E., *Nature*, 179 (1957) 338.
WORK, E., *J. Gen. Microbiol.*, 25 (1961) 167.
WORK, E. and D. L. DEWEY, *J. Gen. Microbiol.*, 9 (1953) 394.
WORK, E. and M. LECADET, *Biochem. J.*, 76 (1960) 39P.
YOSHIDA, N., Y. IZUMI, I. TANI, S. TANAKA, K. TAKAISHI, T. HASHIMOTO and K. FUKUI, *J. Bacteriol.*, 74 (1957) 94.
ZILLIKEN, F., *Fed. Proc.*, 18 (1959) 966.

CHAPTER 5

Structure of cell-wall glycosaminopeptides (mucopeptides) and their sensitivity to enzymic degradation

Since the discovery that the cell walls of Gram-positive bacteria are composed of a small variety of amino acids, amino sugars and sugars (HOLDSWORTH, 1952; SALTON, 1952a) the unusual chemical features of these structures have attracted a great deal of attention. It was apparent that the cell walls of bacteria were composed of chemical entities differing from other known structural compounds such as proteins, polysaccharides and mucopolysaccharides, although closer affinities with the latter class of substances was apparent (HOLDSWORTH, 1952; SALTON, 1952a). MITCHELL AND MOYLE (1951) on the other hand concluded that the 'envelope' of *Staphylococcus aureus* was a glycerophospho-protein complex. The very low amino sugar content (ca.1%) obtained on analysis of the staphylococcal wall by MITCHELL AND MOYLE (1951) is now surprising in retrospect and probably led them to the conclusion that the amino acids of the wall were in the form of a protein.

The earlier concepts of the chemical nature of the cell walls of bacteria were soon to be clarified as further investigations proceeded. SALTON (1952a) concluded that in the *Streptococcus faecalis* wall the polysaccharide was covalently joined to the 'amino-acid residue'. Following the discovery of the 'unknown amino sugar' in spore peptides by STRANGE AND POWELL (1954) and its detection in cell walls (CUMMINS AND HARRIS, 1955; STRANGE AND DARK, 1956; SALTON, 1956a), SALTON (1956b) suggested 'it seems likely that such peptides may be common components of bacterial structures.' The systematic investigations of the wall composition of Gram-positive bacteria (CUMMINS AND HARRIS, 1956a, b) made it abundantly clear that amino sugar-peptides occurred in all groups. WORK (1957) referred to the peptides containing the amino sugars glucosamine and muramic acid, as the 'basal structure'.

The walls of Gram-positive bacteria have become more and more complicated as the variety of constituents detectable on hydrolysis has increased steadily in the past few years. This increase in complexity has followed the discovery of the teichoic acids (ARMSTRONG, BADDILEY, BUCHANAN, CARSS AND GREENBERG, 1958), polysaccharides (IKAWA, 1961; KRAUSE AND MCCARTY, 1961) and teichuronic acid (JANCZURA, PERKINS AND ROGERS, 1961; ROGERS AND PERKINS, 1962). All three types of cell-wall polymer have been extracted from the walls, leaving insoluble residues composed almost entirely of glycosaminopeptide (mucopeptide).

It is apparent that the additional wall compounds are bound to the 'basal structure' by linkages weaker than the internal ones holding the insoluble, rigid component together. Thus ARMSTRONG, BADDILEY, BUCHANAN, CARSS AND GREENBERG (1958) concluded that the teichoic acids are held to the wall mucopeptide by salt linkages. Relatively mild conditions were used in extracting the teichuronic acids (ROGERS, 1962). Removal of the cell-wall polysaccharides seems to require more drastic extraction procedures (hot, saturated picric acid –

The tables are printed together at the end of the book.

HOLDSWORTH, 1952; formamide at 170° – KRAUSE AND MCCARTY, 1961) although some polysaccharide can be solubilized with trichloroacetic acid. The glycosaminopeptide therefore appears to be a very stable complex capable of surviving exposure to a variety of chemical agents and fortunately this property greatly facilitates its separation from other cell-wall polymers. The resistance of the mucopeptides to dissolution has enabled SCHOCHER, BAYLEY AND WATSON (1961) and MANDELSTAM (1962) to purify the mucopeptides from Gram-negative bacteria by extracting the protein, lipopolysaccharide and lipid components in various solvents. In order to solubilize *Micrococcus lysodeikticus* cell walls, SALTON (1957a) found that rather drastic treatment was necessary. Alkaline hypochlorite in the cold was an effective reagent for this purpose and the water soluble material prepared in this way formed viscous solutions and could be degraded further by lysozyme.

The various terms used to describe the 'basal structure' of bacterial walls have already been discussed and of all the names proposed in recent years (muco-complex, mucopeptide, mucopolymer, teichoin – IKAWA, 1961; glycopeptide, peptidopolysaccharide) mucopeptide has been adopted more widely. For the reasons outlined in Chapter 4, the term glycosaminopeptide is chemically more descriptive of this class of substance and its use avoids the inept prefix 'muco' for an insoluble rigid structure.

Interest in the structure of the cell-wall glycosaminopeptide was originally stimulated by the chemical characterization of the PARK (1952) nucleotides and the products released by lysozyme action on isolated bacterial walls. Apart from the substance assumed to be a disaccharide of N-acetylglucosamine and N-acetylmuramic acid (then referred to as an 'unidentified' 'acidic' amino sugar), the products of lysozyme action on walls of *Micrococcus lysodeikticus* possessed the same qualitative composition as the original wall (i.e. glucosamine, muramic acid, alanine, glutamic acid, lysine and glycine together with a higher proportion of glucose in one of the higher molecular weight fractions (SALTON, 1956b). With the identification of the amino sugar in the nucleotides from penicillin-treated *Staphylococcus aureus* as muramic acid (PARK AND STROMINGER, 1957) the main outlines of the glycosaminopeptide structure began to emerge.

Information on the structure of the cell-wall glycosaminopeptides has come from investigations of the products of digestion of isolated walls with the enzyme lysozyme (muramidase, SALTON, 1961a) and from an examination of the compounds in partial acid-hydrolysates of the walls. These studies together with the structures inferred from the chemical characterization of the PARK nucleotides have formed the basis for our present concepts of the nature of glycosaminopeptides.

It should be emphasized that we still do not know whether a 'mucopeptide' or a 'glycosaminopeptide' of the cell wall of a given species, is a single heteropolymer or a mixture of such polymeric substances. Some aspects of this problem will become more apparent as the structure of compounds isolated from enzymic digests and partial acid hydrolysates is discussed.

In the past, determinations of the molar proportions of amino acid and amino sugar constituents usually given in 'round figures', have been taken to indicate a high degree of uniformity of peptide and hence glycosaminopeptide subunit structure. To what extent departures from exact molar ratios of amino acids in walls and isolated glycosaminopeptides infer structural heterogeneity cannot be said at present. Much of this apparent simplicity of the peptides has been inferred from the molar ratios of amino acids observed in the PARK nucleotides but there is increasing evidence to indicate that in some fragments of the wall glycosaminopeptide they may represent only a small part of the whole structure. ROGERS (1962) has pointed out that with *Micrococcus lysodeikticus* walls, the molar ratio of muramic acid to glutamic acid is more than unity, an observation which together with the existence of about 10% of the weight of the wall as oligosaccharide (glucosamine – muramic acid) material with free carboxyl groups, makes the generalization proposed by PARK (1958) a structural over simplification. That the glycosaminopeptide structure is more complicated than originally believed from the examination of the nucleotide 'precursors', is now being recognized, for MANDELSTAM AND STROMINGER (1961) have proposed a covalent linkage of the glycine peptide to N-acetylmuramylpeptides forming the more 'characteristic' part of the glycopeptide. So at present we cannot infer too much structural knowledge from the molar ratios of the various wall constituents unless they are of known molecular weights and specified homogeneity.

Analysis of products in enzymic digests of walls

It was evident from earlier investigations by MEYER, PALMER, THOMPSON AND KHORAZO (1936) and EPSTEIN AND CHAIN (1940) that the characterization of the lysozyme substrate and the products formed by the action of this enzyme should throw a great deal of light on the problem of cell-wall structure. The demonstration that the isolated cell wall of *Micrococcus lysodeikticus* could be used as the 'substrate' for lysozyme (SALTON, 1952) simplified the task of identification and eventual characterization of the digestion products. Apart from the detection of reducing N-acetyl amino sugar groups and the decrease in viscosity on degradation of the soluble 'mucopolysaccharide' substrate (EPSTEIN AND CHAIN, 1940; MEYER AND HAHNEL, 1946) the nature of the compounds produced by lysozyme action was not established until the question was re-examined by SALTON (1956b).

Cell walls of *Micrococcus lysodeikticus*, *Sarcina lutea* and *Bacillus megaterium* were completely degraded to soluble products on digestion with lysozyme and the complex mixture of 'fragments' was separated on the basis of diffusability through dialysis tubing into 'non-dialysable' and 'dialysable' fractions (SALTON, 1956b). An examination of the dialysable fractions from walls of all three organisms showed the presence of a common component which appeared to be a disaccharide of N-acetylglucosamine and N-acetylmuramic acid. In addition to detecting the 'disaccharide' on paper chromatograms, slower-moving components reacting more weakly with aniline phthalate and ammoniacal silver nitrate were also present (SALTON, 1956b). The nature of these substances was unknown until isolated later by GHUYSEN AND SALTON (1960). Thus the principal component detectable on paper chromatograms and also the simplest substance found in lysozyme-digested walls was a compound having the properties of a 'disaccharide' of the two cell-wall amino sugars. It was suggested that the disaccharide formed an important structural unit of the wall (SALTON, 1956b). The presence of N-acetyl amino sugar groups attached to the higher-molecular weight glycosaminopeptide components of the non-dialysable fraction, indicated that the 'disaccharide' units were probably joined to these substances in the original cell wall.

Although the di- and tetrasaccharides isolated from lysozyme-digested walls of *Micrococcus lysodeikticus* by SALTON AND GHUYSEN (1959) and GHUYSEN AND SALTON (1960) are not glycosaminopeptides, it is assumed for the present, that they are part of the native mucopeptide structure of the wall and that they are quite relevant to any discussion of the chemical nature of these heteropolymers.

Confirmation that the disaccharide is the simplest component liberated by lysozyme action (SALTON, 1956b) came from the studies of GHUYSEN AND SALTON (1960) and PERKINS (1960). The structure of the disaccharide was investigated independently by PERKINS (1959; 1960) and SALTON AND GHUYSEN (1959; 1960), both studies reaching the same conclusion that N-acetylglucosamine was joined by a 1–6 linkage to N-acetylmuramic acid. However, the methods employed in the two studies differed.

To determine which of the two amino sugars possessed the free reducing group, PERKINS (1960) carried out oxidation in the cold with alkaline iodine solution (KUHN, GAUHE AND BAER, 1954) and examined the products for the corresponding hexonic acid formed. Since no glucosaminic acid was detected after oxidation and hydrolysis of the product, it was concluded that the C-1 of each molecule of N-acetylglucosamine was bound by a glycosidic linkage. The oxidation product of muramic acid was detected and at the same time free muramic acid was not given on hydrolysis of the oxidized 'oligosaccharide'. PERKINS (1960) concluded that the reducing group of the disaccharide was that of N-acetylmuramic acid. From a study of the chromogens formed under the conditions of the MORGAN-ELSON reaction, PERKINS (1960) obtained evidence indicating the linkage of N-acetylglucosamine to the C-6 position of N-acetylmuramic acid, thus enabling him to conclude that the disaccharide was 6-O-(N-acetylglucosaminyl)-N-acetylmuramic acid.

SALTON AND GHUYSEN (1959; 1960) identified the free reducing group of the disaccharide by reacting with sodium borohydride and examining the hydrolysis products of the reduced compound for the corresponding amino sugar hexitol. No glucosaminitol was detected and accompanying the disappearance of muramic acid, a new compound with the properties of a hexitol of muramic acid (muramicitol) appeared in the products of acid hydrolysis of the reduced disaccharide. Thus muramic acid possessed the free reducing group of the disaccharide. The same technique was used in identifying the reducing group of the tetrasaccharide also present in the digests of cell walls (GHUYSEN AND SALTON, 1960; SALTON AND GHUYSEN, 1960). Indeed muramic acid appeared to be the only reducing group liberated during degradation of the wall by lysozyme. Both non-dialysable and dialysable glycosaminopeptides therefore have reducing end groups of N-acetylmuramic acid.

The nature of the linkage of the disaccharide was investigated also by SALTON AND GHUYSEN (1959; 1960). In this study ^{14}C-disaccharide was prepared and after oxidation with periodate, preparations were examined for [^{14}C]-formaldehyde isolated with carrier formaldehyde by the addition of 5,5-dimethyl-1,3-cyclohexandione (dimedon). The absence of significant amounts of [^{14}C]-formaldehyde from both di- and tetra-saccharides before reduction with borohydride provided evidence for the presence of a 1–6 linkage between N-acetylglucosamine and N-acetylmuramic acid. Preparations of β-glucosidase (Mann Research Labs. Inc., salt-free, almond β-glucosidase) which obviously contained N-acetyl-glucosaminidase activity gave free N-acetylglucosamine and free N-acetylmuramic acid from the di- and tetra-saccharides. It was concluded that the di-saccharide was 6-O-β-N-acetylglucosaminyl-N-acetylmuramic acid and that the probable structure of the tetrasaccharide was O-β-N-acetyl-glucosaminyl-(1 → 6)-O-β-N-acetylmuramyl-(1–4)-O-β-N-acetylglucosaminyl-(1 → 6)-β-N-acetylmuramic acid. The structure of the disaccharide and its reaction with sodium borohydride and β-glucosidase are shown together with the proposed structure of the tetrasaccharide in Fig. 53.

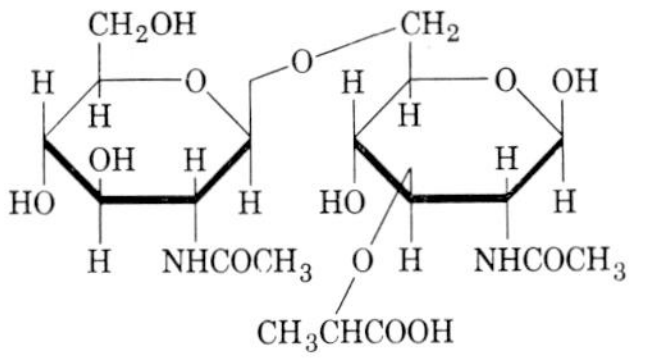

I

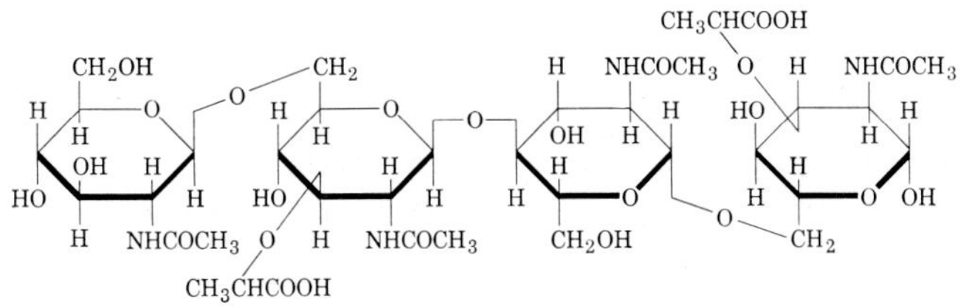

II

FIGURE 53. The proposed structures of amino sugar di- and tetra-saccharides (I and II respectively) isolated from cell walls of *Micrococcus lysodeikticus* digested with the muramidase, egg-white lysozyme (SALTON AND GHUYSEN, 1959).

With the chemical synthesis of a glucosaminyl-muramic acid disaccharide [O-2-acetamido-2-deoxy-β-D-glucopyranosyl-(1→6)-N-acetylmuramic acid] by FLOWERS AND JEANLOZ (1963) there is now some doubt as to the validity of the structure proposed for the naturally occurring disaccharide, tetrasaccharide and wall backbone. The colour reactions of the synthetic disaccharide differ from that of the disaccharide released by muramidase and there is therefore a strong suggestion that the natural compound may indeed be a 1→4 linked compound and not a 1→6 as previously suggested (JEANLOZ, SHARON AND FLOWERS, personal communication). These reservations on the structures of amino sugar oligosaccharides and glycosaminopeptides will now have to be kept in mind when reading subsequent parts of this chapter.

In addition to the di- and tetra-saccharides, the dialysable fractions from lysozyme-digested walls of *Micrococcus lysodeikticus* contained low-molecular weight glycosaminopeptides separating on paper chromatograms. These compounds corresponded to the slower-moving components which reacted weakly with ammoniacal silver nitrate and aniline phthalate (SALTON, 1956b). These glycosaminopeptides were isolated by GHUYSEN AND SALTON (1960) and the molar ratios of the constituent amino acids and amino sugars are shown in Table 51. There were no reasons for suspecting that the amino acid, glycine, was not an integral part of the glycosaminopeptide molecule.

An interesting compound was found in lysozyme digests of the insoluble residues remaining after extracting *Streptococcus pyogenes* group A cell walls with hot formamide (HEYMANN, ZELEZNICK AND MANNIELLO, 1961). This material could be separated from other substances in the digest and on isolation as a homogeneous compound it contained alanine, glucosamine and muramic acid in molar ratios of 1.0:0.94:0.77. Treatment of the compound with β-N-acetylglucosaminidase gave a product behaving as N-acetylglucosamine and a substance yielding alanine and muramic acid on hydrolysis. The evidence suggests that this compound has a single amino acid (alanine) substituent on a disaccharide and it represents the simplest glycosaminopeptide found in enzymic digests. It will be of great interest to see if it can also be isolated from unextracted streptococcal walls.

GHUYSEN'S enzyme, the Streptomyces amidase, has been of great value in studying the structure of the low-molecular weight glycosaminopeptides isolated from *Micrococcus lysodeikticus* walls digested with N-acetylhexosaminidases (GHUYSEN, 1960). Splitting of the amide bond between muramic acid and the peptide was accompanied by the release of a compound having properties identical with the disaccharide of N-acetylglucosamine and N-acetylmuramic acid as well as the appearance of new N-terminal groups of alanine.

GHUYSEN (1961) isolated two glycosaminopeptides from *Micrococcus lysodeikticus* walls, each compound having identical molar proportions of disaccharide 1, alanine 2, glutamic acid 1, glycine 1 and lysine 1. These two compounds were separated on the basis of their different electrophoretic properties. The Streptomyces F_2B amidase liberated the disaccharide (6-O-β-N-acetylglucosaminyl-N-acetylmuramic acid) and N-terminal alanine from each complex, the ratios of α-NH_2-alanine:disaccharide varying from 0.7–0.9. Only one free ε-NH_2 group of lysine was found in each complex. GHUYSEN (1961) presented evidence for the structures of these compounds and he concluded that the slower-moving substance was a dimer of the glycosaminopeptide migrating more rapidly on electrophoresis. The proposed structures for these two compounds are given in Fig. 54. As GHUYSEN (1961) pointed out, the sequence of amino acids was not known, with of course the exception of alanine which occupied the N-terminal position of the peptide liberated from the disaccharide by the amidase.

The electrophoretic properties of the glycosaminopeptides from *Micrococcus lysodeikticus* cell walls suggested that the γ-carboxyl group of glutamic acid and an unidentified α-carboxyl group were free (GHUYSEN, 1961). SALTON (1961b) identified the 'C-terminal' groups in cell walls and the non-dialysable fraction of lysozyme-digested walls of *Micrococcus lysodeikticus* and the findings are pertinent to the structions suggested by GHUYSEN (1961). By hydrazinolysis, glycine was the principal C-terminal group and corrected values indicated 1 glycine with a free carboxyl group/2000 molecular weight for both whole cell walls and non-dialysable fractions. These results, together with those of GHUYSEN'S (1961) have given some tentative suggestions of the subunit size for *Micrococcus lysodeikticus* wall and glycosaminopeptide and the data are presented in Table 52.

Low-molecular weight mucopeptides have also been isolated from lysozyme digests of the rigid portion of *Escherichia coli* walls by PRIMOSIGH, PELZER, MAASS AND WEIDEL (1961). From the complex mixture of compounds detected on paper chromatography of the digests, two components were predominant. These seemed to be of the general type detected in digests of *Micrococcus lysodeikticus* walls i.e. a disaccharide joined to a peptide. Using similar methods to those of SALTON AND GHUYSEN (1959; 1960), the C5 and C6 substances were shown to possess muramic acid with free reducing groups. A snail enzyme preparation liberated free N-acetylglucosamine from the mucopeptide, leaving compounds containing

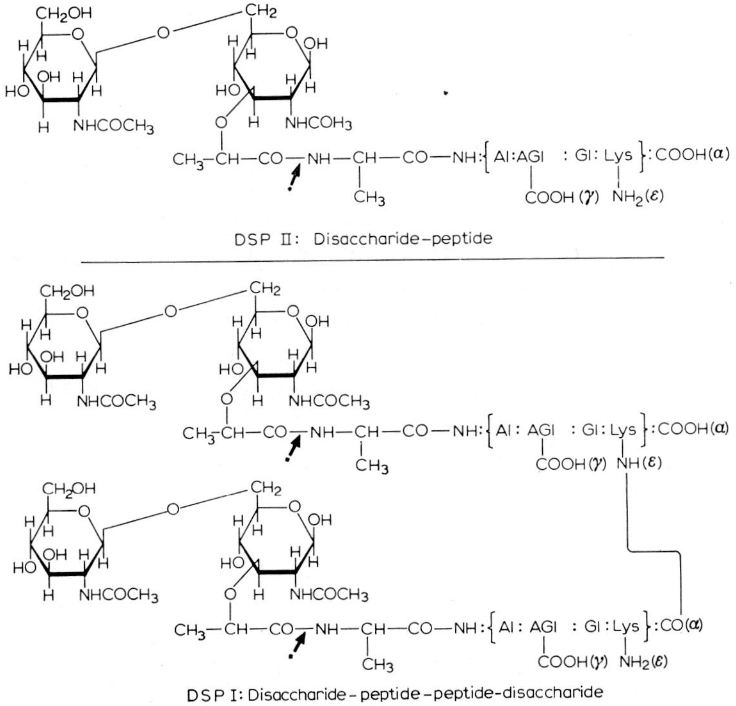

DSP II: Disaccharide-peptide

DSP I: Disaccharide-peptide-peptide-disaccharide

FIGURE 54. The proposed structures of two glycosaminopeptides (a monomer and a dimer) isolated from the cell wall of *Micrococcus lysodeikticus* digested with N-acetyl hexosaminidases (GHUYSEN, 1961).

muramic acid, alanine, glutamic acid and diaminopimelic acid in molar ratios (1:1:1:1 for C5 and 1:2:1:1 for C6) identical to those found in the original starting material. Thus the only difference between the C5 and C6 compounds was an extra alanine residue in the latter.

Molecular weights found for the *Escherichia coli* C5 and C6 compounds were 895 (869 theoretical) and 965 (940 theoretical) respectively. Diaminopimelic acid possessed one of its amino groups free and mono-dinitrophenyl-DAP was obtained on reacting the material with 1-fluoro-2,4-dinitrobenzene. PRIMOSIGH, PELZER, MAASS AND WEIDEL (1961) reported that although C5 and C6 compounds were quantitatively the predominant types of subunits of the mucopolymer, there was a variety of additional mucopeptide compounds in the digests. These authors concluded that one peptide side chain is 'attached to practically every single repeating unit of the polymer'. Such a generalization does not, however, fit in with other observations on walls of various species.

The analysis and structural studies of the C5 and C6 glycosaminopeptides from *Escherichia coli* walls have been carried further by PELZER (1962). The investigations have shown that alanine at the carboxyl end of the mucopeptide C6 is D-alanine, the other alanine joined through its amino group to muramic acid is the L-isomer. Incubation of the C6 mucopeptide with a partially purified D-alanine carboxypeptidase from *Escherichia coli* B yielded the C5 mucopeptide and D-alanine. Partial acid hydrolysis and identification of the peptides enabled the structures of these two glycosaminopeptides to be suggested by PELZER (1962) and these are given in Fig. 55. However, as PELZER (1962) has indicated, no choice can be made at present between the isomeric forms theoretically possible for these two compounds.

GHUYSEN, LEYH-BOUILLE AND DIERICKX (1962) have investigated the structure of *Bacillus megaterium* cell walls and have concluded that two heteropolymers (a 'phosphomucopolysaccharide' complex and a 'mucopeptide') are present in equimolar proportions and are covalently bound through glucosides – muraminyl bridges branched on some of the alanine residues of the basal mucopeptide. From analysis of compounds isolated from enzymic digests GHUYSEN, LEYH-BOUILLE AND DIERICKX (1962) concluded that the mucopeptide fraction has an average composition of N-acetylglucosamine (1), N-acetylmuramic acid (1), alanine (2.2), glutamic acid (1) and diaminopimelic acid (1). Glucose is believed to be part of the mucopep-

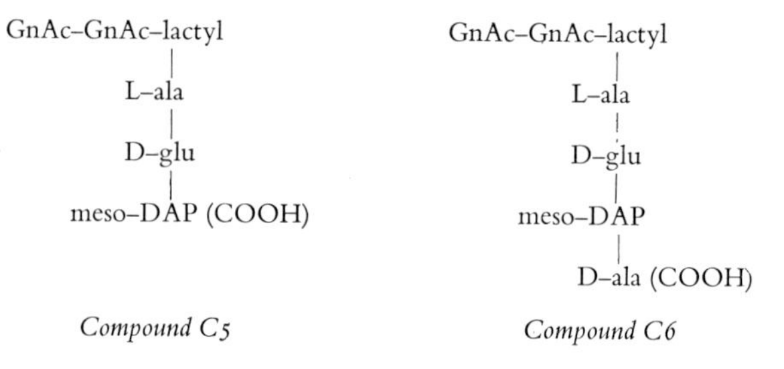

FIGURE 55. Structure of two glycosaminopeptides (compounds C5 and C6) isolated from lysozyme – digested rigid layer of *Escherichia coli* B cell wall (after PELZER, 1962).

tide polymer and they have suggested that four molecules of the above average composition make up a subunit with a molecular weight of about 4,500.

The higher-molecular weight compounds in the non-dialysable fractions of digested walls have not been investigated as extensively as the smaller fragments. However, SALTON (1956b) found that qualitatively, the non-dialysable fraction had the same amino acid and amino sugar

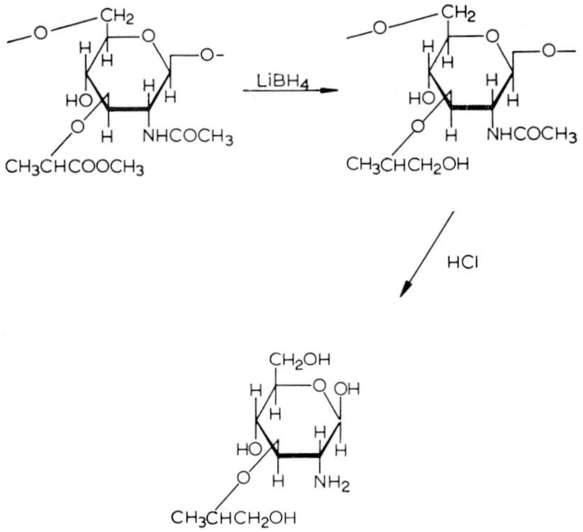

FIGURE 56. The sequence of reactions used in the detection of muramic acid residues (in walls) possessing free carboxyl groups, involves esterification with methanol – HCl followed by reduction with lithium borohydride giving modified 'muramic acid' compound on acid hydrolysis.

composition as the original *Micrococcus lysodeikticus* cell walls. The amino sugar content of the non-dialysable fraction was lower than that of the dialysable portion. This was not unexpected as the dialysable fractions contained disaccharide (and later shown to contain tetrasaccharide). Many investigators have concluded that because the cell wall has a 1:1 ratio of muramic acid to glutamic acid that each muramic acid residue of the wall carries a peptide substituent. There is substantial evidence for the contrary state of affairs (SALTON, 1961a) in the wall of *Micrococcus lysodeikticus*.

By esterifying cell walls of *Micrococcus lysodeikticus* and reducing with $LiBH_4$ in tetrahydrofuran, SALTON (1961a) was able to convert the free –COOH groups of the lactyl side chain of muramic acid to $-CH_2OH$ by the series of reactions shown in Fig. 56. This enables a distinction to be made between N-acetylmuramic acid residues with unsubstituted carboxyl groups and those carrying peptide substituents. Thus in order to preserve the molar ratios observed in the whole cell wall of *Micrococcus lysodeikticus* and at the same time accommodate the unsubstituted di- and tetra-saccharide units, there must be some way of 'duplicating' the peptides in the 10,000–20,000 molecular weight compounds in the non-dialysable fractions.

PARK AND GRIFFITH (1962) have reported that on the basis of methylation studies, 6-methyl muramic acid was the principal derivative obtained from the non-dialysable portion of *Micrococcus lysodeikticus* walls. This very interesting finding suggests that the amino sugars are linked differently in this non-dialysable polymer, the N-acetylglucosamine being attached to the 4 position of muramic acid. It will be recalled that in an earlier study GHUYSEN AND SALTON (1960) found in *Micrococcus lysodeikticus* walls digested with the F_1 N-acetylhexosaminidase, a glycosaminopeptide containing a 10:1 ratio of [N-acetylmuramic acid – N-acetylglucosamine] to peptide, but the reasons for resistance of this large oligosaccharide moiety were not determined.

PARK AND GRIFFITH (1962) also stated that the non-dialysable fraction retained most of the original wall amino acids, suggesting that at least 2 distinct polymers exist. This low diffusibility of compounds containing amino acids is contrasted with the results of GHUYSEN AND SALTON (1960) and PRIMOSIGH, PELZER, MAASS AND WEIDEL (1961), both investigations clearly establishing the presence of substantial amounts of low-molecular weight glycosaminopeptides.

It is clear that the non-diffusible compounds present some interesting structural problems and that further investigations in the future should yield information about the possible mechanism of duplication of peptide units in these glycosaminopeptides. The results of enzymic digestion of walls indicate the complexity of the products and we are still a long way from understanding how all of these fragments are woven together into a rigid wall.

Products of partial acid hydrolysis of walls

Examination of the products of partial acid hydrolysis has long been used in structural studies of peptides and proteins. PERKINS AND ROGERS (1958; 1959) were the first to apply this technique to a study of wall structure. PERKINS AND ROGERS (1959) hydrolysed the cell walls of *Micrococcus lysodeikticus* with $2N$ H_2SO_4 for 20 minutes at 100° and after neutralizing, the diffusible compounds were separated for isolation on chromatography columns. Three fragments were isolated from the acid hydrolysates and the composition of each substance is given in Table 53.

The detection of a compound composed of glucosamine, muramic acid and glycine was of considerable interest. Unfortunately, owing to the small yields, no information was available on the mode of attachment of the glycine. It is therefore difficult to assess the possible relationship between this compound and the simple glycosaminopeptide isolated from lysozyme-digests of the formamide-insoluble fraction of group A streptococcal wall by HEYMANN, ZELEZNICK AND MANNIELLO (1961). The most complicated fragment contained muramic acid, alanine, glutamic acid, lysine and glycine in quite different proportions to those found in the original walls (PERKINS AND ROGERS, 1959) or in lysozyme digest products (GHUYSEN AND SALTON, 1960; cf. Table 51).

An oligosaccharide containing glucosamine and muramic acid was also found in the partial acid hydrolysates examined by PERKINS AND ROGERS (1959). This material was not sensitive to further degradation by lysozyme, a fact which is not surprising as the substance was presumably partially or wholly de-acetylated (N-acetylglucosaminidases are usually inactive towards the substrates containing amino sugars with free amino groups). Apart from the detection of a glucosamine–muramic acid compound, the examination of the products of partial acid hydrolysis did not clarify the structure of the wall mucopeptide and PERKINS AND ROGERS (1959) concluded that 'it is not yet safe to hypothesize about the structure of the whole wall'.

Using an improved method for detecting N-acetyl amino sugar compounds on paper chromatograms, SALTON (1959) examined products of partial acid hydrolysis of *Micrococcus lysodeikticus* walls with $12N$ HCl (1–2 days at 37°; 6 days at 4°) after separating on 'finger prints' by combined electrophoresis and chromatography on paper. Although free amino acids and amino sugars were present, numerous N-acetyl amino sugar compounds were detectable, including free N-acetylmuramic acid and N-acetylglucosamine and a disaccharide of the two compounds corresponding closely to that present in lysozyme digested walls (SALTON, 1959). Thus under the conditions of partial hydrolysis a compound similar to that found in enzymic digests was detected, a finding which added further weight to the suggestion that the disaccharide forms an important structural unit in certain walls (SALTON, 1956b).

In order to reduce de-acetylation to a minimum, SALTON (1961a) used concentrated orthophosphoric acid for partial acid hydrolysis of *Micrococcus lysodeikticus*. This method did indeed yield a variety of N-acetyl amino sugar compounds and appeared to achieve partial hydrolysis of the wall with a minimum of de-acetylation. A compound of similar behaviour on electrophoresis and paper chromatograms to that of the disaccharide liberated by lysozyme action was again found in these phosphoric acid partial hydrolysates. In addition, other oligosaccharides were found together with diffusible mucopeptides. One compound migrating between di- and tetra-saccharides could be a trisaccharide. If in fact this proves to be a trisaccharide, it could be a most useful substrate for lysozyme.

It is evident that apart from confirming the existence of a disaccharide unit of N-acetylglucosamine and N-acetylmuramic acid in the walls of *Micrococcus lysodeikticus*, an examination of the products of partial acid hydrolysis has not contributed a great deal of information to our knowledge of the glycosaminopeptide structure of the wall. For the present, it appears that the most fruitful application of partial acid hydrolysis has been in the determination of the amino acid sequences of the compounds studied by PELZER (1962). There is little doubt however, that the identification of products in the partial acid hydrolysates of the non-diffusible substances isolated from lysozyme digested glycosaminopeptide walls will provide valuable data in the structural analysis of these larger cell-wall polymers.

Proposed structures of glycosaminopeptides

We are now in a position to summarize some aspects of the current status of our knowledge of glycosaminopeptide (mucopeptide) structure which has been based on investigations with whole cell walls assumed to be largely mucopeptides (e.g. *Micrococcus lysodeikticus*) and isolated fragments from enzymic digests and acid hydrolysates of walls. Some early ideas of glycosaminopeptide structure emerged from the study of the nature of the complex mixture of fragments present in lysozyme digested walls. Thus, a compound believed to be a disaccharide was regarded as an important structural unit of the wall (SALTON, 1956b). SALTON (1957b) concluded that 'the cell walls of these lysozyme sensitive bacteria have a highly branched structure in which peptides and peptide-amino sugar complexes may be glycosidically linked through terminal amino sugars to other amino sugars and sugar residues of polysaccharide components.'

The next major advance in understanding the structure of the wall glycosaminopeptide occurred when BRUMFITT, WARDLAW AND PARK (1958) suggested the possible linkages between repeating units of the two amino sugars, N-acetylglucosamine and N-acetylmuramic acid. Based on the results of a differential MORGAN AND ELSON colour reaction used by PARK (1956), it was concluded that carbon atoms 1 and 4 of N-acetylmuramic acid were free (BRUMFITT, WARDLAW AND PARK, 1958). With the identification of the amino sugar of the PARK nucleotides as N-acetylmuramic acid it seemed certain that the wall peptides would be attached through the carboxyl groups of muramic acid as already suggested in the nucleotides (PARK, 1952; PARK AND STROMINGER, 1957).

Upon isolation and characterization of the di- and tetra-saccharides and low-molecular weight glycosaminopeptides (GHUYSEN AND SALTON, 1960; SALTON AND GHUYSEN, 1959; 1960; PERKINS, 1959; 1960), further details of the structure of *Micrococcus lysodeikticus* cell walls were revealed, thus directly confirming the observations of BRUMFITT, WARDLAW AND PARK (1958). The type of structure accommodating the various fragments isolated from the lysozyme digests was therefore proposed by SALTON AND GHUYSEN (1960) and this is illustrated in Fig. 57.

The purification of the *Streptomyces albus* enzyme F_2B by GHUYSEN (1957, 1960) and the discovery of its mode of attack on cell-wall compounds has made a major contribution to the study of the structure of wall glycosaminopeptides. This enzyme possessing amidase activity, breaks the bonds between muramic acid and the peptide. Thus from the low-molecular weight mucopeptides isolated from *Micrococcus lysodeikticus* walls, the Streptomyces amidase liberated the terminal disaccharide from the peptide. Furthermore, alanine was shown to be at the N-terminal end of the peptide (GHUYSEN, 1960) a finding which directly

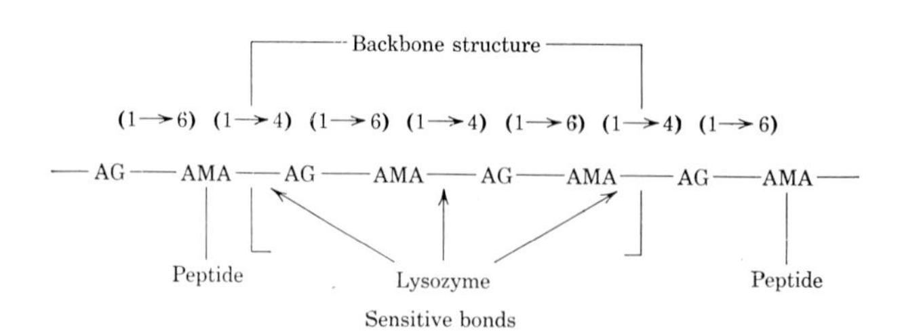

FIGURE 57. The 'backbone structure' proposed for the wall of *Micrococcus lysodeikticus* showing both peptide-substituted and unsubstituted N-acetylmuramic acid (AMA) residues linked to N-acetylglucosamine (AG) and the distribution of lysozyme sensitive bonds.

established the type of structure suspected from the characterization of the *Staphylococcus aureus* nucleotides (PARK AND STROMINGER, 1957).

The location of the amidase sensitive bond is illustrated in Fig. 58 and the action of this enzyme aided the study of the structure of two glycosaminopeptides isolated from *Micrococcus lysodeikticus* walls digested with lysozyme (GHUYSEN, 1961). Liberation of 'N-terminal' alanine by treating walls from several organisms with Streptomyces F_2B enzyme (SALTON, 1961b) confirms the widely accepted idea that alanine is the amino acid usually joined through the carboxyl group of muramic acid to form the glycosaminopeptide.

Based on the compounds isolated and studied by GHUYSEN (1961) and the results of hydrazinolysis experiments, SALTON (1961b) proposed a model for glycosaminopeptide of the type which may occur in the walls of *Micrococcus lysodeikticus*. This hypothetical model shown in Fig. 59 takes into account the observed values of C-terminal glycine and free ε–NH_2 groups of lysine. Such a subunit would consist of two identical halves joined through the ε–NH_2 group of a lysine residue in one half to the COOH of glycine in the other half, thus leaving a free ε–NH_2-lysine and a free C-terminal glycine. Some additional evidence in support of this type of structure was suggested from the increase in ε–NH_2-lysine groups and C-terminal glycine residues on incubating the lysozyme-non-dialysable fraction from *Micrococcus lysodeikticus* walls with Streptomyces F_2B enzyme (incompletely purified fraction). Only weak activity of the F_2B enzyme preparation on synthetic ε–lysyl-glycine peptides was observed (SALTON, unpublished observations) so a re-interpretation of the results may be necessary.

From a detailed analysis of the walls of *Micrococcus lysodeikticus*, CZERKAWSKI, PERKINS AND ROGERS (1963) were able to account for 85% of the wall as constituents of the glycosaminopeptide (N-acetylmuramic acid, N-acetylglucosamine, alanine, glutamic acid, lysine and glycine). They proposed a structure essentially similar to the various ones discussed above and have suggested that there are four glycosaminopeptide units to one disaccharide unit (the muramic acid of which has the carboxyl group unsubstituted). However, as these investigators have pointed out, 'the total titrable acidic and basic groups are too few to be accounted for by a simple structure consisting of a polysaccharide chain with short peptide chains attached to

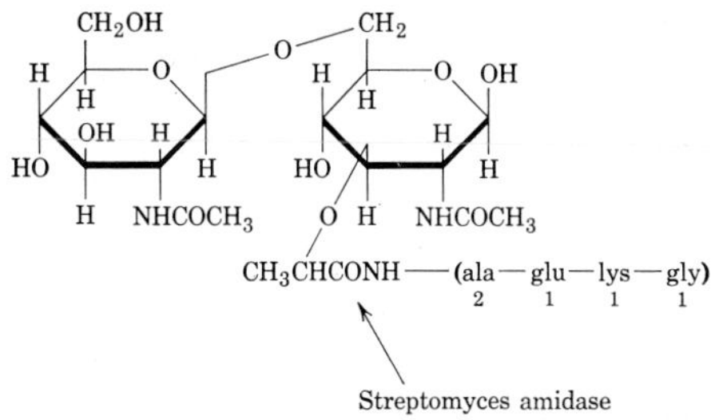

FIGURE 58. Structure of *Micrococcus lysodeikticus* cell-wall glycosaminopeptide and the bond sensitive to streptomyces amidase.

the muramic acid and provide evidence for rather extensive cross-linking in the whole polymer.' The results indicated that about half the ε–NH_2 groups of lysine in the whole mucopeptide are combined, possibly with a similar proportion of the γ-carboxyl groups of the glutamic acid. This suggestion of cross-linking is in accord with various structural studies carried out in other laboratories.

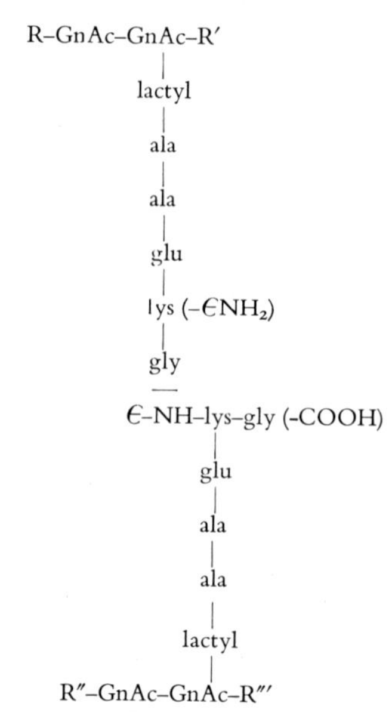

FIGURE 59. A hypothetical 'subunit' of the glycosaminopeptide based on data obtained from *Micrococcus lysodeikticus* walls.

The overall structure of the *Escherichia coli* B mucopeptides has heen established by PELZER (1962) and the main features of these compounds are in accord with those proposed for other wall compounds (i.e. a peptide attached through alanine to the carboxyl group of N-acetylmuramic acid which is in turn glycosidically linked to N-acetylglucosamine). PRIMOSIGH, PELZER, MAASS AND WEIDEL (1961) also indicated the presence of other compounds, the structures of which were more difficult to interpret from the data available.

Purified glycosaminopeptide from *Aerobacter cloacae* wall investigated by SCHOCHER, JUSIC AND WATSON (1961) gave C-terminal groups of alanine and α,ε-diaminopimelic acid (DAP) by the hydrazinolysis method. From the quantitative estimation of amino acid and amino sugar composition (SCHOCHER, BAYLEY AND WATSON, 1962) and the determination of C-terminal residues, these authors suggested a hypothetical subunit for *Aerobacter cloacae* wall glycosaminopeptide. The structure of this subunit is illustrated in Fig. 60.

The elucidation of structure of the cell wall of *Staphylococcus aureus* has been of especial interest, since it was from this organism that the PARK (1952) nucleotides were isolated and fully characterized (PARK AND STROMINGER, 1957). The walls of *Staphylococcus aureus* contain at least two types of polymers, the ribitol teichoic acid and the glycosaminopeptide. Unlike any of the other micrococci, the staphylococcal wall possesses a high proportion of glycine, an amino acid which was not present in the most complicated of the nucleotides isolated from penicillin-treated *Staphylococcus aureus*. There has thus been some doubt as to whether glycine is really part of a glycosaminopeptide or whether it is present as a separate peptide structure.

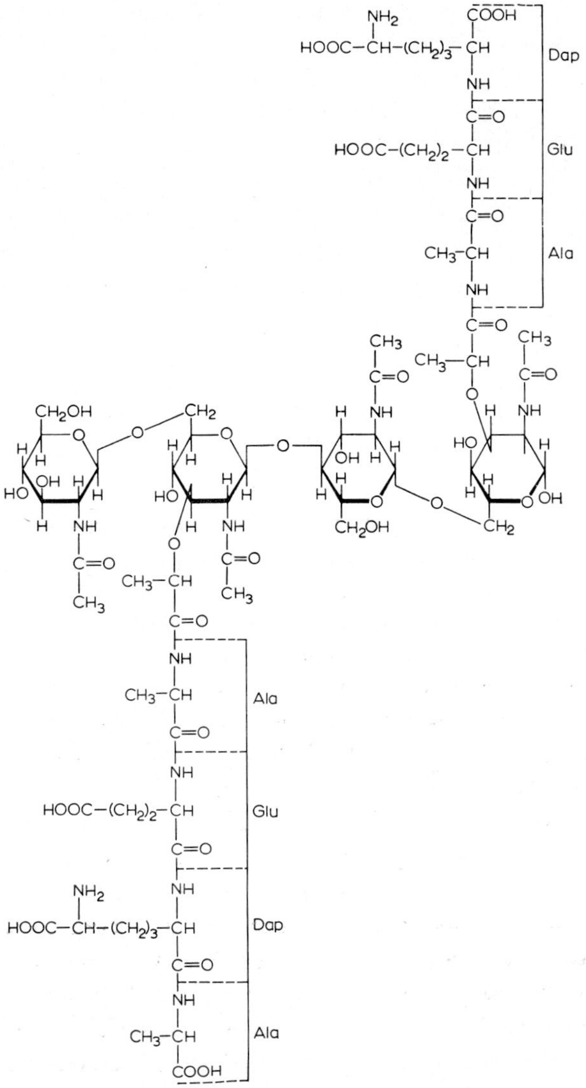

FIGURE 60. Hypothetical glycosaminopeptide subunit of the wall of *Aerobacter cloacae* proposed by SCHOCHER BAYLEY AND WATSON, 1962)

MANDELSTAM AND STROMINGER (1961) have carried out further investigations of the structure of the wall of *Staphylococcus aureus* by studying the removal of teichoic acid and the action of β-N-acetylglucosaminidase and lysozyme on the cell walls. From these studies MANDELSTAM AND STROMINGER (1961) concluded that *Staphylococcus aureus* (Copenhagen) wall has 'three polymeric threads, the glycopeptide backbone, the polyglycine component which appears to cross-link peptide chains through the ε-amino group of lysine; and the ribitol phosphate polymer.' The relatively few free ε-NH_2 groups of lysine and 'C-terminal and N-terminal' glycine residues determined in both Copenhagen and Duncan strains of *Staphylococcus aureus* walls (SALTON, 1961b) is compatible with the cross-linking of the polyglycine part of the wall. The two proposed structures for the *Staphylococcus aureus* (Copenhagen) wall and the possible mode of attachment of the teichoic acids are shown in Fig. 61.

Recent investigations by GHUYSEN AND STROMINGER (1963) have established the presence of terminal disaccharides in the glycopeptides solubilized by enzymic digestion of *Staphylococcus aureus* cell walls. Treatment of the soluble glycopeptide with the Streptomyces amidase yielded a homogeneous peptide containing glycine, alanine, glutamic acid and lysine as well as two types of disaccharide of N-acetylglucosamine and N-acetylmuramic acid. One of the disaccharides differed from the other in that it contained an O-acetyl group (GHUYSEN AND STROMINGER, 1963). The authors concluded that most of the muramic acid residues in the wall of *Staphylococcus aureus* carry peptide substituents. This is contrasted with the wall of *Micrococcus lysodeikticus* which yielded free di- and tetra-saccharides after digestion with lysozyme (SALTON AND GHUYSEN, 1960).

At least one common structural property has been established in all of the studies of the nature of the cell-wall glycosaminopeptides. They possess a glycosidically-linked 'backbone' of alternating groups of the two N-acetylamino sugars, N-acetylglucosamine and N-acetylmuramic acid and the peptides are linked through the amino group of alanine to the carboxyl group of muramic acid. Other features such as the precise sequences of amino acids, the mode

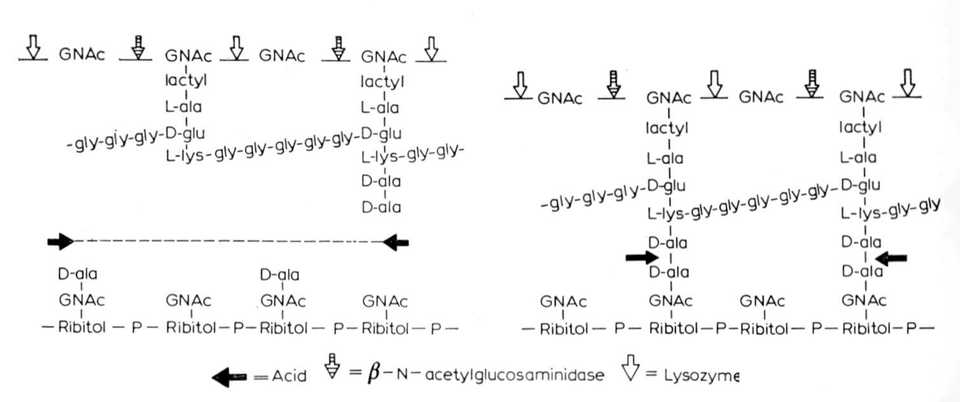

FIGURE 61. The structure proposed by MANDELSTAM AND STROMINGER (1961) for the cell wall of *Staphylococcus aureus* (Copenhagen) with the possible mode of attachment of the teichoic acid polymer shown in structure B.

of linkage of other cell-wall amino acids such as glycine and aspartic acid have yet to be determined. With the attachment of the teichoic acids, oligo- and polysaccharides the final structure of the simpler Gram-positive walls must present a complicated mosaic and much structural work remains before a clear picture of how all the fragments so far characterized, are put together chemically and biosynthetically.

There is at present no experimental information about the macromolecular organization of the glycosaminopeptides in cell walls. ROGERS (1962) has recently suggested several models compatible with the variety of compounds found in enzymic digests. In these models ROGERS (1962) pictured bundles of crossing microfibrils forming a network which may have occasional or many cross-links between closely packed strands. It was suggested that the latter type may be characteristic of the wall of *Staphylococcus aureus*.

Action of muramidases and other enzymes on glycosaminopeptides

From the foregoing discussions it will now be evident that the glycosaminopeptide of bacterial cell walls is the component responsible for the structural rigidity and integrity of the wall. Any enzyme bringing about a dissolution of the wall must involve an attack on the glycosaminopeptide fraction. Investigations of the chemical composition of walls has indicated that in some species of Gram-positive bacteria the glycosaminopeptide may account for as much as 90% of the cell wall (e.g. *Micrococcus lysodeikticus* – PERKINS AND ROGERS, 1959). In other organisms such as *Staphylococcus aureus* about half the wall is glycosaminopeptide and half ribitol teichoic acid. The additional components in the walls (teichoic acids, oligo- and polysaccharides)can usually be extracted, leaving behind the insoluble peptide-amino sugar complexes. Similarly, in Gram-negative bacteria the lipopolysaccharides, protein and lipid compounds can be extracted from the walls leaving an insoluble glycosaminopeptide. All of these structures have now been shown to be sensitive in varying degrees, to digestion with enzymes such as lysozyme (SALTON, 1952b; 1957b; KRAUSE AND MCCARTY, 1961; MANDELSTAM, 1962) and it is now widely recognized that the site of action of this enzyme is in the glycosaminopeptide component.

The problems of the structure of cell-wall peptide-amino sugar polymers and the specificity of lysozyme and related enzymes are so interwoven now that they can hardly be dealt with separately. As already indicated the determination of the chemical structure of the products of lysozyme action has played an important part in elucidating the nature of the wall heteropolymers.

Muramidases (Lysozymes)

It will now be apparent from the discussion of the products found in digested walls that lysozyme attacks the glycosidic bonds between acetyl amino sugars and can therefore be classified as an acetylhexosaminidase. The important discovery that egg-white lysozyme also degrades highly purified chitin (BERGER AND WEISER, 1957) established its $\beta(1 \rightarrow 4)$ N-acetylglucosaminidase activity. Although there is little information about the relative affinities of lysozyme for its substrate in the insoluble cell-wall glycosaminopeptides compared to that of chitin, the speed of dissolution as determined by reduction in turbidity is greater for the wall substrates.

SALTON (1961a) has proposed the term 'muramidase' for the enzyme lysozyme and the related enzymes such as the Streptomyces F1, N-acetylhexosaminidase isolated by GHUYSEN (1957) and studied by SALTON AND GHUYSEN (1960). Another enzyme attacking the amino sugar portion of the wall glycosaminopeptide has been isolated from the *Streptomyces albus* enzymes (GHUYSEN, LEYH-BOUILLE AND DIERICKX, 1962). This enzyme, called Enzyme '32', also acts on the amino sugar 'backbone' of the wall and although it differs from lysozyme and Streptomyces N-acetylhexosaminidase in not liberating free oligosaccharides, it can be classified as a muramidase.

The enzyme lysozyme has been investigated more extensively than any of the other muramidases and some of the substances that are sensitive and resistant to the action of lysozyme are given in Table 54. The simplest 'native' substrate for lysozyme obtained from a bacterial cell wall is a tetrasaccharide, the proposed structure of which is: O-β-N-acetylglucosaminyl-(1–6)-O-β-N-acetylmuraminyl-(1–4)-O-β-N-acetylglucosaminyl-(1–6)-β-N-acetylmuramic acid (SALTON AND GHUYSEN, 1960). PARK, WYNNGATE AND GRIFFITH (1961) have isolated an oligosaccharide almost freed of amino acids by prolonged treatment with alkali and on N-acetylation this material is degraded by lysozyme.

In addition to lysozyme, the specificity of the Streptomyces N-acetylhexosaminidase has also been studied. This latter enzyme shows a number of properties in common with lysozyme. It degrades the walls of *Micrococcus lysodeikticus* liberating di-and tetra-saccharides (structures shown in Fig. 53), no action is observed on phenyl-N-acetylglucosaminides, and it hydrolyses dichitobiose (DIERICKX AND GHUYSEN, 1962). Although the F1 enzyme preparations studied by SALTON AND GHUYSEN (1960) were capable of degrading the tetrasaccharide liberated from *Micrococcus lysodeikticus* walls by lysozyme, further purification of the enzyme by zone electrophoresis gave preparations no longer active on this substrate (DIERICKX AND GHUYSEN, 1962). Lysozyme on the other hand degrades the tetrasaccharide further, giving disaccharide which is identical in behaviour to that found in the original lysozyme digests.

The discovery that chitin is indeed susceptible to egg-white lysozyme as well as the inducible microbial chitinases (REYNOLDS, 1954; BERGER AND REYNOLDS, 1958), clearly established the β (1→4) N-acetylglucosaminidase activity of this muramidase. The action of lysozyme on soluble oligosaccharides isolated from chitin was first reported by MEYER (1959) and later confirmed by SALTON AND GHUYSEN (1959). The products of lysozyme action on chitin (BERGER AND WEISER, 1957) and chitodextrin (MEYER, 1959) were not characterized. The chitodextrin used by MEYER (1959) had an avarage chain length of 6–7 units. The viscosity of solutions of glycol chitin was reduced by incubation with egg-white lysozyme (HAMAGUCHI AND FUNATSU, 1959) but characterization of the products was not reported.

The action of lysozyme on di-N-acetyl-chitobiose and tetra-N-acetyl-chitotetraose was studied by examining the products on paper chromatograms (SALTON AND GHUYSEN, 1959;

1960). It was concluded that free N-acetylglucosamine and chitobiose were present in the digests of tetrasaccharide and that free N-acetylglucosamine was also found in the lysozyme-treated disaccharide (di-N-acetyl-chitobiose). However, in a later investigation DIERICKX AND GHUYSEN (1962) were unable to confirm the liberation of free N-acetylglucosamine from these substrates. Chitobiose was not attacked by lysozyme or the more highly purified Streptomyces N-acetylhexosaminidase; the tetra-N-acetyl-chitotetraose yielded only chitobiose after incubation with these enzymes (DIERICKX AND GHUYSEN, 1962). On the other hand HATTON (personal communication) has found free N-acetylglucosamine liberated from lysozyme digested chitodextrin. The reasons for these discrepancies are not known at present. WENZEL, LENK AND SCHÜTTE (1961) have studied the action of lysozyme on tritium-labelled tri-N-acetyl-chitotriose and reported the hydrolysis of this substrate to a mixture of di-N-acetyl-chitobiose, N-acetylglucosamine and an unknown compound. Unchanged trisaccharide was also present. Unfortunately there was no quantitative information which would have indicated whether or not there was an excess of N-acetylglucosamine liberated when the trisaccharide was used as a substrate, so the action of lysozyme on di-N-acetyl-chitobiose is still in doubt.

The cleavage of the tetrasaccharide, O-β-N- acetylglucosaminyl-(1–6)-O-β-N-acetylmuraminyl-(1–4)-O-β-N-acetylglucosaminyl-(1–6)-β-N-acetylmuramic acid, to the disaccharide, β-(1–6)-N-acetylglucosaminyl-N-acetylmuramic acid, has established that lysozyme splits 1–4 linkages and that 1–6 linkages are not attacked (SALTON AND GHYUSEN, 1960). The action on the chitin oligosaccharides suggests that the specificity of several of the muramidases (lysozyme and Streptomyces N-acetylhexosaminidase) is towards 1–4 linkages of the β configuration. It will be of interest to see what structural requirements determine the specificity of Enzyme '32' (GHUYSEN, LEYH-BOUILLE AND DIERICKX, 1962).

From these considerations of the structure of the cell-wall glycosaminopeptides, there are obviously many features which may determine their sensitivity to lysozyme and related muramidases. Following the discovery of O-acetyl groups in cell walls of *Streptococcus faecalis* and several other bacteria (ABRAMS, 1958) the importance of these substituents in relation to lysozyme sensitivity also became conspicuous (BRUMFITT, WARDLAW AND PARK, 1958). Resistance to lysozyme by selection of mutants was shown to be due to the production of walls containing about 1 group of O-acetyl per muramic acid residue. Cells and walls could also be rendered completely resistant to lysozyme by direct chemical O-acetylation with acetic anhydride (BRUMFITT, WARDLAW AND PARK, 1958). These investigations indicated that a particular hydroxyl group in the amino sugar backbone of the wall, had to be free for splitting to occur.

BRUMFITT (1959) selected lysozyme resistant mutants of *Bacillus megaterium* and these too were shown to be related to acetylation of cell-wall hydroxyl groups. However, *Streptococcus faecalis* and *Staphylococcus aureus* cell walls were apparently resistant for other reasons as deacetylation did not increase the sensitivity to lysozyme. SALTON AND PAVLIK (1960) found only one species *(Lactobacillus arabinosus)* of the six organisms they investigated, which showed greater lysozyme sensitivity after treatments removing O-ester groups. PARK, WYNNGATE AND GRIFFITH (1961) reported that walls of *Staphylococcus aureus* strain H were rendered sensitive to lysozyme after alkali treatment which removed primarily ester-linked alanine.

In discussing the various factors affecting sensitivity of the glycosaminopeptide of cell wall to lysozyme, SALTON AND PAVLIK (1960) and SALTON (1961a) suggested that the variations in he types of linkages between amino sugars may be of some importance, especially the bonds between carbon atom 1 of muramic acid and the adjacent sugar (e.g. $1 \rightarrow 3$, $1 \rightarrow 4$, $1 \rightarrow 6$ linkages). PARK AND GRIFFITH (1962) have recently obtained evidence that the non-dialysable fraction of *Micrococcus lysodeikticus* contains a polymer in which the N-acetylglucosamine is linked to position 4 of muramic acid. This was established by methylation studies which gave 6-methyl muramic acid as the principal product in this non-dialysable fraction. PARK AND GRIFFITH (1962) suggest that position 4 of muramic acid must be free to permit the lysozyme to act, a suggestion also in accord with the inhibitory action of blocking the hydroxyl groups by O-acetylation.

The extent to which the frequency of peptides on the amino sugar backbone contributes to lysozyme resistance cannot be judged at present. There is no doubt that muramidases such as lysozyme and Streptomyces N-acetylhexosaminidase can degrade the substrate adjacent to the peptides, leaving single peptide chains attached through muramic acid to disaccharide units (GHUYSEN AND SALTON, 1960; GHUYSEN, 1961). PARK AND GRIFFITH (1962) believe that the amino acids in the substrate play no role in determining the number of bonds split by lysozyme. It may well be true that the peptides only bring about a minimum of hindrance to lysozyme action on highly branched and cross-linked structures and it could be that dissolution of the wall is a poor index of the number of bonds split in such a wall. In this context the cell wall of *Bacillus cereus* would be a particularly interesting one, as it has a high amino sugar content, low O-acetyl values (SALTON AND PAVLIK, 1960) and yields no free amino sugar compounds (e.g. disaccharide) on incubation with lysozyme despite a reduction in turbidity varying from 24–50% (SALTON AND PAVLIK, 1960).

Factors other than the structure of the substrate can also contribute to the sensitivity or resistance of intact cells and isolated walls to muramidases and these have been discussed fully by SALTON (1957, 1958, 1961), LITWACK (1960) and REPASKE (1960). Perhaps one of the commonest reasons for 'apparent resistance' is the occlusion of the substrate, especially in the walls of Gram-negative bacteria. However, it is quite clear that the glycosaminopeptides of the Gram-negative bacteria are not intrinsically resistant to lysozyme, for digestion of the substrates has been amply demonstrated (EPSTEIN AND CHAIN, 1940; SALTON, 1958; WEIDEL, FRANK AND MARTIN, 1960; MANDELSTAM, 1962).

The mechanism of resistance of the glycosaminopeptide to the action of lysozyme on intact cells of Gram-negative bacteria has been investigated by REPASKE and the influence of ethylenediaminetetraacetic acid (EDTA) in rendering the substrate accessible and sensitive to the enzyme has been discussed by REPASKE (1960). There seems little doubt that EDTA acts by chelating with bivalent cations in the wall, but the site at which it acts to enable the lysozyme to degrade the glycosaminopeptide is still obscure. Whether the bivalent cations are involved in cross-linking glycosaminopeptide to protein or bonds between carboxyl groups of muramic acid in adjacent chains cannot be stated at present. It appears to the author that the most likely explanation of the EDTA effect is to render the substrate accessible by altering the conformation of the overlying protein structures, especially as pretreatment of the bacteria with 4% *n*-butanol also renders them sensitive to lysozyme and this latter reagent is more likely to involve the wall lipid and protein components. Moreover, as lysozyme is capable of breaking substrate bonds irrespective of the substitution of muramic acid with peptide side-chains, it would seem unlikely that bivalent cations attached to carboxyl groups of muramic acid would alter the sensitivity to enzymic attack.

Streptomyces amidase

The only other enzyme attacking glycosaminopeptides which has been studied in any detail is the amidase discovered and isolated by GHUYSEN (1957; 1960; 1961). Although this enzyme does not bring about dissolution of isolated cell walls on its own, it does aid the digestion of walls when allowed to act with other enzymes e.g. muramidases.

The bond sensitive to the Streptomyces amidase has already been indicated in Fig. 58. It seems likely that L-alanine is joined to muramic acid in most of the wall glycosaminopeptides but beyond this there is very little information, at the present time, about the specificity requirements for this amidase.

Cell-wall degrading enzymes

Many cell-wall degrading enzymes have been studied now and it can be concluded that those enzymes bringing about a dissolution of the walls of Gram-positive bacteria do so by splitting the linkages of the glycosaminopeptide polymers. The action of some of these enzymes on isolated walls has been reviewed by SALTON (1960). Just how many of these enzymes could be classified as muramidases cannot be said at this stage as most of these systems have not been purified and the nature of the products of their activities has not been established.

Some of the cell-wall lytic enzymes may prove of great value in further studies of glycosaminopeptide structure. KOTANI, KATO, MATSUBARA, HIRANO AND HIGASHIGAWA (1959) found culture filtrates of a *Flavobacterium* sp. active in lysing cell walls of lysozyme-resistant organisms such as *Staphylococcus aureus*, *Mycobacterium tuberculosis* BCG, *Corynebacterium diphtheriae* and *Streptococcus pyogenes*. An enzyme from a *Streptomyces* sp. brought about dissolution of the walls of *Corynebacterium diphtheriae* and *Bacillus megaterium;* it was inactive towards some of the more lysozyme-sensitive organisms (MORI, KATO, MATSUBARA AND KOTANI, 1960). No doubt many of these enzymes, autolytic enzymes and phage-induced enzymes will attract attention in the future and their purification and characterization should add a great deal to our knowledge of the chemical structure of the rigid components of bacterial cell walls.

REFERENCES

ABRAMS, A., *J. Biol. Chem.*, 230 (1958) 949.

ARMSTRONG, J. J., J. BADDILEY, J. G. BUCHANAN, B. CARSS and G. R. GREENBERG, *J. Chem. Soc.*, (1958) 4344.

BERGER, L. R. and D. M. REYNOLDS, *Biochim. Biophys. Acta*, 29 (1958) 522.

BERGER, L. R. and R. S. WEISER, *Biochim. Biophys. Acta*, 26 (1957) 517.

BRUMFITT, W., *Brit. J. Exptl. Pathol.*, 40 (1959) 441.

BRUMFITT, W., A. C. WARDLAW and J. T. PARK, *Nature*, 181 (1958) 1783.

CUMMINS, C. S. and H. HARRIS, *J. Gen. Microbiol.*, 13 (1955) iii.

CUMMINS, C. S. and H. HARRIS, *J. Gen. Microbiol.*, 14 (1956a) 583.

CUMMINS, C. S. and H. HARRIS, *Intern. Bull. Bacteriol. Nomen.*, 6 (1956b) 111.

CZERKAWSKI, J. W., H. R. PERKINS and H. J. ROGERS, *Biochem. J.*, 86 (1963) 468.

DIERICKX, L. and J. M. GHUYSEN, *Biochim. Biophys. Acta*, 58 (1962) 7.

EPSTEIN, L. A. and E. CHAIN, *Brit. J. Exptl. Pathol.*, 21 (1940) 339.

FLOWERS, H. M. and R. W. JEANLOZ, *J. Org. Chem.*, 28 (1963) 1564.

GHUYSEN, J. M., *Arch. int. Physiol.*, 65 (1957) 174.

GHUYSEN, J. M., *Biochim. Biophys. Acta*, 40 (1960) 473.

GHUYSEN, J. M., *Biochim. Biophys. Acta*, 47 (1961) 561.

GHUYSEN, J. M., M. LEYH-BOUILLE and L. DIERICKX, *Biochim. Biophys. Acta*, 63 (1962) 286.

GHUYSEN, J. M. and M. R. J. SALTON, *Biochim. Biophys. Acta*, 40 (1960) 462.

GHUYSEN, J. M. and J. T. STROMINGER, *J. Biol. Chem.* (1963) In the press.

HAMAGUCHI, K. and M. FUNATSU, *J. Biochem. (Japan)*, 46 (1959) 1659.

HEYMANN, H., L. D. ZELEZNICK and J. A. MANNIELLO, *J. Am. Chem. Soc.*, 83 (1961) 4859.

HOLDSWORTH, E. S., *Biochim. Biophys. Acta*, 9 (1952) 19.

IKAWA, M., *J. Biol. Chem.*, 236 (1961) 1087.

JANCZURA, E., H. R. PERKINS and H. J. ROGERS, *Biochem. J.*, 80 (1961) 82.

KOTANI, S., K. KATO, T. MATSUBARA T. HIRANO and M. HIGASHIGAWA, *Biken's Journal*, 2 (1959) 211.

KRAUSE, R. M. and M. MCCARTY, *J. Exptl. Med.*, 114 (1961) 127.

KUHN, R., A. GAUHE and H. H. BAER, *Chem. Ber.*, 87 (1954) 289.

LITWACK, G., Symposium, *Lysozyme as Related to Problems in Microbiology*, Convened at Soc. Amer. Bacteriologists, Philadelphia, 1960.

MANDELSTAM, J., *Biochem. J.*, 84 (1962) 294.

MANDELSTAM, J. and J. L. STROMINGER, *Biochem. Biophys. Res. Comm.*, 5 (1961) 466.

MEYER, K., *Polysaccharides in Biology*, 5th Confr. Josiah Macy, Jr. Foundation, 1959, p. 78.

MEYER, K. and E. HAHNEL, *J. Biol. Chem.*, 163 (1946) 723.

MEYER, K., J. W. PALMER, R. THOMPSON and D. KHORAZO, *J. Biol. Chem.*, 113 (1936) 479.

MITCHELL, P. and J. MOYLE, *J. Gen. Microbiol.*, 5 (1951) 981.

MORI, Y., K. KATO, T. MATSUBARA and S. KOTANI, *Biken's Journal*, 3 (1960) 139.

PARK, J. T., *J. Biol. Chem.*, 194 (1952) 877, 885, 897.

PARK, J. T., *Abstr. Am. Chem. Soc.*, (1956).

PARK, J. T., *Soc. Gen. Microbiol. Symposium* No. 8, 1958, p. 49.

PARK, J. T. and M. E. GRIFFITH, *Abstracts 8th Intern. Congr. Microbiol.*, Montreal (1962).

PARK, J. T. and J. L. STROMINGER, *Science*, 125 (1957) 99.

PARK, J. T., A. WYNNGATE and M. E. GRIFFITH, *Bacteriol. Proc.*, (1961) 689.

PELZER, H., *Biochim. Biophys. Acta*, 63 (1962) 229.

PERKINS, H. R., *Biochem. J.*, 73 (1959) 33P.

PERKINS, H. R., *Biochem. J.*, 74 (1960) 182.

PERKINS, H. R. and H. J. ROGERS., *Biochem. J.*, 69 (1958) 15P.
PERKINS, H. R. and H. J. ROGERS, *Biochem. J.*, 72 (1959) 647.
PRIMOSIGH, J., H. PELZER, D. MAASS and W. WEIDEL, *Biochim. Biophys. Acta*, (1961).
REPASKE, R., Symposium, *Lysozyme as Related to Problems in Microbiology*, Convened at Soc. Amer. Bacteriologists, Philadelphia, 1960.
REYNOLDS, D. M., *J. Gen. Microbiol.*, 11 (1954) 150.
ROGERS, H. J., *Biochem. Soc. Symposium* No. 22 (1962) p. 55.
ROGERS, H. J. and H. R. PERKINS, *Biochem. J.*, (1962) 35P.
SALTON, M. R. J., *Biochim. Biophys. Acta*, 8 (1952a) 510.
SALTON, M. R. J., *Nature*, 170 (1952b) 746.
SALTON, M. R. J., *Soc. Gen. Microbiol. Symposium* No. 6 (1956a).
SALTON, M. R. J., *Biochim. Biophys. Acta.*, 22 (1956b) 495.
SALTON, M. R. J. in *The Nature of Viruses*, Eds. G. E. W. Wolstenholme and E. C. P. Millar, Churchill, London, 1957a.
SALTON, M. R. J., *Bacteriol. Rev.*, 21 (1957b) 82.
SALTON, M. R. J., *J. Gen. Microbiol.*, 18 (1958) 481.
SALTON, M. R. J., *Biochim. Biophys. Acta*, 34 (1959) 308.
SALTON, M. R. J. in *The Bacteria*, Vol. I, p. 97, Academic Press, New York, 1960.
SALTON, M. R. J., *Proc. 2nd Symposium on Fleming's Lysozyme*, Milan, 1961a, Vol. II, p. 1/xii.
SALTON, M. R. J., *Biochim. Biophys. Acta*, 52 (1961b) 329.
SALTON, M. R. J. and J. M. GHUYSEN, *Biochim. Biophys. Acta*, 36 (1959) 552.
SALTON, M. R. J. and J. M. GHUYSEN, *Biochim. Biophys. Acta*, 45 (1960) 355.
SALTON, M. R. J. and J. G. PAVLIK, *Biochim. Biophys. Acta*, 39 (1960) 398.
SCHOCHER, A. J., S. T. BAYLEY and R. W. WATSON, (1961).
SCHOCHER, A. J., D. JUSIC and R. W. WATSON, *Biochem. Biophys. Res. Comm.*, 6 (1961) 16.
STRANGE, R. E. and F. A. DARK, *Nature*, 177 (1956) 186.
STRANGE, R. E. and J. F. POWELL, *Biochem. J.*, 58 (1954) 80.
WEIDEL, W., H. FRANK and H. H. MARTIN, *J. Gen. Microbiol.*, 22 (1960) 158.
WENZEL, M., H. P. LENK and E. SCHÜTTE, *Hoppe-Seyler's Z. Physiol. Chem.*, 327 (1961) 13.
WORK, E., *Nature*, 179 (1957) 338.

CHAPTER 6

The occurrence and structure of teichoic acids

It is a curious fact that most of the unusual compounds now known to be localized in bacterial cell walls were not discovered by direct examination of walls but were first found in other cell compounds. Thus muramic acid was initially detected in spore peptides (STRANGE AND POWELL, 1954), α, ε-diaminopimelic acid in hydrolysates of whole bacterial cells (WORK, 1951), D-alanine and D-glutamic acid in the uridine nucleotides isolated by PARK (1952). It came as no surprise then, that the discovery of the last major group of 'unusual' cell-wall substances originated from the identification of ribitol in a cytidine nucleotide by BADDILEY AND MATHIAS (1954) and BADDILEY, BUCHANAN, CARSS, MATHIAS AND SANDERSON (1956). MITCHELL AND MOYLE (1951) did in fact find a polyol phosphate compound in the 'envelope' of *Staphylococcus aureus;* they concluded that it was primarily a glycerophosphate polymer and it is now recognized as the 'intracellular teichoic acid' (BADDILEY, 1961).

Following the isolation of cytidine diphosphate ribitol (CDP-ribitol) and cytidine diphosphate glycerol (CDP-glycerol) from *Lactobacillus arabinosus* and the confirmation of their structures by chemical synthesis (BADDILEY, BUCHANAN AND SANDERSON, 1958; BADDILEY, BUCHANAN AND FAWCETT, 1959) the role these nucleotides may play in polymer biosynthesis was contemplated. The CDP-ribitol contained a D-ribitol 5-phosphate residue and the CDP-glycerol was found to be a derivative of the naturally occurring L-glycerol 3-phosphate. The biosynthetic functions of the nucleotides became even more probable when polymers containing ribitol phosphate and glycerol phosphate were isolated from trichloroacetic acid extracts of *Lactobacillus arabinosus* and other bacteria (ARMSTRONG, BADDILEY, BUCHANAN, CARSS AND GREENBERG, 1958).

An examination of the isolated cell walls of *Lactobacillus arabinosus* showed the presence of the ribitol phosphate polymer, but none of the glycerol phosphate compound was detected in the wall fraction (BADDILEY, BUCHANAN AND CARSS, 1958). Walls of *Bacillus subtilis,* known to be very rich in phosphorus (SALTON, 1953) were also examined and they too contained appreciable amounts of the ribitol phosphate polymer. The localization of these compounds in bacterial walls led to the description of this new class of polyol phosphate substances as the 'teichoic acids' (ARMSTRONG, BADDILEY, BUCHANAN, CARSS AND GREENBERG, 1958). These 'teichoic acids' do not appear to have any role in maintaining the rigidity of the wall structure, for it has already been pointed out that their removal from the cell wall leaves the rigid glycosaminopeptide component (ARCHIBALD, ARMSTRONG, BADDILEY AND HAY, 1961).

The chemical structure of the teichoic acids has been investigated intensively by BADDILEY and his colleagues and there is now more precise information about the structure of these wall polymers than there is for any of the other compounds (e.g. glycosaminopeptides,

The tables are printed together at the end of the book.

oligosaccharides). Moreover, some of the structures proposed for the teichoic acids have been confirmed by direct chemical synthesis.

Detection and occurrence of teichoic acids

The detection of the teichoic acids in bacterial walls rests largely on the phosphorus contents of the cell walls and an examination of the products of acid hydrolysis for the presence of characteristic constituents. The spray reaction for identifying polyols, developed by BUCHANAN, DEKKER AND LONG (1950) has been of great value in studying the teichoic acids.

Hydrolysis of the teichoic acids from the walls of *Lactobacillus arabinosus*, *Bacillus subtilis* and *Staphylococcus aureus* with $2N$ HCl for several hours yielded a mixture of substances detectable on paper chromatography of the hydrolysates. The products found in the various teichoic acids from these organisms are shown in Table 55. One of the conspicuous components detected by the BUCHANAN, DEKKER AND LONG (1950) spray reagent is 1,4-anhydroribitol formed as a decomposition product from ribitol phosphates during acid hydrolysis of the polymer. SALTON AND PAVLIK (1960) also detected another fast-moving product in $6N$ HCl hydrolysates which they suggested may possibly be a dianhydroribitol. However, BADDILEY has pointed out that a dianhydroribitol would be sufficiently stable to resist oxidation with periodate and would not be detected by the usual method; the fast-moving component is probably a chloro-derivative formed in strong acid and BADDILEY, BUCHANAN, MARTIN AND RAJBHANDARY (1962) have recently concluded that this compound is 5-chloro-5-deoxy-1,4-anhydroribitol.

Teichoic acid of the glycerol type may also occur in bacterial walls and on hydrolysis the products detected are glycerol, phosphate, alanine and usually a sugar (BADDILEY, 1961; BADDILEY AND DAVISON, 1961).

More Gram-positive bacteria have been tested for the presence of the teichoic acids in their walls than Gram-negative organisms. The occurrence of ribitol and glycerol teichoic acids in various bacterial walls was investigated by ARMSTRONG, BADDILEY, BUCHANAN, DAVISON, KELEMEN AND NEUHAUS (1959); SALTON AND PAVLIK (1960); BADDILEY (1961) and IKAWA (1961) and the results of these studies are summarized in Table 56. It can be seen that cell walls usually contain either the ribitol or the glycerol type of teichoic acid. However, the wall of *Streptococcus faecalis* contains both ribitol and glycerol and probably possesses the two types of polymer (ARMSTRONG *et al.*, 1959; IKAWA, 1961).

Traces of a glycerol type of polymer were detected in the isolated walls of *Escherichia coli* B. Very few Gram-negative bacteria have been tested for the presence of teichoic acids and the status of the material in *Escherichia coli* B wall does not appear to have been investigated further. However, LILLY (1962) has recently reported the presence of trichloroacetic acid extractable polymer containing ribitol, phosphate, glucose and mucopeptide from *Escherichia coli*, strain 26–26, cell walls. So far as the author is aware this is the first suggestion of the presence of a ribitol teichoic acid in the wall of a Gram-negative organism. The teichoic acid in this instance was believed to be linked in some way to the mucopeptide component. Moreover, both cytidine diphosphate glycerol and cytidine diphosphate ribitol were found in the nucleotides which had accumulated in the growth medium of *Escherichia coli* 26–26 (LILLY, 1962).

The teichoic acid of *Bacillus megaterium* deserves special mention, since it appears to differ markedly from those found so far in other bacterial walls. Although SALTON AND MILHAUD (1959) detected a compound in hydrolysates of uniformly ^{14}C labelled *Bacillus megaterium* cell walls which corresponded to anhydroribitol (as revealed by spraying the chromatograms with the Schiff's reagent), GHUYSEN (1961) found only traces of anhydroribitol and even then only after hydrolysis with 6*N* HCl. The principal compound revealed as a polyol in the walls and isolated compounds from enzymic digests, could not be identified as belonging to the ribitol or glycerol class (GHUYSEN, 1961). BADDILEY (1961) has also recorded the presence of 'Ribitol?' teichoic acid in two strains of *Bacillus megaterium*. It will be of considerable interest to see the elucidation of the nature of the polyol of *Bacillus megaterium* wall and to clarify these conflicting results.

The teichoic acid contents of bacterial walls cannot be determined by direct chemical means, although approximations can be made from total phosphorus values (assuming it is all in the form of teichoic acid polymer). The absence of a satisfactory chemical method is not surprising, considering the complexity of the products of acid hydrolysis (Table 55), the lability of ester-linked D-alanine residues and variations in chain length with different proportions of phosphomonoester groups. In addition to values calculated from phosphorus determinations, the contribution of the teichoic acids can also be estimated by weighing the residues after removal of the polymers by extraction with trichloroacetic acid.

The only other sources of compounds related to the teichoic acids are the pneumococcal type specific substances. REBERS AND HEIDELBERGER (1959) were the first to report the presence of ribitol phosphate in a pneumococcal capsular polysaccharide. SHABAROVA, BUCHANAN AND BADDILEY (1962) examined 23 pneumococcal type specific substances and found ribitol or glycerol together with other sugars and amino sugars and in some instances small amounts of amino acids. The absence of alanine as a major component helps to distinguish these substances from the wall teichoic acids.

Following the detection of the two types (ribitol and glycerol) of teichoic acid in the cell walls of various bacteria the 'intracellular teichoic acid' came to light and it has now been found in a number of bacteria. BADDILEY (1961) and BADDILEY AND DAVISON (1961) examined various species of lactobacilli and other bacteria to determine the distribution of teichoic acids in the walls and cells and the results are summarized in Table 57.

The functions of the cell-wall and intracellular teichoic acids of bacteria are not clear at present. Structurally, the wall teichoic acids do not appear to be concerned with the mechanical rigidity of the cell wall and indeed they may be regarded as non-essential components (*Bacillus subtilis* has been reported to be devoid of teichoic acid when grown in media containing 2% glucose – ARMSTRONG, BADDILEY AND BUCHANAN, 1960). Recent investigations are in accord with the view that the teichoic acids are on the outer surface of the cell wall where they constitute the important antigenic components of the cells possessing these polymers (HAUKENES, ELLWOOD, BADDILEY AND OEDING, 1961; SANDERSON, JUERGENS AND STROMINGER, 1961). On the other hand it is possible that because of their ionic nature the teichoic acids may contribute to the passage of ions across the cell surface, thereby possessing some regulatory function in the bacterial cell (ARCHIBALD, ARMSTRONG, BADDILEY AND HAY, 1961). The ability of organisms possessing major teichoic acid components in their walls to grow in acid solutions and in fairly high salt concentrations has been noted by these authors.

The functions of the intracellular glycerol 'teichoic acid' are not known, although it is tempting to speculate that they may be concerned with some aspect of the transfer of D-alanine residues into the wall peptides or the locking of peptide chains of the glycosaminopeptides together. The fact that none of the other wall amino acids has been found associated with the intracellular glycerol teichoic acid, would argue against a general role (comparable to that of S-RNA) in wall peptide synthesis and it suggests a fairly specific function. No doubt this problem will attract much attention in the future and this will lead to a general solution to the problem of the functions of such an interesting type of polymer. There is already some suggestion that this material is located between the wall and protoplast membrane (HAY, WICKEN AND BADDILEY, 1963).

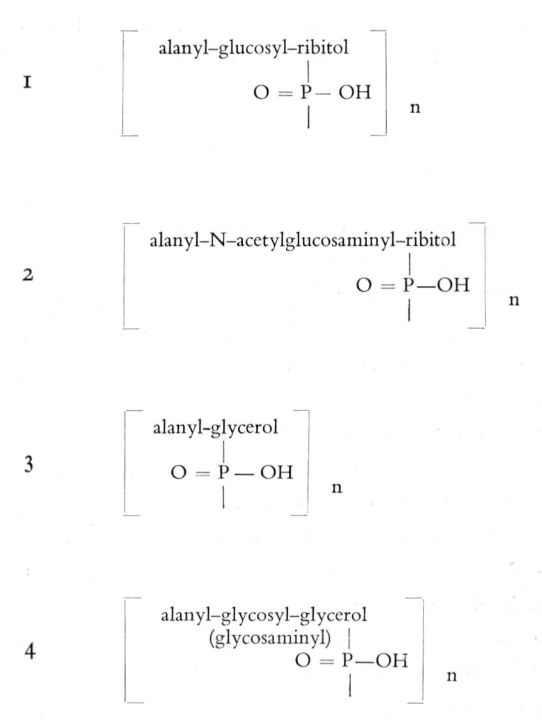

FIGURE 62. The four general types of teichoic acid represented in formulas 1, 2, 3 and 4.

Structure of the teichoic acids

Several characteristics of the structure of teichoic acids emerged from the early studies by ARMSTRONG, BADDILEY, BUCHANAN, CARSS AND GREENBERG (1958) and ARMSTRONG, BADDILEY, BUCHANAN, DAVISON, KELEMEN AND NEUHAUS (1959). The presence of anhydroribitol phosphate in acid hydrolysates and the detection of ribitol diphosphate derivatives on alkali hydrolysis was consistent with a general structure of ribitol units joined together by phosphodiester linkages.

One of the other distinguishing features of the teichoic acids was the presence of ester-linked D-alanine. The glycerol teichoic acid also showed these structural features (glycerol linked by phosphodiester bonds and ester linked D-alanine). Two general types of cell-wall teichoic acid can thus be distinguished as represented in the general formulae in Fig. 62 which also shows that the alanyl-glycosyl-ribitol phosphate type of polymer can be sub-divided further depending on whether the glycosyl residues are glucose or N-acetylamino sugars. The intracellular glycerol teichoic acid prepared from *Lactobacillus casei* by KELEMEN AND BADDILEY

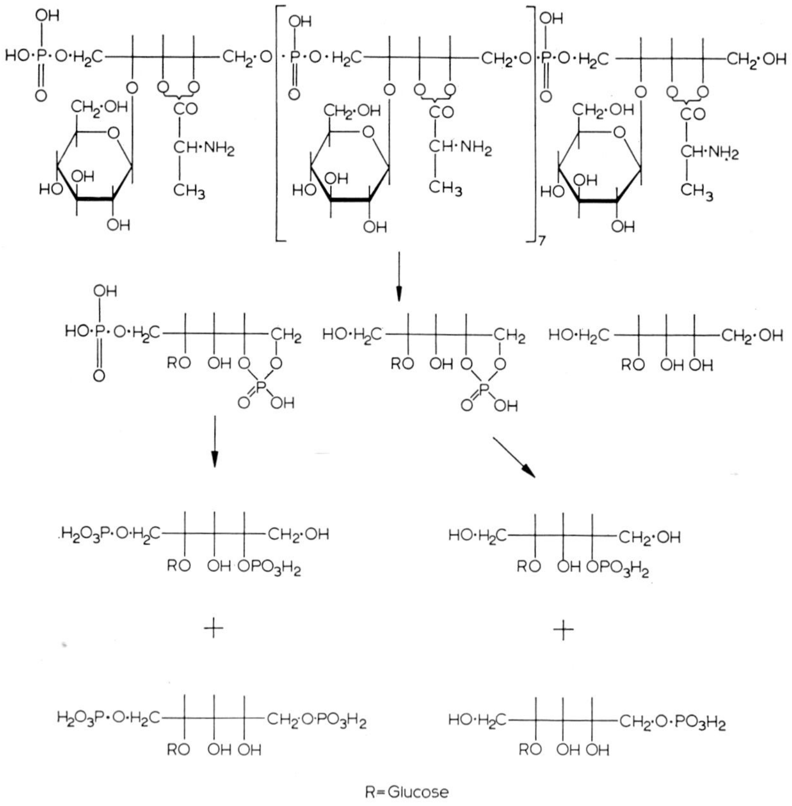

FIGURE 63. Structure of the cell-wall ribitol teichoic acid from *Bacillus subtilis*.

(1961) was of the general type shown in Fig. 62. BADDILEY (1962) has recently mentioned further details of the structure of both wall and intracellular glycerol teichoic acids and it is evident that these too may be of the general alanyl-glycosyl-glycerol phosphate type although there is probably no overall uniformity in the number of glycosyl residues/glycerol of different types of teichoic acid.

Bacillus subtilis teichoic acid

The structure of the teichoic acid of *Bacillus subtilis* was the first one to be established in full detail. By degrading the teichoic acid with alkali, after removal of the alanine residues the hydrolysis of the polymer proceeded through the formation of intermediate cyclic phosphates yielding isomeric glucosyl ribitol monophosphates as the main products, together with small amounts of glucosylribitol and its diphosphates, the latter compounds arising from the ends of chains (ARMSTRONG, BADDILEY AND BUCHANAN, 1960). The structure of *Bacillus subtilis* teichoic acid was thus established by degradation with alkali, followed by treatment with phosphatase and determining the structure of the dephosphorylated compounds by periodate oxidation and treatment with β-glucosidase. By these various techniques the bonds to which the glucose residues were attached and the structures of the teichoic acid were established (ARMSTRONG, BADDILEY AND BUCHANAN, 1960; BADDILEY, BUCHANAN AND HARDY, 1961; SARGENT, BUCHANAN AND BADDILEY, 1962). Fig. 63 presents the structure of *Bacillus subtilis* teichoic acid and the course of its degradation by alkali.

The exact location of glucose on the ribitol phosphate 'backbone' was established from periodate oxidation studies which indicated that it must be on either the 2- or 4- position of D-ribitol. The product of periodate oxidation was optically active and yielded D-glyceric acid on further oxidation with bromine water and hydrolysis as shown in Fig. 64. An inactive (meso)oxidation product from the 2-isomer would have given DL-glyceric acid on further oxidation and hydrolysis. It followed from these experiments that the glucose is on the 4-position of D-ribitol (ARMSTRONG, BADDILEY AND BUCHANAN, 1960). The structure of the

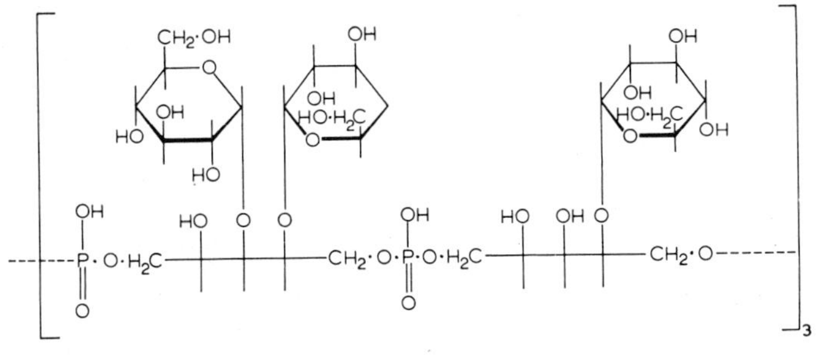

Lactobacillus arabinosus wall teichoic acid (alanine omitted)

FIGURE 64. Structure of the cell-wall ribitol teichoic acid from *Lactobacillus arabinosus*.

glucosylribitol was confirmed by total synthesis of both α and β forms of 4-O-(D-glucopyranosyl)-D-ribitol (BADDILEY, BUCHANAN AND HARDY, 1961).

The location of the alanine in the teichoic acid was indicated by the stability of the ribitol residues to periodate before, but not after, removal of the alanine. It was concluded that the alanine was attached to a hydroxyl group at either the 2- or 3- position in the ribitol residues ARMSTRONG, BADDILEY AND BUCHANAN, 1961a); at present it has not been possible to distinguish between these alternatives.

The size of the teichoic acid chain has been determined from the amount of formaldehyde produced from periodate oxidation of alanine-free polymer. The formaldehyde arises from a glycol group at one end of the polymer. This evidence together with titration data indicated a chain of 9 alanyl-glucosylribitol-5-phosphate units joined to give the structure shown in Fig. 63.

Lactobacillus arabinosus teichoic acid

The cell-wall teichoic acid from *Lactobacillus arabinosus* has shown a greater variability in composition than that encountered in preparations of *Bacillus subtilis* polymer (ARCHIBALD, BADDILEY AND BUCHANAN, 1961b). Some of the earlier preparations of *Lactobacillus arabinosus* teichoic acid were polymers containing mono- and diglucosylribitol as well as ribitol without sugar residues (ARMSTRONG, BADDILEY, BUCHANAN, CARSS AND GREENBERG, 1958). In later preparations, sugar residues were found on each ribitol (ARCHIBALD, BADDILEY AND BUCHANAN, 1961b). The monoglucosylribitol is the α anomer of that found in *Bacillus subtilis* teichoic acid. The structure of this α anomer has been confirmed by chemical synthesis. The diglucosylribitol bears an additional α-glucosyl residue at position 3. This teichoic acid is believed to be made up of about 6–7 ribitol units linked at the 1,5-positions by phosphodiester linkages. The suggested repeating structure for *Lactobacillus arabinosus* is given in Fig. 64 which omits the location of alanine in the polymer (BADDILEY, 1962a). The teichoic acid of this organism differs from that of *Bacillus subtilis* in possessing mono- and di-glucoside residues joined exclusively by α glycosidic bonds.

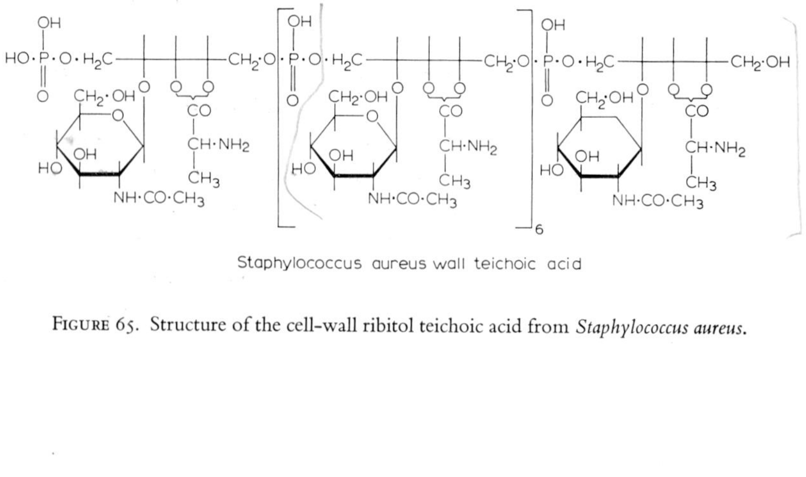

Staphylococcus aureus wall teichoic acid

FIGURE 65. Structure of the cell-wall ribitol teichoic acid from *Staphylococcus aureus*.

Staphylococcal teichoic acids

The teichoic acid of *Staphylococcus aureus*, strain H, cell wall was shown by ARMSTRONG, BADDILEY, BUCHANAN, CARSS AND GREENBERG (1958) to be a ribitol phosphate polymer with attached acetylglucosamine and alanine residues. The alanine of this teichoic acid also had the D-configuration and as in all other cases it was ester linked through hydroxyl groups in the polymer.

Degradation of the teichoic acid gave products identified as: ribitol, anhydroribitol, glucosamine, 2,5-anhydromannose, ribitol-1-phosphate, ribitol-2-phosphate, glucosaminyl-ribitol-1-phosphate, glucosaminylribitol-2-phosphate, glucosaminylribitol diphosphates and ribitol diphosphates.

The detailed structure of the *Staphylococcus aureus* teichoic acid was determined by the methods employed for the other teichoic acids. Alkali hydrolysis gave a mixture of isomeric mono-and di-phosphates of the 4-glucosaminylribitol or its N-acetyl derivative were obtained. Both α and β glycosidic linkages were detected but the proportions varied with different batches of cells. Periodate oxidation after removal of alanine, and determination of the formaldehyde produced, indicated that the 4-O-(N-acetylglucosaminyl)-D-ribitol units were joined through 1,5-phosphodiester linkages. Formaldehyde yield and titration suggested that 8 units were present. As with the other teichoic acids the ester-linked D-alanine was assigned to hydroxyl groups at the 2 or 3 positions of ribitol. The evidence from this study by BADDILEY, BUCHANAN, MARTIN AND RAJBHANDARY (1962) enabled them to give the structure in Fig. 65 to this teichoic acid.

SANDERSON, JUERGENS AND STROMINGER (1961) investigated the structure of the teichoic acid from another strain of *Staphylococcus aureus*, the strain 'Copenhagen'. In common with other cell-wall teichoic acids, this staphylococcal polymer was believed to be a 1,5-phosphodiester linked polymer of ribitol with the N-acetylglucosamine residues joined at positions 2 or 4 on the ribitol. Purified β-acetylglucosaminidase from pig epididymis was used in showing that about 85% of the N-acetylglucosaminyl residues were β-linked. After removal of acetylamino sugar with the pig epididymis enzyme, the resistant amino sugar was slowly liberated when treated with an extract of rat epididymis, which contains both α as well as β-acetylglucosaminidase.

The teichoic acid from *Staphylococcus aureus* Copenhagen contained ratios of ribitol phosphate: acetylglucosamine: D-alanine of 1:0.99:0.49. These analytical data together with evidence from periodate oxidation led SANDERSON, JUERGENS AND STROMINGER (1961) to suggest that the D-alanine residues were on the 6-position of statistically every other acetylglucosamine residue. If this suggestion is confirmed, the *Staphylococcus aureus* Copenhagen. teichoic acid differs from the other three cell-wall teichoic acids so far chemically characterized Some difference in antigenic properties of the Copenhagen strain may reflect differences in chemical structure although it should be remembered that these properties do not seem to be dependent on the ester-linked alanine.

In contrast to the *Staphylococcus aureus* ribitol type of teichoic acid, a glycerol teichoic acid was isolated from the wall of the closely related species *Staphylococcus albus*. The composition of the walls of *Staphylococcus albus* NCTC 7944 was first examined by SALTON AND PAVLIK (1960) and was shown to be very similar indeed to other staphylococcal species. Our wall preparations of this organism were examined by ARMSTRONG, BADDILEY, BUCHANAN, DAVISON, KELEMEN AND NEUHAUS (1959) and a glycerol polymer was detected. That this glycerol teichoic acid was in the wall was confirmed later during the structural investigations carried out by ELLWOOD, KELEMEN AND BADDILEY (1963).

The products of degradation of the *Staphylococcus albus* teichoic acid included glycerol and its mono- and di-phosphates, alanine, galactosamine and under milder conditions, N-acetylgalactosamine. The alanine has the D-configuration and is ester-linked as in all other teichoic acids. 2-O-D-galactosaminylglycerol was formed by hydrolysis with alkali and the structure of this compound was established by hydrolysis, periodate oxidation and reaction with nitrous acid. The amino sugar was shown to be of the D-series.

The glycosidic linkages between galactosamine and glycerol have mostly the α-configuration but a small proportion of β-linkages may occur. The structure of the teichoic acid has typical 1:3 phosphodiester linked glycerol with a chain length of about 18 glycerol phosphate units. An average of every third glycerol bears a N-acetylgalactosaminyl residue. A preliminary examination of the intracellular teichoic acid of this organism has shown that it is qualitatively similar, probably with a lower proportion of amino sugar (ELLWOOD, KELEMEN AND BADDILEY, 1963).

Cell-wall teichoic acids from other bacteria

BADDILEY (1962b) has reported that cell walls of *Streptococcus faecalis* contained a ribitol compound containing both N-acetylgalactosamine and glucose.

GHUYSEN (1961) has described material in the cell wall of *Bacillus megaterium* KM giving a periodate-Schiff positive reaction. However, no glycerol or ribitol or the compounds characteristic of degraded teichoic acid from other sources was found in this product. The nature of the periodate-Schiff positive material remains unknown; the ratio of phosphorus: hydroxyl groups for one of the substances separable from other wall compounds was 1:3.5. Whether this material is analogous to the teichoc acids of other bacterial walls will depend to a large extent, on its identification and the nature of the structure containing the unknown compound.

SALTON AND MILHAUD (1959) reported the presence of a compound corresponding to anhydroribitol in paper chromatograms of hydrolysed cell walls and material released by phage adsorption to the wall of *Bacillus megaterium*. It seems likely that some of these conflicting reports could well be due to differences in media for if it can be shown that glucose generally suppresses the formation of teichoic acids as reported for *Bacillus subtilis* by ARMSTRONG, BADDILEY AND BUCHANAN (1960) it may account for the absence of teichoic acids under some conditions. In commenting on the 'teichoic acid' of *Bacillus megaterium* KM, BADDILEY, BUCHANAN, MARTIN AND RAJBHANDARY (1962) suggested that the small amount of phosphorus in the wall of this strain may belong to mucopeptide rather than teichoic acid. Apart from *Bacillus megaterium* walls no new polyols other than ribitol or glycerol have so far been indicated as teichoic acid components. Recent evidence, obtained by GHUYSEN (1964), has shown that the polyol in the complex isolated from *Bacillus megaterium* is glycerol.

Intracellular 'teichoic acids'

With the discovery of intracellular glycerol phosphate polymers in bacteria we are now faced with the contradicting terminology of 'intracellular teichoic acids.' However, these polymers are so closely related chemically to the ribitol and glycerol teichoic acids of cell walls that it would be confusing to give them a new trivial name before the broad features of the intracellular compounds from a variety of bacteria have been fully established.

Intracellular glycerol teichoic acids were found in all of the lactobacilli examined by BADDILEY AND DAVISON (1961) as well as being detected in other bacteria (BADDILEY, 1961). It is probable that they are identical to the polyglycerophosphate antigens isolated from Group A streptococci and other organisms by MCCARTY (1959), although these preparations appeared to be devoid of the ester-linked D-alanine.

The walls of *Lactobacillus casei* ATCC 7469 contained no detectable teichoic acid but an intracellular compound was present and this discovery provided a suitable organism for the isolation of the material. KELEMEN AND BADDILEY (1961) isolated the glycerol teichoic acid from cell contents or whole defatted cells of *Lactobacillus casei* with cold dilute trichloroacetic acid. The purification of this teichoic acid naturally proved more difficult than the task of isolating the cell-wall polymers. However, the trichloroacetic acid extracts could be treated with ethanol to precipitate the glycerol teichoic acid. By reprecipitation several times with ethanol the intracellular teichoic acid was purified sufficiently for structural studies.

Acid hydrolysis of this compound gave glycerol its mono- and di-phosphates, inorganic phosphate and alanine. The products of hydrolysis with alkali included the alkali-stable diglycerol triphosphate. Diglycerol phosphate was isolated from the latter after treatment with prostatic phosphomonoesterase and this compound was then hydrolysed by alkali to glycerol and its monophosphates. It was concluded that this teichoic acid was a polymer in which the glycerol was joined by phosphodiester linkages at positions 1 and 3 in the glycerol units. The alanine was shown to have the D-configuration and to be joined at the 2-position

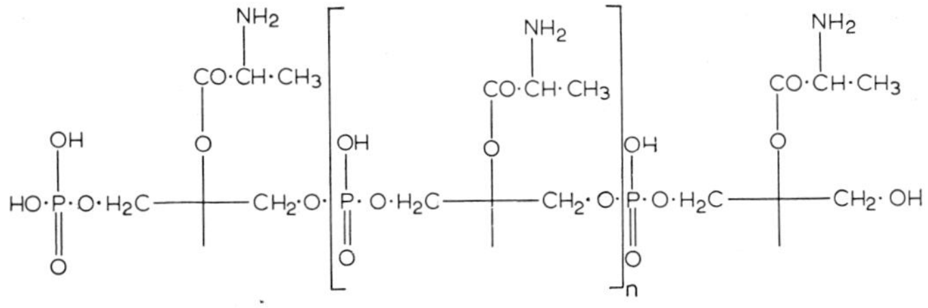

FIGURE 66. Structure of the intracellular glycerol teichoic acid isolated from *Lactobacillus casei*.

of each glycerol residue by ester linkages (KELEMEN AND BADDILEY, 1961). The structure assigned to this intracellular teichoic acid by KELEMEN AND BADDILEY (1961) is presented in Fig. 66 and shows the characteristic features of other cell-wall teichoic acids.

The intracellular teichoic acid from *Lactobacillus arabinosus* is similarly constituted, being a 1,3-polymer of about 18 alanyl glycerophosphate units. This teichoic acid also contained about two α-glucosyl residues at the 2-position of glycerols in each molecule. Quantitative analysis of formaldehyde and titration of acidic groups indicated a polymer of 18 units (BADDILEY, 1962a).

A more complex intracellular teichoic acid from *Streptococcus faecalis* has been studied by Dr. A. J. WICKEN (BADDILEY, 1962a) and he has found that each glycerol unit bears statistically three glucose residues, probably as a trisaccharide. In this teichoic acid the alanine must be attached to the sugar rather than to the glycerol residues.

Capsular polysaccharides containing ribitol phosphate

For some time it appeared that the ribitol teichoic acids were found exclusively in bacterial cell walls. However, the detection of ribitol phosphate and sugars in the capsular polysaccharide of type VI pneumococci (S.6) indicated that ribitol may be found in other bacterial products (REBERS AND HEIDELBERGER, 1959). On hydrolysis the constituents of the S.6 polysaccharide indicated that it was a polymer of ribitol phosphate, galactose, glucose and rhamnose. The structure of the S.6 polysaccharide has been established by REBERS AND HEIDELBERGER (1961) as a polymer containing galactosyl-glucosyl-rhamnosyl-ribitol units with phosphodiester bonds joining the ribitol of one unit to the galactose of its neighbour. This mode of linkage clearly distinguishes this polysaccharide from the teichoic acids and it is evident from the absence of any phosphates of anhydroribitol that the phosphodiester bonds do not join ribitol units directly to one another.

Similar structural properties of the S.34 polysaccharide have also been reported recently by ROBERTS, BUCHANAN AND BADDILEY (1962).

Mode of attachment of teichoic acids

From their extractibility in cold, dilute trichloroacetic acid, ARMSTRONG, BADDILEY, BUCHANAN, CARSS AND GREENBERG (1958) concluded that the cell-wall teichoic acids were joined to the other wall components by salt linkages or by hydrogen bonding. This view that the teichoic acids are attached by salt linkages or weak bonds, has been challenged by MANDELSTAM AND STROMINGER (1961) and by GHUYSEN (1961).

The material classified as a teichoic acid by GHUYSEN (1961) could not be extracted by trichloroacetic acid and he presented evidence for the attachment of the unidentified polyol to amino sugars which in turn were linked to the glycosaminopeptide. GHUYSEN (1961) therefore concluded that the *Bacillus megaterium* polymer was in covalent linkage as a 'teichoic acid – mucopeptide' complex.

MANDELSTAM AND STROMINGER (1961) found that a considerable period of extraction with

10% trichloroacetic acid was needed for the removal of the teichoic acid from *Staphylococcus aureus* Copenhagen walls. Even after three weeks some 5% of the total phosphorus remained in the insoluble fraction. MANDELSTAM AND STROMINGER (1961) considered the possibility that the teichoic acid may be joined to the glycopeptide through a peptide bond between D-alanine linked to the teichoic acid by an ester linkage. However, as BADDILEY, BUCHANAN, MARTIN AND RAJBHANDARY (1962) have pointed out it seems unlikely that trichloroacetic acid hydrolysis of peptide bonds would occur whilst the highly labile ester linkages were still retained. The latter investigators have suggested that residual phosphorus in cell walls is not associated with teichoic acid components and probably represents occasional phosphate groups of the glycosaminopeptides.

SALTON (1961) reported an increase in C-terminal and N-terminal groups (except for alanine) following extraction of the teichoic acids in hot trichloroacetic acid. However, such conditions of extraction could not be regarded as optimal for quantitative removal of the teichoic acid and minimal for damage to the other cell-wall components.

It is evident that to resolve this problem an enzyme capable of degrading the native teichoic acid or alternatively a muramidase to degrade the glycosaminopeptide to release the teichoic acid, is badly needed. STROMINGER AND GHUYSEN (1963) have recently examined enzymic digests of staphylococcal walls and they have found a small amount of glycosaminopeptide attached to the teichoic acid. As a consequence of the isolation of a homogeneous compound of teichoic acid and glycosaminopeptide they have concluded that the two are covalently linked. One curious feature of this study is the report that the Streptomyces amidase results in a complete separation of teichoic acid from the glycosaminopeptide. The nature of the bond in such a complex is still unknown.

REFERENCES

ARCHIBALD, A. R., J. J. ARMSTRONG, J. BADDILEY and J. B. HAY, *Nature*, 191 (1961) 570.
ARCHIBALD, A. R., BADDILEY, J. and J. G. BUCHANAN, *Biochem. J.*, 81 (1961) 124.
ARMSTRONG, J. J., J. BADDILEY and J. G. BUCHANAN, *Biochem. J.*, 76 (1960) 610.
ARMSTRONG, J. J., J. BADDILEY and J. G. BUCHANAN, *Biochem. J.*, 80 (1961) 254.
ARMSTRONG, J. J., J. BADDILEY, J. G. BUCHANAN, B. CARSS and G. R. GREENBERG, *J. Chem. Soc.*, (1958) 4344.
ARMSTRONG, J. J., J. BADDILEY, J. G. BUCHANAN, A. L. DAVISON, M. V. KELEMEN and F. C. NEUHAUS, *Nature*, 184 (1959) 247.
BADDILEY, J., In *Immunochemical Approaches to Problems in Microbiology*, Eds. M. Heidelberger, O. J. Plescia and R. A. Day, Rutgers Univ. Press, N.J., 1961, p. 91.
BADDILEY, J., *J. Roy. Instit. Chem. (London)*, (1962a) 366.
BADDILEY, J., *Biochem. J.*, 82 (1962b) 36P.
BADDILEY, J., J. G. BUCHANAN and B. CARSS, *Biochim. Biophys. Acta*, 27 (1958) 220.
BADDILEY, J., J. G. BUCHANAN, B. CARRS, A. P. MATHIAS and A. R. SANDERSON, *Biochem. J.*, 64 (1956) 599.
BADDILEY, J., J. G. BUCHANAN and C. P. FAWCETT, *J. Chem. Soc.*, (1959) 2192.
BADDILEY, J., J. G. BUCHANAN and F. E. HARDY, *J. Chem. Soc.*, (1961) 2180.
BADDILEY, J., J. G. BUCHANAN, R. O. MARTIN and U. L. RAJBHANDARY, *Biochem. J.*, 85 (1962) 49.

BADDILEY, J., J. G. BUCHANAN and A. R. SANDERSON, *J. Chem. Soc.*, (1958) 3107.
BADDILEY, J. and A. L. DAVISON, *J. Gen. Microbiol.*, 24 (1961) 295.
BADDILEY, J. and A. P. MATHIAS, *J. Chem. Soc.*, (1954) 2723.
BUCHANAN, J. G., C. A. DEKKER and A. G. LONG, *J. Chem. Soc.*, (1950) 3162.
ELLWOOD, D. C., M. V. KELEMEN and J. BADDILEY, *Biochem. J.*, 86 (1963) 213.
GHUYSEN, J. M., *Biochim. Biophys. Acta*, 50 (1961) 413.
GHUYSEN, J. M., *Biochim, Biophys. Acta*, 83 (1964), in the press.
HAUKENES, G., D. C. ELLWOOD, J. BADDILEY and P. OEDING, *Biochim. Biophys. Acta*, 53 (1961) 425.
HAY, J. B., A. J. WICKEN and J. BADDILEY, *Biochim. Biophys. Acta*, 71 (1963) 188.
IKAWA, M., *J. Biol. Chem.*, 236 (1961) 1087.
KELEMEN, M. V. and J. BADDILEY, *Biochem. J.*, 80 (1961) 246.
LILLY, M. D., *J. Gen. Microbiol.*, 28 (1962) ii–iii.
MANDELSTAM, M. H. and J. L. STROMINGER, *Biochem. Biophys. Res. Comm.*, 5 (1961) 466.
MCCARTY, M., *J. Exptl. Med.*, 109 (1959) 361.
MITCHELL, P. and J. MOYLE, *J. Gen. Microbiol.*, 5 (1951) 981.
PARK, J. T., *J. Biol. Chem.*, 194 (1952) 877, 885, 897.
REBERS, P. A. and M. HEIDELBERGER, *J. Am. Chem. Soc.*, 81 (1959) 2415.
REBERS, P. A. and M. HEIDELBERGER, *J. Am. Chem. Soc.*, 83 (1961) 3056.
ROBERTS, W. K., J. G. BUCHANAN and J. BADDILEY, *Biochem. J.*, 82 (1962) 42P.
SALTON, M. R. J., *Biochim. Biophys. Acta*, 10 (1953) 512.
SALTON, M. R. J., *Biochim. Biophys. Acta*, 52 (1961) 329.
SALTON, M. R. J. and G. MILHAUD, *Biochim. Biophys. Acta*, 35 (1959) 254.
SALTON, M. R. J. and J. G. PAVLIK, *Biochim. Biophys. Acta*, 39 (1960) 398.
SANDERSON, A. R., W. G. JUERGENS and J. L. STROMINGER, *Biochem. Biophys. Res Comm.*, 5 (1961) 472.
SARGENT, L. J., J. G. BUCHANAN and J. BADDILEY, *J. Chem. Soc.*, (1962) 2184.
SHABAROVA, Z. A., J. G. BUCHANAN and J. BADDILEY, *Biochim. Biophys. Acta*, 57 (1962) 146.
STRANGE, R.E. and J. F. POWELL, *Biochem. J.*, 58 (1954) 80.
STROMINGER, J. L. and J. M. GHUYSEN, *Biochem. Biophys. Res. Comm.*, 12 (1963) 418.
WORK, E., *Biochem. J.*, 49 (1951) 17.

CHAPTER 7

Cell-wall antigens and bacteriophage receptors

Prior to the isolation and chemical characterization of bacterial cell walls, it had long been assumed that antigenic components and bacteriophage receptors were in the cell surface structures of those organisms not producing capsules. The importance of capsular polysaccharides as the dominant antigenic compounds of the bacterial surface emerged from the classical studies of the pneumococcal type-specific substances. The nature of group specific substances, somatic O antigens and bacteriophage receptors had been explored at the chemical level but the topography of these antigenic complexes and phage adsorption centres and their anatomical location in the bacterial cell was not clearly established until homogeneous wall and membrane structures were isolated.

This chapter will be devoted to a brief discussion of the principal antigenic substances of the cell walls of Gram-positive and Gram-negative bacteria and the nature of bacteriophage receptors. This latter topic has not been investigated very fully and apart from the earlier work of MILLER AND GOEBEL (1949) and WEIDEL (1951, 1958) with phage receptors of Gram-negative bacteria, only several studies relating to the problem in Gram-positive bacteria have appeared.

Although it has been known for some time (CUMMINS, 1954; TOMCSIK AND GUEX-HOLZER, 1954) that cell-wall preparations can elicit a satisfactory antibody response, there are surprisingly few studies on the antigenic and immunochemical properties of isolated walls. However, this aspect of the problem of defining cell-wall structure and function has received more attention in the past couple of years and it should not be long before the principal antigenic determinants in cell walls of a wide variety of bacteria and other microorganisms will be known. For the Gram-negative bacteria a great deal of information is available on the antigenic specificity of the cell surface, most of this having come from studies of the isolated lipopolysaccharides rather than from investigations with cell-wall preparations.

As the 'carriers' of antigenic determinants the cell walls of both Gram-positive and Gram-negative bacteria are of considerable immunological importance. In addition to bearing the surface antigens they have now been shown to possess other immunobiological components such as protective antigens, the endotoxins and pyrogenic substances of Gram-negative bacteria, the histamine-sensitizing factor of *Bordetella pertussis* and the capacity to induce tuberculin hypersensitivity. Cell walls of *Listeria monocytogenes* containing 5 amino acids (alanine, glutamic acid, diaminopimelic acid, aspartic acid and leucine), hexosamine and sugars were found by KEELER AND GRAY (1960) to elecit monocyte production. Thus the monocyte-producing factor appears to be localized in the wall of *Listeria monocytogenes* thereby adding a further biological property to the bacterial wall.

The tables are printed together at the end of the book.

Cell-wall antigens of Gram-positive bacteria

As our knowledge of the chemistry of the cell walls of Gram-positive bacteria has advanced it has become more obvious that very few walls are composed entirely of glycosaminopeptide and that the vast majority have some other polymer or molecular species in close association with the structure. There is an increasing body of evidence to show that it is the 'extra' component in or on the rigid glycosaminopeptide structure which is the principal antigenic determinant of the isolated wall. Thus polysaccharides attached to the wall are the important antigenic substances for most of the Lancefield groups of streptococci. Similarly, the teichoic acids are the principal antigens of staphylococcal cell walls. Indeed, it is now difficult to say whether the glycosaminopeptide structure of the wall contributes at all to the antigenic properties. So far as the author is aware no direct attempts have been made to determine the antigenicity of a 'pure' glycosaminopeptide, although CUMMINS (1962a) has assumed that the 'mucocomplex' antigenic specificities were dominant when suspensions of cell-wall fragments were used in agglutination reactions. Perhaps there are no valid reasons for suspecting that the glycosaminopeptide should not be antigenic but their ability to stimulate antibody production must be established experimentally before any assumption can be made. The walls of *Micrococcus lysodeikticus* are the closest of any species to 'pure' glycosaminopeptide but even with this organism the detection of a cell-wall polymer of glucose and an aminouronic acid (PERKINS, 1962) could complicate the antigenic potentialities of the wall.

Streptococci

The antigenic properties of the cell walls of streptococci have been investigated in greater detail than those of other groups of Gram-positive bacteria and will accordingly be discussed first. The separation of the streptococci into well-difined groups by the LANCEFIELD's (1933) serological classification has provided a stimulus to attempt to relate cell-wall composition and antigenic specificity.

In early studies of the properties of isolated walls it became apparent that certain antigenic components of the cells were associated with or were part of the cell-wall structure. Thus SALTON (1952b; 1953) found that the M-protein antigen of a group A streptococcus was retained on the cell-wall surface throughout the isolation procedures. As may have been expected from the earlier work of LANCEFIELD (1943) the M-protein was removed from the wall when the preparations were digested with trypsin. The detection of protein antigens bound to the streptococcal walls was also confirmed by BARKULIS AND JONES (1957) and FREIMER, KRAUSE AND MCCARTY (1959). The mode of attachment of the M-protein antigen to the wall is not clearly understood. It is evident that its production is independent of cell-wall biosynthesis for L-forms and protoplasts of group A streptococci are capable of synthesizing M-protein (SHARP, HIJMANS AND DIENES, 1957; FREIMER, KRAUSE AND MCCARTY, 1959; GOODER AND MAXTED, 1961). Thus, a glycosaminopeptide structure is not a prerequisite for the production of the M-proteins. However in the L-phase cells the M-protein appeared in the medium, there being no wall structure on which to anchor the protein.

BARKULIS AND JONES (1957) found that the M-protein represented only a part of the trypsin-sensitive protein associated with the cell wall. They concluded that the isolated M-protein is not representative of the total trypsin-labile protein of the wall. TEPPER, HAYASHI AND BARKULIS (1960) investigated the amino acid composition of cell-wall proteins from virulent and avirulent group A streptococci and they concluded that changes in configuration of the cell-wall protein which were associated with the acquisition of type-specific antigenicity and virulence were accompanied by relatively small changes in amino acid composition.

As pointed out by LANCEFIELD (1962) it is now well established that the M-protein is the cellular component of group A streptococci responsible for their immunological type specificity. No completely purified preparations of M-protein have been isolated so its chemical properties and the mode of attachment of this and other proteins to the cell-wall surface still remain unknown.

In addition to the substances conferring type-specificity in streptococci, the 'carbohydrates' associated with the serological grouping have been of direct interest to those investigating cell-wall structure and antigenicity. Following MAXTED's (1948) discovery that the C substance from β-haemolytic streptococci was released into solution when cells were treated with an enzyme preparation from *Streptomyces albus*, McCARTY (1952b) and SALTON (1952) reported the lytic action of the enzyme upon isolated group A streptococcal walls. These investigations indicated that the group specific substance was part of the cell-wall structure.

With the purified enzymes from *Streptomyces albus*, McCARTY (1952a, b) was able to obtain the group A carbohydrate in solution and purify it further by removing other products of the lytic action of the enzyme on the isolated cell walls. Although the group A carbohydrate isolated from enzymically digested walls was predominantly polysaccharide material containing rhamnose and glucosamine, some 'peptide' with the typical wall constitution (muramic acid, alanine, glutamic acid and lysine) persisted in these preparations (McCARTY, 1952b; KRAUSE AND McCARTY, 1961).

Thus streptococcal cell walls became recognized as the anatomical site of certain group-specific antigens. The reactions of group antisera with walls in cell-wall agglutination tests have been reported by CUMMINS (1962a) and these results are presented in Table 58. It is of interest to note that there is a number of cross-reactions, at considerably lower titres however. The group antigen of group D streptococci on the other hand is not localized in the wall. JONES AND SHATTOCK (1960) reported that the group D-specific substance was in the cytoplasmic membrane fraction. The group D antigen has now been identified as the intracellular teichoic acid (ELLIOTT, 1962; WICKEN, ELLIOTT AND BADDILEY, 1963). Neither cell-wall antigen nor group D antigen could be detected in the L-phase variants of group D streptococci (HIJMANS, 1962).

SLADE AND SLAMP (1962) have also reported the agglutination of streptococcal walls by grouping antisera. Antisera for groups A, B, C, D, E, F, G, H, K, L, M, N, O, and Q were tested and only group D failed to give a positive cell-wall agglutination although low titres (1/20) were also observed with some strains of other groups.

The immunochemical basis of the different group specificities has not been investigated fully. Following the detection of rhamnose in various streptococcal walls (McCarty, 1952b; Salton, 1952a, b; Cummins and Harris, 1956) is seemed likely that this monosaccharide may be involved in the antigenic specificity. Recent studies by Slade and Slamp (1962) have shown that rhamnose is absent from one group (O) and variable in three others. The principal constituents of streptococcal cell walls have been identified by Cummins and Harris (1956), Hayashi and Barkulis (1959), Roberts and Stewart (1961) and Slade and Slamp (1962) and the typical glycosaminopeptide constituents (glucosamine, muramic acid, alanine, glutamic acid and lysine) are common to all groups investigated. Galactose occurs less frequently than rhamnose or glucose. Galactosamine is also present in a number of groups. Beyond this, it is impossible to say which of the chemical constituents found in the walls determines the group antigenic specificity. Apart from the presence of polysaccharides, the distribution of teichoic acids throughout the various streptococcal groups is unknown at present. Although galactosamine has been detected in group specific polysaccharide it may also occur in the teichoic acid polymers.

However, the investigations of McCarty (1952b; 1956), McCarty and Lancefield (1955) and Krause and McCarty (1961) on the group A and group A variant streptococci do throw some light on the nature of the antigenic determinants in these group-specific polysaccharides. Variant strains of group A streptococci were isolated after mouse passage and the cell-wall carbohydrates of these and normal group A strains were examined by McCarty and Lancefield (1955). The loss of group reactivity on mouse passage (Wilson, 1945) was found to be due to an alteration in the chemical structure and serological specificity of the wall carbohydrate. The carbohydrate of the variant strains contained both rhamnose and glucosamine as in the group A polysaccharide but the proportion of these differed. Variants with intermediate properties were also encountered. The ratios of rhamnose: glucosamine ranged from 1.5–2.0 for the group A substances, 2.4–3.4 for the intermediate carbohydrates and 4.0–6.0 for the Variant polysaccharides (McCarty and Lancefield, 1955).

The immunochemistry of the group specific substances was advanced further when McCarty (1956) isolated soil organisms producing inducible enzymes capable of degrading group A and variant (V) streptococcal polysaccharides. The V polysaccharide was extensively degraded by the enzyme and rhamnose oligosaccharide obtained in the dialysable products inhibited the reaction between intact V carbohydrate and its homologous antiserum. McCarty (1956) therefore concluded that the serological specificity of the V carbohydrate appeared to be primarily dependent on a rhamnose-rhamnose linkage.

The action of the enzyme on group A carbohydrate was characterized by the removal of 50–70% of the total glucosamine in the form of free N-acetylglucosamine. The bacterial enzyme was believed to be a β-N-acetylglucosaminidase, though its specificity apparently differed from that of other glucosaminidases (e.g. emulsin, and mammalian enzymes) which did not attack the linkage joining the N-acetylglucosamine to the rest of the group A carbohydrate. Although the specific bond was not determined, the investigations have clarified the nature of the determinant groups, for the removal of N-acetylglucosamine was accompanied by a loss of group A reactivity and at the same time a markedly increased cross-reactivity with V-antisera was observed. McCarty (1956) was able to suggest then, that the specificity of group A carbohydrate was determined to a large extent, by side chains of N-acetylglucosamine. The latter also served to mask the underlying rhamnose – rhamnose linkages with V specificity.

In addition to the solubilization of the group specific carbohydrates by wall-degrading enzymes (*Streptomyces albus* enzyme and the phage-associated lysin developed by Krause, 1958), Krause and McCarty (1961) discovered that the polysaccharides could be extracted with hot formamide. The conclusion on the nature of serological specificity, based on the group A and V polysaccharides isolated by enzymic digestion of the walls was further substantiated on examination of the materials solubilized by the hot formamide method. The variant carbohydrate contained even smaller amounts of glucosamine (1–3%) and as much as 80–85% rhamnose (Krause and McCarty, 1961).

A further contribution to our understanding of the specificity of other streptococcal groups has come from the more recent studies of Krause and McCarty, (1962a, b) Group C strains differ from group A in possessing N-acetylgalactosamine as the determinant amino sugar compared to N-acetylglucosamine in Group A. This difference in the group A and C carbohydrates is of considerable interest, in view of the fact that the highest level of cross-reactivity occurred between these two groups in the cell-wall agglutination tests (Cummins, 1962a) shown in Table 58. Moreover, Krause and McCarty (1962b) found that certain group C-intermediate strains contained an antigen giving a precipitin cross-reaction with A-variant antiserum. The polysaccharides of these strains possessed rhamnose: hexosamine ratios of 2.4–2.6 compared to 1.1–1.7 for the typical group C strains. The high concentration of rhamnose in the C-intermediate carbohydrate suggested that some of the rhamnose residues in the oligosaccharide side chains are devoid of terminal N-acetylgalactosamine and thus react with group A-variant antiserum.

Polyglycerophosphate antigens found in culture supernatants and from the cytoplasm of certain Gram-positive organisms including streptococci (McCarty, 1959; Stewart, 1961) are probably not cell-wall antigens but may be more closely related to the intracellular teichoic acids.

Staphylococci

The importance of cell-wall teichoic acids as antigenic components of staphylococci has only recently been realized and in the past couple of years several papers have appeared on this problem. With the detailed knowledge of the structure of the teichoic acids provided by the studies of Baddiley and his colleagues (Baddiley, 1962) the chemical basis of the antigenicity of these polymers has and will continue to be established with great rapidity.

JULIANELLE AND WIEGHARD (1934) described an antigenic 'polysaccharide A' from pathogenic staphylococci. The problem of elucidating the nature of this antigen was taken up again by HAUKENES (1962a), who attempted to purify the polysaccharide by isolating it from disrupted bacteria, extracting and precipitating the crude serologically active material with ethanol. After purification of the 'polysaccharide A' by column chromatography (HAUKENES, 1962b, c, d) and identification of the products of acid hydrolysis, HAUKENES (1962d) concluded that the material contained two components, a mucopeptide and a ribitol teichoic acid, both resembling the corresponding compounds of the staphylococcal cell wall. The teichoic acid constituted the greater part of the purified 'polysaccharide A'. These interesting investigations clearly indicated the nature of the polysaccharide A which had not been studied since the mid-thirties.

A more direct approach to the study of the nature of the antigens of *Staphylococcus aureus* cell walls was made by SANDERSON, JUERGENS AND STROMINGER (1961) and the results of their investigations preceeded the publication of the series of papers on polysaccharide A by HAUKENES (1962a, b, c, d). The inhibition of the agglutination of cell walls of *Staphylococcus aureus* Copenhagen by rabbit antiserum was investigated by adding teichoic acid haptenes and known parts of the teichoic acid structure. The results of these studies are summarized in Table 59 and indicated a specific inhibition of the agglutination reaction by α-N-acetylglucosaminides either in the free state or attached to the ribitol teichoic acid polymer (e.g. compound in Table 59). SANDERSON, JUERGENS AND STROMINGER (1961) therefore concluded that the immunologically determinant group in the wall of the Copenhagen strain of *Staphylococcus aureus* is an α-N-acetylglucosaminyl-ribitol residue of the teichoic acid. The D-alanine groups of the teichoic acids do not appear to play any part in the serological reactions of these polymers.

HAUKENES (1962e) also concluded that the serologically active groupings of polysaccharide A were the N-acetylglucosaminyl-ribitol residues and although one preparation of teichoic acid contained no detectable α-linked residues, it gave specific precipitin ring tests. Differences in strains are to be expected and it should be remembered that the antisera used by SANDERSON, JUERGENS AND STROMINGER (1961) did not agglutinate walls of another *Staphylococcus aureus* strain.

Lactobacilli

BADDILEY AND DAVISON (1961) have suggested that there is a relationship between the serological properties of certain *Lactobacillus* spp. and the teichoic acids in their walls. It is clear of course that the antigenic behaviour of the lactobacilli is not determined solely by the teichoic acids. Organisms of the group D lactobacilli possessed ribitol teichoic acid in the walls, whereas the glycerol type was found in groups A and E. No cell-wall teichoic acid was detected in groups B, C, and F. It will be of great interest to establish the nature of the determinant groups of the teichoic acid antigens of the lactobacilli.

The cell walls of *Lactobacillus casei* belonging to group B (according to the classification of SHARPE AND WHEATER, 1952) were studied by KNOX AND BRANDSEN (1962). There is no teichoic acid in the wall of this organism (Baddiley and Davison, 1961) and the serological specificity of the wall antigens resides in the polysaccharide component. Knox and Brandsen (1962) obtained soluble products from the walls of *Lactobacillus casei* by allowing an autolytic enzyme to degrade the isolated walls. This enabled these investigators to test the inhibition of precipitin reactions between soluble cell-wall material and antiserum by monosaccharides. Different chemical fractions were obtained and for one in which the major sugar component was rhamnose it was shown that L-rhamnose was very effective in inhibiting the precipitin reaction. With another fraction containing less rhamnose the inhibitory ability of L-rhamnose was correspondingly lower. The results indicated a heterogeneous wall structure and galactose and glucose also contributed to the serological specificity (Knox and Brandsen, 1962).

Corynebacteria, Mycobacteria and Nocardias

Interest in the cell-wall antigens of the corynebacteria, mycobacteria and nocardias originated from the earlier studies of Cummins (1954) with *Corynebacterium diphtheriae* and the demonstration of the immunological activity associated with mycobacterial walls (Ribi, Larsen, List and Wicht, 1958).

Cummins (1954) suggested that the surface antigen responsible for the agglutination of intact cells of the 'mitis' strain of *Corynebacterium diphtheriae* was a protein and that a periodate sensitive polysaccharide represented a common group antigen in strains of this organism.

As the pattern of cell-wall composition was so similar for a variety of corynebacteria, mycodacteria and nocardias, Cummins (1962b) examined this group for possible relationships between their immunological reactions and wall constituents. Indeed, it was concluded that a common antigenic component occurred in all those strains of these three groups of organisms having arabinose and galactose as their principal cell-wall monosaccharides (Cummins, 1962b). Strains of corynebacteria and 3 strains of *Nocardia pelletieri* had a different pattern of cell-wall constituents and appeared to be devoid of the common cell-wall antigen. At present there is no evidence to indicate the determinant group of the antigen common to these organisms.

Cummins (1962b) has emphasized that it is doubtful whether the cell-wall agglutination tests performed with the isolated walls can be compared with the agglutination reactions using intact bacteria. As pointed out earlier the method used in 'cleaning up' the walls of organisms rich in surface lipids, waxes and glycolipids probably involves removal of some antigenically active components. Cell-wall preparations of *Mycobacterium butyricum* (Ribi, Larsen, List and Wicht, 1958) and *Mycobacterium tuberculosis* BCG (Kotani, Kitaura, Hirano and Tanaka, 1959; Kotani, Kitaura, Higashigawa, Kato, Mori, Matsubara and Tsujimoto, 1960) were probably less degraded and were undoubtedly chemically and antigenically more complex. The study by Kotani *et al.* (1960) established that the walls of the BCG strain were the most active of the cell fractions in inducing tuberculin hypersensitivity and in enhancing the resistance of mice to tuberculosis infection.

KANAI, YOUMANS AND YOUMANS (1960) have also reported the high degree of tuberculin sensitivity induced by cell-wall fractions of *Mycobacterium tuberculosis* strain H37 Ra. The relationship of the immunogenic particles of YOUMANS, MILLMAN AND YOUMANS (1955) to the cell wall or other structures is not clear and at present their results appear to conflict somewhat with those of KOTANI *et al.* (1960).

It is evident that the complex antigenic structure of the surface and cell walls of some members of the corynebacteria, mycobacteria and nocardias will require extensive investigation for these organisms abound in a great variety of lipids, waxes, glycolipids and peptidoglycolipids (LEDERER, 1961).

Cell-wall antigens of Gram-negative bacteria

The detection of monosaccharides characteristic of antigenic polysaccharides, in isolated wall preparations of Gram-negative bacteria (SALTON, 1953; WEIDEL, 1955) indicated that the walls would be of considerable immunological importance. However, at the time these early studies were performed, the anatomical status of the polysaccharide material constituting the typical O antigens of Gram-negative bacteria was uncertain. Indeed, as pointed out in Chapter 1 the surface protein-lipid-polysaccharide complexes have been classified as 'microcapsular' components (WILKINSON, 1958; SALTON, 1960) a term which is now regarded as misleading for there is no evidence of any layer resembling a capsular layer outside the multiple-layered wall or membrane of these organisms.

YOSHIDA *et al.* (1955) were the first to study the immunological properties of isolated cell-wall preparations of the Gram-negative organism, *Bordetella pertussis*. They demonstrated the presence of the protective and agglutinating antigens on the cell walls of the organism. In more precise and extensive tests MUNOZ, RIBI AND LARSON (1959) established that the protective antigen and the histamine-sensitizing factor were located principally in the cell wall of *Bordetella pertussis*.

The antigenic reactions of spheroplast membranes of *Echerichia coli* B were investigated by HOLME, MALMBORG AND COTA-ROBLES (1960) and were found to be very similar to those of intact cells. There was some evidence that additional new antigenic sites were exposed as a result of spheroplast formation.

SHAFA AND SALTON (unpublished observations) studied the agglutination of O-suspensions of cells, isolated cell walls, penicillin spheroplast walls and lysozyme-treated walls of *Salmonella gallinarum* by homologous and cross-reacting rabbit antisera. These cell and cell-wall agglutination tests are presented in Table 60 and indicate that isolated walls and O suspensions have very similar serological properties. The spheroplast walls were agglutinated at rather lower titres by O and cell-wall antisera. Although treatment of the walls with lysozyme was believed to remove principally the glycosaminopeptide constituents (SALTON, 1958) the lowered agglutination titres suggested a loss of surface antigenic components as well.

In addition to examining *Bordetella pertussis* cell walls, RIBI and his colleagues have examined wall and protoplasmic fractions from *Bacterium tularense* (SHEPARD, RIBI

AND LARSON, 1955), *Salmonella enteritidis* (RIBI, MILNER AND PERRINE, 1959) and *Brucella abortus* (FOSTER AND RIBI, 1962). In all cases the cell wall was the site of antigenic substances capable of immunizing against infection. However, there are not enough chemical data at present to indicate the nature of the determinant groups in these cell-wall antigens. The relationship between chemical composition and host-reactive properties usually referred to as 'endotoxic' have been considered by RIBI, HASKINS, LANDY AND MILNER (1961) but no definite conclusion about the nature of the substance responsible for the toxic activity of *Salmonella enteritidis* endotoxin could be made.

Antigenic specificity of cell-wall lipopolysaccharides

The lipopolysaccharides have attracted a great deal of attention in the past ten years and both the chemistry and immunochemistry of these surface antigenic substances have been studied intensively by WESTPHAL, STAUB AND DAVIES, and their colleagues. The isolation and chemical characterization of these lipopolysaccharides has not only revealed the occurrence of dideoxysugars and heptoses in Nature but has also provided a chemical basis for the Kauffmann-White serological classification of the Enterobacteriaceae (KAUFFMANN, 1954). Without the knowledge of the chemistry of these antigenic lipopolysaccharides, the Kauffmann scheme would remain a purely determinitive serological classification and it seems inappropriate to suggest that the chemical investigations represent merely 'a logical extension of the work of KAUFFMANN AND WHITE' (CUMMINS, 1962a). As pointed out by STAUB, TINELLI, LÜDERITZ AND WESTPHAL (1959) the immunochemical investigations provide a chemical basis for the Kauffmann-White scheme established only on serological data and moreover offer an explanation for the mechanisms of some of the cross-reactions between Salmonellas.

Although these lipopolysaccharides are derived from the wall or membrane structures of Gram-negative bacteria, their functions in the 'non-rigid' layers of the envelope are not known. It is clear from the investigations of NIKAIDO (1962) that organisms such as *Salmonella enteritidis* can tolerate the loss of certain sugars of the polysaccharide without impairing the ability of the cell to grow and survive. Whether there is any functional need for a polysaccharide in the walls of Gram-negative organisms cannot be stated at present, but at least in some mutants of salmonellas this component can be 'simplified' to the extent that only heptose is present (HORECKER, personal communication).

The composition of the lipopolysaccharides of a wide variety of salmonellas of the different serological groups and antigenic specificities of the KAUFFMANN (1954) scheme has been determined and some of the data from KAUFFMANN, LÜDERITZ, STIERLIN AND WESTPHAL (1960) are summarized in Table 61. On the basis of the monosaccharides present in the cell-wall lipopolysaccharide complexes, the various antigens of the salmonellas could be separated into 16 'chemotypes' and these results are given in Table 62. Thus the O antigens of species belonging to one and the same O group or O subgroup, have identical qualitative chemical constitutions and accordingly belong to the same chemotype. The simplest chemotype of the series studied was represented by a polysaccharide composed of glucosamine, heptose, galactose and glucose. The most complicated structures contained 3 different sugars with the relatively lipophilic 6-deoxy- and 3,6-dideoxy-hexoses occurring frequently.

The role of the 6-deoxy- and 3,6-dideoxy-hexoses as terminal, determinant groups in the antigens has been investigated by the inhibitory effects of single sugars on the specific precipitation of antigen and antibody (STAUB, 1959). Greatest inhibitory activity is believed to occur with sugars identical to the terminal ones of the lipopolysaccharide (STAUB, 1959; STAUB, TINELLI, LÜDERITZ AND WESTPHAL, 1959) although variations in antisera from one animal species to another have been observed.

From these studies it was concluded that tyvelose (3,6-dideoxy-D-mannose) was responsible for the specificity of group D, antigen 9 and colitose (3,6-dideoxy-L-galactose) determined the specificity of group P antigen 35. The immunological complexity of antigen 12 was related to two side chains, one terminating in rhamnose, the other in glucose. The cross reactions of anti- *Salmonella paratyphi* B sera with galacto-mannans suggested the presence of an oligosaccharide with the sequence: abequose-galactose-mannose in O antigen 4 (STAUB *et al.*, 1959).

The role of the terminal sugars of determinant groups in certain of the O antigenic polysaccharides has been elegantly demonstrated with the preparation of an artificial antigen containing colitose (3,6-dideoxy-L-galactose) by LÜDERITZ, WESTPHAL, STAUB AND LE MINOR (1960). Colitose phenylazo-proteins were prepared and they gave strong precipitation reactions with horse antiserum to *Escherichia coli* 0111. The results obtained with this synthetic antigen gave good agreement with previous inhibition studies indicating that colitose accounted, to a high degree, for the antigenic specificity of *Escherichia coli* 0111, and Salmonella 035 antigenic specificities.

The fact that the polysaccharide O antigens of many of the serotypes of *Salmonella* spp. frequently carry more than one immunological specificity is an intrigueing one. This has raised the question as to whether the 'antigen formula' of the Kauffmann-White scheme is based on a heterogeneous mixture of oligo- or polysaccharides. LÜDERITZ, O'NEILL AND WESTPHAL (1960) have largely resolved this problem by using selectively adsorbed and suitable cross-reacting antisera and analysing specific precipitates with purified Salmonella O antigens for protein and known sugar constituents (e.g. rhamnose quantitatively, dideoxy-hexoses qualitatively). The sugar analyses gave identical results both before and after cross-precipitation irrespective of the antigenic factor involved in the reaction. LÜDERITZ, O'NEILL AND WESTPHAL (1960) concluded that the various antigenic factors of the O antigen complex of a particular serotype, are linked together in the one polysaccharide molecule and that for the species investigated only one single O-antigenic polysaccharide is produced as a constituent of their cell walls.

Similar immunochemical studies have now been performed with the O antigens of *Escherichia coli* strains, Arizona and Citrobacter strains (WESTPHAL, KAUFFMANN, LÜDERITZ AND STIERLIN, 1960; KAUFFMANN, BRAUN, LÜDERITZ, STIERLIN AND WESTPHAL, 1960). For the coli strains examined, 14 chemotypes have been recognized, the simplest ones containing 4 monosaccharides (glucosamine, heptose, galactose or mannose and glucose). The 6-deoxy-hexoses, fucose and rhamnose and the 3,6-dideoxy-hexose, colitose, were present singly with the other sugars in the more complicated chemotypes. These studies greatly facilitate an understanding of the cross reactions between *Salmonella* spp. and *Escherichia coli* and other groups, at the chemical level.

There seems little doubt that in the years to come the full chemical structure of many of these lipopolysaccharides will be elucidated and the spacings of the side chains on a basal polysaccharide backbone determined. These studies will lead to a full understanding of the immunochemical reactions of these fascinating polysaccharides. The work of NIKAIDO (1962) has shown how the structure and biosynthesis of these polysaccharides can be studied by the selection of suitable mutants and these investigations will undoubtedly stimulate much interest in the genetic aspects of these wall components. At the present time the great potentialities of bacterial conjugation have not been exploited in this field; but this state of affairs is not likely to persist for long. Finally at the more complex biological level, the wealth of chemical and immunochemical knowledge should greatly assist our understanding of the mechanisms of pathogenicity and the effects of the endotoxins upon the hosts.

Smooth – rough variation and polysaccharide structure

The smooth (S) to rough (R) variation in *Salmonella* spp. appears to be fairly straightforward but in other Gram-negative organisms the change is more complex (DAVIES, 1960). From the various *Salmonella* spp. so far examined a loss of smoothness is accompanied by the loss of certain sugars of the polysaccharide O-antigens. Thus the polysaccharides of the rough form yield glucose and galactose having lost the mannose, rhamnose and dideoxyhexose of the corresponding smooth strains (WESTPHAL, LÜDERITZ, FROMME AND JOSEPH, 1953; MIKULASZEK, 1956). More detailed studies with *Salmonella paratyphi* B and *Salmonella typhimurium* established that the rough strains contained glucosamine, heptose, galactose and glucose in their polysaccharides and the smooth forms contained in addition mannose, rhamnose and abequose (LÜDERITZ, KAUFFMANN, STIERLIN AND WESTPHAL, 1960). In *Escherichia coli* 08 the rough polysaccharide contained glucose and in the smooth, galactose, glucose and rhamnose were found (WESTPHAL, LÜDERITZ, FROMME AND JOSEPH, 1953).

DAVIES (1957) investigated the nature of the lipopolysaccharides in rough and smooth forms of *Shigella dysenteriae* and found that the rough polysaccharide contained glucosamine, galactose and an aldoheptose in contrast to the glucosamine, galactose and rhamnose in the smooth strain lipopolysaccharide. No evidence could be obtained with this organism that indicated the coexistence of rough and smooth polysaccharides in the smooth strain.

In Pasteurellas the S→R change may involve the loss of a protein, an effect which can be superimposed on a second smooth to rough variation which involves changes in the specificity of the polysaccharide (CRUMPTON AND DAVIES, 1958). Aldoheptoses are present in both rough and smooth forms. The change to rough cultures in *Pasteurella pseudotuberculosis* involved a loss of galactose, mannose, fucose, paratose, abequose, tyvelose, ascarylose and galactosamine depending on the serological group of the corresponding smooth strain (DAVIES, 1958; CRUMPTON, DAVIES AND HUTCHISON, 1958).

Bacteriophage receptors of cell walls

Just as it had been assumed that antibodies brought about agglutination of bacteria by reaction with specific surface components, so with bacteriophage infection it has long been believed that the initial phase involved the attachment of the phage to a receptor on the outside of the bacterial cell. Where the phage-sensitive bacteria are non-encapsulated organisms it follows that the receptor substance or area is anatomically located on or in the wall. WEIDEL (1951) observed the direct attachment of bacteriophages to the isolated walls of *Escherichia coli* by examining the preparations in the electron microscope. A clearer picture of the mode of attachment of the phage to the cell wall was obtained with isolated walls of *Staphylococcus aureus* and one of its phages by HOTCHIN, DAWSON AND ELFORD (1952). Since these early investigations, the binding of bacteriophages to isolated walls of sensitive hosts has been observed for a variety of Gram-positive and Gram-negative bacteria.

Two independent studies on the nature of the specific surface substances responsible for the affinities between host and infecting bacteriophage were initiated by GOEBEL and his colleagues and by WEIDEL.

MILLER AND GOEBEL (1949) suggested that the type-specific 'somatic antigen' (O antigen) acted as the virus receptor in the *Shigella sonnei* Phase II – T phage system. The isolated Phase II antigen inactivated, *in vitro*, all of the T phages to which the organism was susceptible. GOEBEL AND JESAITIS (1952), moreover, obtained a mutant strain which was susceptible to only T2 and T6 phages. The variant Phase II antigen was a powerful inactivating agent for T2 phage. The lipopolysaccharide antigens of the parent and variant strains differed chemically and serologically. Glucose, galactose, glucosamine and an aldoheptose were identified as the sugar components of the Phase II lipocarbohydrate, while the variant II/3,4,7 contained only a hexosamine.

In subsequent studies JESAITIS AND GOEBEL (1953) and GOEBEL AND JESAITIS (1953) demonstrated that the soluble lipocarbohydrate triggered the release of nucleic acid from the bacteriophage. Extraction of the lipocarbohydrate with 70% ethanol removed lipid from the complex and the product no longer caused lysis of the phage. By mixing the extracted lipid or other saturated fatty acids such as myristic, palmitic and stearic acids the residual lipopolysaccharide regained its ability to inactivate the phage. JESAITIS AND GOEBEL (1953) concluded that lipocarbohydrate, a component of the antigenic complex on the cell surface was the substance responsible for inducing the release of genetically active material from the virus.

The investigation of the virus receptor in *Escherichia coli* B led WEIDEL (1950) to the isolation of a fraction which possessed the anatomical features of the wall or membrane of this organism. This material prepared by autolysis and digestion with trypsin and differential centrifugation was found to be composed of protein and lipid (WEIDEL, 1951). Its ability to interact with the phages paved the way for a series of investigations which have given us a detailed understanding of the chemical anatomy of the cell wall of *Escherichia coli* B as well as the nature of the phage receptors.

The cell-wall preparations were capable of specific and irreversible binding of T2, T4 and T6 bacteriophages; T3 and T7 were less adsorbed (WEIDEL, 1951). The early investigations by WEIDEL (1953) suggested a mosaic-like structure of phage receptors. Two chemically distinct layers were recognized on the basis of solubility in 90% phenol, one layer composed of lipoprotein and some free lipid and another with 'building blocks of the rigid inside layer' giving glucose, L-gala-D-manno-heptose, glucosamine, muramic acid, bound lipid and a few characteristic amino acids (WEIDEL, 1955; WEIDEL AND PRIMOSIGH, 1958). The 'plastic' lipoprotein layer accounted for 80% of the dry weight of the wall and the rigid component, 20% of the wall mass.

The receptor activity indicated by inactivation tests, showed the location of T3, T4 and T7 receptors in the '20% layer'. WEIDEL, KOCH AND LOHSS (1954) found that the mutant strain *Escherichia coli* B/3,4,7 was resistant to the phages T3, T4 and T7 and moreover it was unable to bind these phages. It appeared that some difference in the properties of the '20%' or rigid layer would be involved and this indeed was the case; the heptose was missing from the 20% layer in the resistant mutant B 3,4,7. This loss of heptose from the phage receptor component is reminiscent of a similar finding by GOEBEL AND JESAITIS (1952, 1953) with the *Shigella sonnei* phase II/3,4,7 phage resistant variant which contained an unidentified hexosamine and none of the sugars in the lipocarbohydrate.

Although the isolated cell wall of *Escherichia coli* B adsorbs T2 and T6 phages, the localization of their receptors has not been fully determined. The receptors are removed from the wall on extraction with 90% phenol and it was concluded that the activity for these two phages resided in two components of the lipoprotein layer (WEIDEL, KOCH AND LOHSS, 1954).

Of the two remaining phages of the *Escherichia coli* B, T series only T5 has been amenable to direct investigation. Nothing is known of the T1 receptor, which can only function fully to the point of irreversibly attaching the phage when the cell is living (WEIDEL, 1958). The T5 receptor on the other hand can be isolated in particle form by extracting the cells with dilute alkali (WEIDEL, KOCH AND BOBOSCH, 1954). Direct attachment of one receptor particle to the tail of each T5 phage was demonstrated by WEIDEL AND KELLENBERGER (1955).

The chemical composition of the T5 receptor particle has been interpreted as a two layered 'mirror' of the wall structure. Each particle contained a core of lipopolysaccharide, the surface of which was covered with lipoprotein, rendering the receptor activity exclusive for T5 phage. Removal of the outer lipoprotein layers uncovered the receptor activity of the lipopolysaccharide core for the T3, T4 and T7 phages.

WEIDEL (1958) thus pictures the phage receptors in the *Escherichia coli* B cell walls in the following way: 'The rigid 20% layer, carrying the receptor sites for T3, T4 and T7 is buried under a layer of plastic lipoprotein, but in a way which leaves part of these sites exposed. In other words, the lipoprotein layer, which represents an ocean of T2-sites with considerably less sites for T6 admixed in the form of a separate lipoprotein component, has holes or rather discontinuities in it through which many of the potential sites for T3, T4 and T7 remain accessible. In addition to the holes, it is dotted with comparatively few, loosely bound particles of their own individual make-up, carrying the sites for T5'.

The problem of determining the number of receptors in the intact wall has been considered by WEIDEL (1958). Some of the difficulties encountered in estimating the numbers include steric hindrances that may be encountered with saturation experiments and structural changes in the wall following irreversible attachment. The latter could induce the exposure of new phage receptor sites or may result in loss of activities due to deformation of the structures.

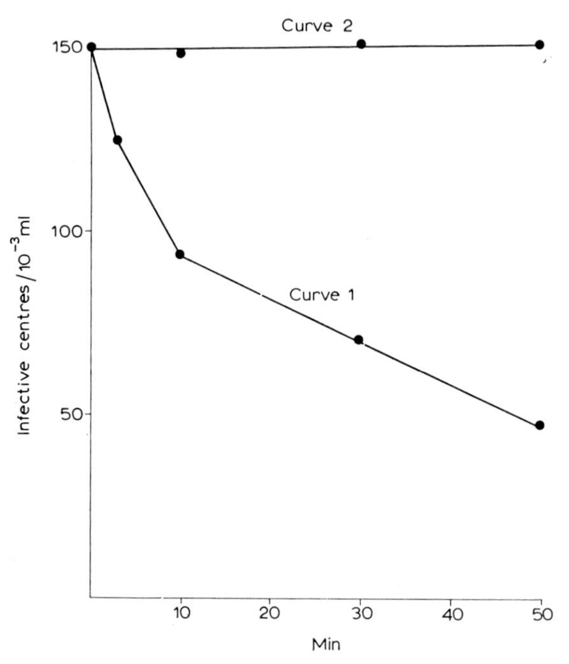

FIGURE 67. The inactivation of bacteriophage by isolated cell walls of *Bacillus megaterium* KM (curve 1). The 'receptor' activity of the walls is destroyed by digesting the walls with lysozyme (curve 2). Data from SALTON, 1956.

Isolated cell walls of Gram-positive bacteria bring about an inactivation of their specific bacteriophages in a similar way to that observed for the systems studied with the Gram-negative organisms. The irreversible attachment of bacteriophages to walls with the consequent loss in phage titre over a period of time is shown for the *Bacillus megaterium* KM cell walls and phage C in Fig. 67 (SALTON, 1956). Similar results have also been obtained with walls from *Micrococcus lysodeikticus* and its bacteriophages by SALTON (1957) AND BRUMFITT (1960).

The chemical nature of the bacteriophage receptors in Gram-positive bacteria is not very well understood, except perhaps in the case of the strains of *Micrococcus lysodeikticus* studied by

BRUMFITT (1960). The lysozyme-resistant strain, (LR_1), possessing a high content of O-acetyl groups was completely resistant to the bacteriophages for this organism, although there was no significant difference in the adsorption of phage by the parent lysozyme-sensitive strain and the lysozyme-resistant mutant. The results suggest that O-acetyl groups block the infection without preventing initial attachment.

SALTON (1956, 1957) showed that if the cell walls of *Bacillus megaterium* and *Micrococcus lysodeikticus* were digested with egg-white lysozyme prior to the addition of bacteriophage, no inactivation of the phage occurred. BRUMFITT (1960) confirmed the effect with *Micrococcus lysodeikticus*. It was therefore evident that the receptor area of the cell wall no longer presented the correct configuration for phage attachment and inactivation. The results infer that in these organisms the phage receptor is more in the form of a molecular mosaic which is not preserved when the wall is degraded into fragments varying in molecular weights from about 500–20,000. Reaction of free amino groups in the walls of *Bacillus megaterium* neither destroyed the receptor activity nor sensitivity to lysozyme (SALTON, 1957).

In walls that are essentially glycosaminopeptide, the nature of the phage receptor area still remains obscure. Whether the additional structures of the walls of Gram-positive bacteria (e.g. teichoic acids, teichuronic acids, polysaccharides) are phage receptors has not been determined as yet by direct experimentation. It would be surprising if staphylococcal teichoic acids with their surface antigenic activities were not interrelated with phage types and the chemical structure of the teichoic acids.

Thus with the Gram-positive bacteria the nature of the groups responsible for the binding and triggering the release of the phage deoxyribonucleic acid is not known at present, but the results of BRUMFITT (1960) strongly suggest a correlation with part of the lysozyme substrate.

One final interesting facet of the investigations of the interaction between bacteriophages and cell walls has been the discovery of the rather widespread activity of muramidases (lysozymes, etc.) associated with bacteriophages and phage lysis. The disintegration of *Escherichia coli* walls after interaction with T2 phage was observed by WEIDEL (1951) and led to the isolation of the phage enzyme with very similar enzymic properties to those of egg-white lysozyme (KOCH AND DREYER, 1958; WEIDEL AND PRIMOSIGH, 1958; PRIMOSIGH, PELZER, MAASS AND WEIDEL, 1961).

The release of material from *Escherichia coli* walls on attachment of bacteriophage particles was reported by BARRINGTON AND KOZLOFF (1956) and KOZLOFF AND LUTE (1957) and the nature of the products released was eventually shown to be of the glycosaminopeptide type (WEIDEL AND PRIMOSIGH, 1958). Thus the T2 phage enzyme degrades the glycosaminopeptide rigid layer of the Gram-negative cell wall. The evidence is suggestive that the phage 'drills' a hole through the most rigid structure of all bacterial walls with an enzyme located somewhere in the phage tail area, thereby permitting the injection of the viral deoxyribonucleic acid. The purified bacteriophage muramidase seems to be a broad spectrum enzyme and is by no means species or indeed strain specific, differing markedly in this respect to the phage attachment mechanism which appears to be highly specific.

Enzymes of the muramidase type have also been reported for other bacteriophage-host systems. MURPHY (1957, 1960) investigated the lytic enzyme associated with phage release from *Bacillus megaterium* and suggested the 'muralytic' nature of this lysin. SALTON AND MILHAUD (1959) also found low molecular weight products released from the wall of *Bacillus megaterium* during the course of phage attachment. These products had the same overall qualitative composition as the original wall but with different proportions of certain constituents.

BRUMFITT (1960) also prepared an enzyme from *Micrococcus lysodeikticus* phage by freezing and thawing concentrated suspensions of the virus. The enzyme was found to be very similar to egg-white lysozyme and it too responded to the blocking action of O-acetyl groups in the cell-wall substrate. BRUMFITT (1960) concluded that the enzyme could be regarded as a phage lysozyme.

Unfortunately none of these enzymes have been examined fully for the nature of the products they release from glycosaminopeptide substrates, nor has their activity been determined upon defined oligosaccharides such as those used in characterizing the muramidases (see Chapter 5). The release of the C–1 reducing group of muramic acid has been confirmed for the action of the T2 phage enzyme as well as the release of low molecular weight glycosaminopeptides (PRIMOSTGH, PELZER, MAASS AND WEIDEL, 1961).

The lytic enzymes released after phage multiplication have also attracted attention and KRAUSE AND MCCARTY (1962a,b) have used the cell-wall degrading capacity of the streptococcal group C phage-enzyme system (KRAUSE, 1958) in their studies. The phage-induced lysin from group C streptococci was purified by DOUGHTY AND HAYASHI (1962) but only partial lysis of isolated walls of group A streptococci was observed and these investigators concluded that lysis of walls was not identical to the more complex process involved in the lysis of intact cells. The virolysins of phage infected *Staphylococcus aureus* have been studied by RALSTON, BAER, LIEBERMAN AND KRUEGER (1961) and these authors indicate that the phage-induced lysin has a direct action on staphylococcal cell walls although no experimental evidence was included in this particular paper.

Whether the enzymes released after lysis of the cells following bacteriophage multiplication are identical to the enzymes found in the phage particles themselves has yet to be determined. It is of interest to note that REITER AND ORAM (1962) observed that suramin inhibited the lysis of *Streptococcus lactis* by preventing the initial step of phage adsorption; the same compound also inhibited the action of lysozyme on *Micrococcus lysodeikticus*. These results could imply a close relationship between phage-tail enzyme and lytic enzymes of the lysozyme type. No doubt antibodies to purified enzymes will greatly assist the task of solving this problem.

REFERENCES

BADDILEY, J., *J. Roy. Instit. Chem.*, 86 (1962) 366.
BADDILEY, J. and A. L. DAVISON, *J. Gen. Microbiol.*, 24 (1961) 295.
BARKULIS, S. S. and M. F. JONES, *J. Bacteriol.*, 74 (1957) 207.
BARRINGTON, L. F. and L. M. KOZLOFF, *J. Biol. Chem.*, 223 (1956) 615.
BRUMFITT, W., *J. Pathol. Bacteriol.*, 79 (1960) 1.
CRUMPTON, M. J. and D. A. L. DAVIES, *Intern. Congr. Microbiol., 7th Congr.*, Stockholm, 7b (1958).
CRUMPTON, M. J., D. A. L. DAVIES and A. M. HUTCHISON, *J. Gen. Microbiol.*, 18 (1958) 129.
CUMMINS, C. S., *Brit. J. Exptl. Pathol.*, 35 (1954) 166.
CUMMINS, C. S., *Soc. Gen. Microbiol. Symposium*, No. 12, 1962a, p. 212.
CUMMINS, C. S., *J. Gen Microbiol.*, 28 (1962b) 35.
CUMMINS, C. S. and H. HARRIS, *J. Gen. Microbiol.*, 14 (1956) 583.
DAVIES, D. A. L., *Biochim. Biophys. Acta*, 26 (1957) 151.
DAVIES, D. A. L., *J. Gen. Microbiol.*, 18 (1958) 118.
DAVIES, D. A. L., *Adv. Carbohydrate Chem.* Vol. 15, 1960 p. 271.
DOUGHTY, C. C. and J. A. HAYASHI, *J. Bacteriol.*, 83 (1962) 1058.
ELLIOTT, S. D., *Nature*, 193 (1962) 1105.
FOSTER, J. W. and E. RIBI, *J. Bacteriol.*, 84 (1962) 258.
FREIMER, E. H., R. M. KRAUSE and M. MCCARTY, *J. Exptl. Med.*, 110 (1959) 853.
GOEBEL, W. F. and M. A. JESAITIS, *J. Exptl. Med.*, 96 (1952) 425.
GOEBEL, W. F. and M. A. JESAITIS, *Ann. Instit. Pasteur*, 84 (1953) 66.
GOODER, H. and W. R. MAXTED, *Soc. Gen. Microbiol. Symposium*, No. 11, 1961, p. 151.
HAUKENES, G., *Acta Pathol. Microbiol. Scand.*, 55 (1962a) 110.
HAUKENES, G., *Acta Pathol. Microbiol. Scand.*, 55 (1962b) 117.
HAUKENES, G., *Acta Pathol. Microbiol. Scand.*, 55 (1962c) 289.
HAUKENES, G., *Acta Pathol. Microbiol. Scand.*, 55 (1962d) 299.
HAUKENES, G., *Acta Pathol. Microbiol. Scand.*, 55 (1962e) 463.
HAYASHI, J. A. and S. S. BARKULIS, *J. Bacteriol.*, 77 (1959) 177.
HIJMANS, W., *J. Gen. Microbiol.*, 28 (1962) 177.
HOLME, T., A. S. MALMBORG and E. COTA-ROBLES, *Nature*, 185 (1960) 57.
HOTCHIN, J. E., I. M. DAWSON and W. J. ELFORD, *Brit. J. Exptl. Pathol.*, 33 (1952) 177.
JESAITIS, M. A. and W. F. GOEBEL, *Nature*, 172 (1953) 622.
JONES, D. and P. M. F. SHATTOCK, *J. Gen. Microbiol.*, 23 (1960) 335.
JULIANELLE, L. A. and C. M. WIEGHARD, *Proc. Soc. Exptl. Biol. Med.*, 31 (1934) 947.
KANAI, K., G. P. YOUMANS and A. S. YOUMANS, *J. Bacteriol.*, 80 (1960) 615.
KAUFFMANN, F., *Enterobacteriaceae*, 2nd ed., E. Munksgaard, Copenhagen, 1954.
KAUFFMANN, F., O. H. BRAUN, O. LÜDERITZ, H. STIERLIN and O. WESTPHAL, *Z. Bakt. Parasitenk.*, 180 (1960) 180.
KAUFFMANN, F., O. LÜDERITZ, H. STIERLIN AND O. WESTPHAL, *Z. Bakt. Parasistenk.* (Orig.), 178 (1960) 442.
KEELER, R. F. and M. L. GRAY, *J. Bacteriol.*, 80 (1960) 683.
KNOX, K. W. and J. BRANDSEN, *Biochem. J.*, 85 (1962) 15.
KOCH, G. and W. J. DREYER, *Virology*, 6 (1958) 291.
KOTANI, S., T. KITAURA, M, HIGASHIGAWA, K. KATO, Y. MORI, T. MATSUBARA and T. TSUJIMOTO, *Biken's J.*, 3 (1960) 159.
KOTANI, S., T. KITAURA, T. HIRANO and A. TANAKA, *Biken's J.*, 2 (1959) 129.
KOZLOFF, L. M. and M. LUTE, *J. Biol. Chem.*, 228 (1957) 529.
KRAUSE, R. M., *J. Exptl. Med.*, 108 (1958) 803.
KRAUSE, R. M. and M. MCCARTY, *J. Exptl. Med.*, 114 (1961) 127.

KRAUSE, R. M. and M. MCCARTY, *J. Exptl. Med.*, 115 (1962a) 49.
KRAUSE, R. M. and M. MCCARTY, *J. Exptl. Med.*, 116 (1962b) 131.
LANCEFIELD, R. C., *J. Exptl. Med.*, 57 (1933) 571.
LANCEFIELD, R. C., *J. Exptl. Med.*, 78 (1943) 465.
LANCEFIELD, R. C., *J. Immunol.*, 89 (1962) 307.
LEDERER, E., *Pure and Applied Chemistry*, Vol. 2, p. 587, Butterworths, London, 1961.
LÜDERITZ, O., F. KAUFFMANN, H. STIERLIN and O. WESTPHAL, *Z. Bakt. Parasitenk.* (*Orig.*), 179 (1960) 180.
LÜDERITZ, O., G. O'NEILL and O. WESTPHAL, *Biochem. Z.*, 333 (1960) 136.
LÜDERITZ, O., O. WESTPHAL, A. M. STAUB and L. LE MINOR, *Nature*, 188 (1960) 556.
MAXTED, W. R., *Lancet*, ii (1948) 255.
MCCARTY, M., *J. Exptl. Med.*, 96 (1952a) 555.
MCCARTY, M., *J. Exptl. Med.*, 96 (1952b) 569.
MCCARTY, M., *J. Exptl. Med.*, 104 (1956) 629.
MCCARTY, M., *J. Exptl. Med.*, 109 (1959) 361.
MCCARTY, M. and R. C. LANCEFIELD, *J. Exptl. Med.*, 102 (1955) 11.
MIKULASZEK, E., *Ann. Instit. Pasteur*, 91 Suppl. (1956) 40.
MILLER, E. and W. F. GOEBEL, *J. Exptl. Med.*, 90 (1949) 255.
MUNOZ, J., E. RIBI and C. L. LARSON, *J. Immunol.*, 83 (1959) 496.
MURPHY, J. S., *Virology*, 4 (1957) 563.
MURPHY, J. S., *Virology*, 11 (1960) 510.
NIKAIDO, H., *Proc. Nat. Acad. Sci.* U.S., 48 (1962) 1337.
PERKINS, H. R., *Biochem. J.*, 83 (1962) 5P.
PRIMOSIGH, J., H. PELZER, D. MAASS and W. WEIDEL, *Biochim. Biophys. Acta*, 46 (1961) 68.
RALSTON, D. J., B. BAER, M. LIEBERMAN and A. P. KRUEGER, *J. Gen. Microbiol.*, 24 (1961) 313.
REITER, B. and J. D. ORAM, *Nature*, 193 (1962) 651.
RIBI, E., W. T. HASKINS, M. LANDY and K. C. MILNER, *Bacteriol. Rev.*, 25 (1961) 427.
RIBI, E., C. L. LARSEN, R. LIST and W. WICHT, *Proc. Soc. Exptl. Biol. Med.*, 98 (1958) 263.
RIBI, E., K. C. MILNER and T. D. PERRINE, *J. Immunol.*, 82 (1959) 75.
ROBERTS, W. S. L. and F. S. STEWART, *J.* Gen. *Microbiol.*, 24 (1961) 253.
SALTON, M. R. J., *Biochim. Biophys. Acta*, 8 (1952a) 510.
SALTON, M. R. J., *Biochim. Biophys. Acta*, 9 (1952b) 334.
SALTON, M. R. J., *Biochim. Biophys. Acta*, 10 (1953) 512.
SALTON, M. R. J., *Soc. Gen. Microbiol. Symposium*, No. 6, 1956.
SALTON, M. R. J. in *The Nature of Viruses*, Eds. G. E. W. Wolstenholme and E. C. P. Millar, Churchill, London, 1957, p. 263
SALTON, M. R. J., *J. Gen. Microbiol.*, 18 (1958) 481.
SALTON, M. R. J. in '*The Bacteria*', Ed. I. C. Gunsalus and R. Y. Stanier, Academic Press, New York, 1960, Vol, I, p. 97.
SALTON, M. R. J. and G. MILHAUD, *Biochim. Biophys. Acta*, 35 (1959) 254.
SANDERSON, A. R., W. G. JUERGENS and J. L. STROMINGER, *Biochem. Biophys Res. Comm.*, 5 (1961) 472.
SHARP, J. T., W. HIJMANS and L. DIENES, *J. Exptl. Med.*, 105 (1957) 153.
SHARPE, M. E. and D. M. WHEATER, *J. Gen. Microbiol.*, 16 (1957) 676.
SHEPARD, C. C., E. RIBI and C. L. LARSON, *J. Immunol.*, 75 (1955) 7.
SLADE, H. D. and W. C. SLAMP, *J. Bacteriol.*, 84 (1962) 345.
STAUB, A. M. in *Polysaccharide in Biology, 5th Confr. Josiah Macy, Jr. Foundation*, 1959, p. 139.
STAUB, A. M., R. TINELLI, O. LÜDERITZ and O. WESTPHAL, *Ann. Instit. Pasteur*, 96 (1959) 303.
STEWART, F. S., *Nature*, 190 (1961) 464.
TEPPER, B. S., J. A. HAYASHI and S. S. BARKULIS, *J. Bacteriol.*, 79 (1960) 33.

TOMCSIK, J. and S. GUEX-HOLZER, *J. Gen. Microbiol.*, 10 (1954) 317.
WEIDEL, W. in *Viruses*, 1950, Ed. M. Delbrück, California Institute of Technology, 1950, p. 119.
WEIDEL, W., *Z. Naturforsch.*, 6b (1951) 251.
WEIDEL, W., *Ann. Instit. Pasteur*, 84 (1953) 60.
WEIDEL, W., *Hoppe-Seyler's Z. Physiol. Chem.*, 299 (1955) 253.
WEIDEL, W., *Ann. Rev. Microbiol.*, 12 (1958) 27.
WEIDEL, W. and E. KELLENBERGER, *Biochim. Biophys. Acta*, 17 (1955) 1.
WEIDEL, W., G. KOCH and K. BOBOSCH, *Z. Naturforsch.*, 9b (1954) 573.
WEIDEL, W., G. KOCH and F. LOHSS, *Z. Naturforsch.*, 9b (1954) 398.
WEIDEL, W. and J. PRIMOSIGH, *J. Gen. Microbiol.*, 18 (1958) 513.
WESTPHAL, O., F. KAUFFMANN, O. LÜDERITZ and H. STIERLIN, *Z. Bakt. Parasitenk.*, 179 (1960) 336.
WESTPHAL, O., O. LÜDERITZ, I. FROMME and N. JOSEPH, *Angew. Chem.*, 65 (1953) 555.
WICKEN, A. J., S. D. ELLIOTT and J. BADDILEY, *J. Gen. Microbiol.*, 31 (1963) 231.
WILKINSON, J. F., *Bacteriol. Rev.*, 22 (1958) 46.
WILSON, A. T., *J. Exptl. Med.*, 81 (1945) 593.
YOSHIDA, N., S. TANAKA, K. TAKAISHI, I. FUKUYA, K. NISHINO, I.KAKATANI, S. INCI, K. FUKUI, A. KONO and T. HASHIMOTO, *Tokushima J. Exptl. Med.*, 2 (1955) 11.
YOUMANS, G. P., I. MILLMAN and A. S. YOUMANS, *J. Bacteriol.*, 70 (1955) 557.

CHAPTER 8

Biochemistry of the bacterial cell wall

In preceding chapters we have been concerned principally with the physical and chemical properties of the bacterial cell wall. And now we must turn to the 'biochemistry' of the cell wall and here our main interest will be in the functions of these structures and their biosynthesis. It will already be apparent from the discussion in the first chapter of this book, that as an 'organelle', the wall of a Gram-positive bacterium does not appear to be of great biochemical importance. Indeed, it is not absolutely indispensable for survival, as upon the loss of the cell wall the organism can continue its existence in a somewhat different 'guise' as a L-phase culture (SHARP, HIJMANS AND DIENES, 1957; FREIMER, KRAUSE AND MCCARTY, 1959; PANOS AND BARKULIS, 1959; GOODER AND MAXTED, 1961). As indicated in Chapter 1, it is likely that the wall or outer compound membrane of Gram-negative bacteria is of considerable biochemical import, but unfortunately at present it has not been possible to disentangle the enzymic activities of the outer multiple-layered 'wall' from those of the inner multiple-layered plasma membrane.

The contribution of the cell wall to the biochemical functioning of Gram-positive bacteria seems fairly clear from the results of investigations with isolated protoplasts (WEIBULL, 1953; 1958; MCQUILLEN, 1960). The loss of the cell wall in Nature undoubtedly results in the loss of biochemical activity by dilution and disorganization when the cell lyses, consequent 'death' of the species occurring except in special environments favouring protoplast formation. As pointed out by SALTON (1956) the wall confers considerable survival value upon an organism. Bacterial protoplasts produced *in vitro* in an osmotically suitable environment can continue the biosynthesis of proteins, nucleic acids, lipids, polysaccharides as well as supporting the synthesis to completion, of complex structures such as bacteriophages and endospores (WEIBULL, 1958; MCQUILLEN, 1960; MARKOVITZ AND DORFMAN, 1961; HILL, 1962). Thus under conditions of protoplast formation the walls of the Gram-positive organisms which are amenable to biochemical dissection, do not appear to play a role in the biosynthesis of most of the major molecular species of the cell.

The experiments of MITCHELL AND MOYLE (1956) have shown that the high solute concentrations in certain Gram-positive bacteria can exert osmotic pressures of up to 20 atmospheres e.g. *Staphylococus aureus* and other micrococci. The wall of the Gram-positive organism must therefore be strong enough to provide the cell with a rigid 'corset' to prevent osmotic explosion when the cells encounter an environment of low tonicity. The principal function of the walls inferred from osmotic and protoplast studies would seem to be to provide the cell with a structure of sufficient tensile strength to contain and protect the bacterial protoplast.

Perhaps the most important 'biochemical function' of the cell wall of the Gram-positive

The tables are printed together at the end of the book.

bacteria is to provide acceptor sites for the addition of the low-molecular weight precursors during wall biosynthesis. Protoplasts prepared by dissolution of the wall with egg-white lysozyme or with related muramidases do not readily revert to the original cells, although FREIMER, KRAUSE AND MCCARTY (1959) reported reversion to streptococci, of L-phase forms induced and stabilized in the presence of penicillin. To what extent the loss of the cell wall during protoplast formation means permanent loss of ability to synthesize a wall cannot be answered with certainty at the moment. The growth of protoplasts of *Bacillus megaterium* occurred in liquid media without any new wall being formed and without detectable incorporation of ^{14}C-radioactive compounds into wall constituents such as diaminopimelic acid (MCQUILLEN, 1955; 1960; FITZ-JAMES, 1958). It is not clear whether incorporation into substances associated with the protoplasts only was examined by MCQUILLEN (1955, 1960). ROGERS AND PERKINS (1962) have stated that on removal of the cell wall from cocci, 'the remaining protoplasts do not form new cell walls, but we have now shown that they do form large soluble mucopeptides when prepared and incubated in the correct manner.' Full experimental details of this report and the recent observations by LANDMAN AND HALLE (1962) that L-bodies of *Bacillus subtilis* give rise to revertants within 24 hours when transferred to media containing 2% agar, have not yet appeared.

A more comprehensive investigation of the biochemical properties of protoplasts of Gram-positive organisms and derived L-forms in relation to wall biosynthesis is obviously necessary. It now seems evident that under appropriate conditions, protoplasts can synthesize soluble glycosaminopeptides (ROGERS AND PERKINS, 1962) and that a certain population of L-form cells still possesses the capacity for complete wall biosynthesis when placed in the right physiological *milieu* (LANDMAN AND HALLE, 1962). The permanent loss of wall synthesizing ability which seems to occur with some of the protoplasts and stabilized L-phase cells (from either protoplasts formed enzymically or from penicillin induced forms) is intrigueing and suggests the loss of more than just the cell-wall structure. The discovery of the nature of any additional components that are lost during protoplast formation may reveal new facts about the mechanism of wall biosynthesis. It is tempting to speculate that the specific 'intracellular teichoic acids' may be involved in wall biosynthesis and that this material constitutes the 'gap substance' between wall and membrane and its loss during protoplast formation deprives the cell of the steps necessary for the final stages of wall formation. Indeed, SALTON (1956) suggested that sites of biosynthesis of the wall may be external to the membrane and their loss on conversion to protoplasts could explain the inability to form a new wall. It is perhaps highly significant that the group D streptococcal L-phase had lost both its cell-wall substance and the group antigen (HIJMANS, 1962) now recognized as an 'intracellular' teichoic acid (WICKEN, ELLIOTT AND BADDILEY, 1963). There seems little doubt that the next few years will see intensive activities with protoplasts and L-forms in an effort to resolve some of the problems of wall formation.

Whether cell walls play any active part in the passage of solutes from the outside environment to the surface of the plasma membrane has not been established experimentally. The retention of amino acids on cell walls (BUTLER, CRATHORN AND HUNTER, 1958; BRITT AND GERHARDT, 1958) and the highly charged groups of the teichoic acids and their possible role in salt tolerance (ARCHIBALD, ARMSTRONG, BADDILEY AND HAY, 1961) suggest that walls may have some functions in selective passage of substances across the cell envelope (wall-membrane). However, from the work performed with bacterial protoplasts it would seem that its role is a minor one and that it does not exert permeability control by limiting the passage of low-molecular weight materials into the cell. It is not known whether or not the size of pores in the cell wall controls the entrance and exit of high molecular weight substances into the bacterial cell. If it is assumed that high-molecular weight substances do not leave the cell through mechanical breaches, then it follows that the wall offers no barrier to the outward passage of large molecules.

The functions of the 'wall' of Gram-negative organisms where both multilayered 'wall' and plasma membrane occur together have already been discussed and little further need be added here. When methods are available for the selective removal of the outer multiple-layered structure then direct answers to the question of wall function will be possible. It can be anticipated that the structures will be of considerable biochemical importance to the cell. Although the Gram-negative bacteria can persist in the L-form (LEDERBERG AND ST. CLAIR, 1958) this discovery has not assisted us any further in deciding upon the functions of the 'wall' in this group of organisms. One of the principal changes seems to be in the total amount of glycosaminopeptide formed, rather than its complete absence. All of the stable L-phase cells examined by WEIBULL (1958) and MORRISON AND WEIBULL (1962) contained α, ε-diaminopimelic acid and hexosamine. If these constituents are present in the form of glycosaminopeptides, then it is evident that L-phase cells of Gram-negative bacteria differ from those obtained from Gram-positive organisms in that those of the former group contain the components of the rigid wall structure, whereas in the latter group they disappear altogether.

It would be of great interest to know whether the outer layers of the wall of the Gram-negative organism are concerned with the biosynthesis of the rigid glycosaminopeptide structure or whether this is formed by the plasma membrane underlying this layer. At least there are some suggestions that the wall may be the site of biosynthesis of the polysaccharide moiety of the lipopolysaccharide component (OSBORN, ROSEN, ROTHFIELD AND HORECKER, 1962). With a mutant strain of *Salmonella typhimurium* lacking the uridine diphosphate-galactose-4-epimerase, these investigators found that the heavy particulate fractions of sonically disrupted cells contained the enzymic activity responsible for the transfer of galactose into the wall polysaccharide. About 75% of the activity was recovered in the fraction sedimenting between 1,200 $\times$ g and 12,000 $\times$ g. It will be recalled that walls and envelopes are usually deposited over this range of gravitational fields. Attempts to solubilize the enzyme or the bound acceptor were unsuccessful (OSBORN *et al.*, 1962).

The demonstration that capsular hyaluronate is synthesized by streptococcal membrane fractions (MARKOVITZ AND DORFMAN, 1961) in addition to the above observations suggest that it would be profitable to examine the ability of wall and envelope fractions of Gram-negative bacteria to biosynthesize their specific capsular polysaccharides and extracellular gums.

Interest in the biochemistry of the bacterial cell wall has not been confined merely to direct studies of the functions and enzymic reactions intimately associated with the wall structures. Indeed, this aspect now forms a minor part of our biochemical knowledge of this field because attention has been focussed on the fascinating problem of the biosynthesis of this complex entity of the bacterial cell. Much of this work orginated from the discovery by PARK AND JOHNSON (1949), of the imbalance in nucleotide metabolism brought about by the action of penicillin on *Staphylococcus aureus*. The significance of these observations was not fully appreciated until the amino sugar constituent of the PARK nucleotide was identified as muramic acid. This important finding thus gave the first clues about the mechanisms involved in cell-wall biosynthesis (PARK AND STROMINGER, 1957). Despite the fact that this aspect of the biochemistry of the wall is still quite young, new and exceedingly interesting knowledge has been gained in the past 7 years.

Although the pathways for biosynthesis of bacterial cell walls still remain largely hypothetical, a considerable body of basic information has been built up steadily over the past 7 years. The biosynthesis of some of the 'unique' wall compounds has been investigated, the incorporation of radioactive 'wall' compounds into the cell walls has received a lot of attention and synthesis of glycosaminopeptides by washed suspensions and the effects of inhibitions on the process have been studied. Perhaps one of the most important contributions has been the isolation and characterization of a wide variety of nucleotide wall precursors. However, attempts to demonstrate the transfer of the glycosaminopeptide components of the nucleotides into wall have not been very successful and this final aspect of wall biosynthesis awaits the development of suitable systems for further investigations.

In the following discussion of wall biosynthesis only compounds bearing a direct relationship to those known to occur in the wall will be dealt with. No attempt will be made to discuss the origins of many of the commoner constituents, the biosynthetic pathways of which have been established for general cell substances. It will become apparent that there are many gaps in our knowledge of the synthesis of the simple monomeric constituents and their assembly into intermediate stages prior to the formation of insoluble cell wall. To avoid the impression that all has been solved, no attempt will be made to fill these gaps by assuming that the substances are made by existing pathways, no matter how likely this may be.

BIOSYNTHESIS OF BACTERIAL CELL WALLS

1. *Biochemistry of cell-wall amino acids, amino sugars and monosaccharides*

Amino acids

The amino acids that may have 'special' enzymes concerned with their formation for the specific purpose of wall biosynthesis are D-alanine, D-glutamic acid, D-aspartic acid and α, ε-diaminopimelic acid. There are no reasons for suspecting that any of the other L-amino acids occurring in the protein components of Gram-negative cell walls or in peptides of the walls of Gram-positive bacteria, are produced by special mechanisms.

Alanine occurs in the wall as the D and L forms, both isomers being present in the peptide moiety, but only the D-isomer is found in the teichoic acid. It has been known for some time that bacteria contain an alanine racemase. This enzyme was detected in *Streptococcus faecalis* by WOOD AND GUNSALUS (1951) who also showed that a variety of bacteria had active racemases. It seems likely that the alanine racemase fulfils the function of providing the bacteria with D-alanine for cell-wall precursors (nucleotides, alanyl-alanine peptides) and for the teichoic acids.

For the origin of D-glutamic acid there are two possibilities. The series of steps catalysing the formation of D-glutamic acid have been investigated mainly in relation to the synthesis of the D-glutamyl-polypeptides of *Bacillus* spp. by THORNE (1956) and his colleagues. The following enzymic reactions have been detected in *Bacillus subtilis* and *Bacillus anthracis* by THORNE, GOMEZ AND HOUSEWRIGHT (1955) and THORNE AND MOLNAR (1955) and these could account for the conversion of L-glutamic acid to the D-isomer or its origin from D-alanine:

L-glutamic acid + pyruvic acid → α-ketoglutaric acid + L-alanine
L-alanine → DL-alanine
D-alanine + α-ketoglutaric acid → pyruvic acid + D-glutamic acid

From these investigations it was concluded that the D-glutamic acid could arise in these bacteria by an indirect conversion or by the D-amino acid transamination. Attempts to demonstrate a glutamic acid racemase were unsuccessful (THORNE, 1956). Whether the cell-wall glutamic acid of the *Bacillus* spp. is formed by the above reactions cannot be said at present. So far as the author is aware no experiments have been performed to determine the labelling of cell-wall glutamic acid.

NARROD AND WOOD (1952) presented evidence indicating the possibility of a glutamic acid racemase in *Lactobacillus arabinosus* but little further progress in the characterization of this enzyme was made until GLASER (1960) purified a glutamate racemase. A specific racemase has also been detected in another species, *Lactobacillus fermenti;* the enzyme has been purified from extracts of this organism by TANAKA, KATO AND KINOSHITA (1961). While it seems likely that the glutamate racemase would provide a direct mechanism for D-glutamic acid formation prior to incorporation into wall compounds, the assessment of which of the two reactions (racemase or D-amino acid transaminase) is operative for wall biosynthesis will have to await the results of further experiments.

The isolation of mutants of *Bacillus subtilis* requiring D-glutamic acid by MOMOSE (1961a) has led to a clarification of the most probable pathway for cell-wall D-glutamate formation in this organism. In the mutant strain '3d', the alanine racemase activity was about one tenth of that found in the wild-type strain (MOMOSE AND IKEDA, 1961). The mutant required D-alanine D-aspartic acid or D-glutamic acid for growth and it therefore appeared that the genetic block is between L-alanine and D-alanine. Another mutant (4a) possessed a low activity of

D-alanine – D-glutamic acid transaminase and responded to D-aspartic acid, D-glutamic acid, L-aspartic acid and L-glutamic acid. The dependence of the D-amino acids was understandable but the requirements for the L-isomers remained unexplained (MOMOSE AND IKEDA, 1961).

That the requirement of D-glutamic acid and the reactions for its biosynthesis are really intimately concerned with wall formation was indicated by the unbalanced growth and death of the '3d' mutant of *Bacillus subtilis* when it was placed in a medium devoid of D-glutamic acid (MOMOSE, 1961b). This response to limitation of one of the 'building-blocks' for the cell-wall is very similar to the phenomenon observed with diaminopimelic acid-requiring mutants of *Escherichia coli* (BAUMAN AND DAVIS, 1957; MEADOW, HOARE AND WORK, 1957; RHULAND, 1957). In both cases the amino acids are required for the synthesis of the wall, and growth in limiting amounts of the amino acid results in death and lysis when the exogenous and endogenous supplies are exhausted. From the enzymic activities of the wild-type and mutant strains of *Bacillus subtilis* and the obvious dependence of the mutant strain on D-glutamic acid for cell-wall biosynthesis, it appears that in this organism the wall glutamic acid is normally synthesized by a D-amino acid transamination reaction.

At the present time there is no information on the mechanism of formation of the cell-wall D-aspartic acid and it will be of interest to see if it can be formed by the two types of mechanisms found for D-glutamic acid.

In addition to being required as wall components, a number of D-amino acids including D-alanine, D-aspartic acid and D-glutamic acid, can induce the formation of 'protoplasts' of *Alcaligenes faecalis* (LARK AND LARK, 1958) and bizarre-shaped cells of *Rhodospirillum rubrum* (TUTTLE AND GEST, 1960) when they are present in the culture media. Many of the morphological changes occurring in bacterial cells have now been traced to deformation of the rigid structure of the wall and it is tempting to conclude that the effects of the D-isomers may be due to altered equilibria of reactions involved in the biosynthesis of wall precursors. However, TUTTLE AND GEST (1960) also observed morphological changes in *Rhodospirillum rubrum* induced by 'non-wall' amino acids such as D-serine, D-valine, D-histidine, D-leucine, D-lysine and D-methionine. The mechanism of action of these latter amino acids would require further investigation but it is conceivable that their effects may be on the proteins involved in, or at the site of wall formation.

Several other D-amino acids have been found in products closely associated with bacterial walls. The peptido-glycolipids of mycobacteria contain D-phenylalanine, D-allo-threonine (IKAWA, SNELL AND LEDERER, 1960) and of these two amino acids only the reactions leading to D-phenylalanine in bacteria have been described. Extracts of *Bacillus anthracis* can convert the L-isomer of phenylalanine to the D-form by transamination reactions and alanine racemase (THORNE AND MOLNAR, 1955). The final step in the reaction is given below:

$$\text{D-alanine} + \text{phenylpyruvic acid} \rightarrow \text{pyruvic acid} + \text{D-phenylalanine.}$$

Although the mechanism for the formation of D-phenylalanine has been demonstrated in

Bacillus anthracis, it is not known whether the D-isomer of the mycobacterial mycosides is produced by the same reactions.

IKAWA AND SNELL (1962) have found D-allo-isoleucine in *Nocardia asteroides* and D-leucine in the mycoside C fraction of *Mycobacterium avium* but there is no information about the manner of formation of these D-isomers.

The occurrence of α, ε-diaminopimelic acid (DAP) in Nature is largely restricted to a variety of bacteria and related blue-green algae. Its distribution is thus unlike that of the other amino acids discussed above, as their L-isomers are universally present in proteins of plant, animal and microbial origin. The biosynthesis of DAP as a cell-wall amino acid would involve the possession of a complement of enzymes not found in other cells and tissues and therefore merits special attention. The metabolism of DAP is also of interest, especially as an intermediate for the origin of lysine in bacterial species. The role of DAP in the economy and metabolism of bacteria has been reviewed by WORK (1959, 1961) and RHULAND (1960).

Following the demonstration by ABELSON, BOLTON, BRITTEN, COWIE AND ROBERTS (1953) that four of the carbon atoms of α, ε-diaminopimelic acid and lysine were derived from aspartic acid, GILVARG AND RHULAND carried out independent studies on the pathways for DAP biosynthesis using selected mutants of *Escherichia coli*. With strain 26–26 of *Escherichia coli*, RHULAND AND BANNISTER (1956) found a substance in the culture medium which was capable of supporting the growth of *Escherichia coli* 81–29 in the absence of DAP. This compound was identified as succinic acid and it was found to be necessary for the maximal synthesis of DAP by cell-free extracts of *Escherichia coli* 26–26. These investigators reported that succinic acid, aspartate, pyruvate, adenosine triphosphate and triphosphopyridine nucleotide were required for this synthesis.

GILVARG (1956) also reported the synthesis of α, ε-diaminopimelic acid by cell-free extracts and in subsequent studies he isolated the important intermediate N-succinyl-L-diaminopimelic acid from *Escherichia coli* D–1 (GILVARG, 1957). The succinyl side-chain was later shown to be attached to the aspartate portion of N-succinyl-L-diaminopimelic acid (GILVARG, EDELMAN AND KINDLER, 1958). The enzymic hydrolysis of the N-succinyl-L-diaminopimelic acid to succinic acid and α, ε-diaminopimelic acid was effected by extracts of *Escherichia coli* 26–26 and wild-type, but not by strain D–1. These experiments established the role of the N-succinyl compound as an intermediate and explained the effects observed with succinic acid.

The accumulation of another intermediate, 2-N-succinyl-6-ketopimelic acid, by *Escherichia coli* D–1 was reported by GILVARG (1958) and subsequent studies by PETERKOFSKY AND GILVARG (1959) indicated the presence in *Escherichia coli* of a transaminase converting the N-succinyl-keto-pimelic acid to N-succinyl-L-diaminopimelic acid. The purified transaminase showed a specificity for L-glutamic acid as the amino donor. The structure of the intermediate N-succinyl-α-amino-ε-ketopimelate was confirmed by GILVARG (1961).

Isotope experiments on the aspartic acid family of amino acids performed by ROBERTS, ABELSON, COWIE, BOLTON AND BRITTEN (1955) eliminated homoserine or threonine as direct precursors of diaminopimelic acid and lysine. This showed that the branch point for diaminopimelic acid and lysine biosynthesis could occur at any of the three preceding intermediates, aspartic acid itself, aspartyl phosphate or aspartic semialdehyde. GILVARG (1962) has used an auxotrophic mutant to determine the branch point. *Escherichia coli* M-145 had an absolute requirement for diaminopimelic acid, threonine and methionine and a relative requirement for lysine and isoleucine. This mutant lacked the ability to form aspartic semialdehyde dehydrogenase and placed the branch point at aspartic semialdehyde. Replacement of aspartic acid by aspartic semialdehyde enabled diaminopimelic acid synthesis to occur in the blocked mutant. Moreover, further confirmation came from the restoration of diaminopimelic acid synthesis by the addition of purified aspartic semialdehyde dehydrogenase to extracts of *Escherichia coli* M-145 utilizing aspartic acid as the carbon source.

Most of the principal steps in the biosynthesis of α, ε-diaminopimelic acid and lysine can now be specified as shown below. Triphosphopyridine nucleotide reductions occur in the conversion of aspartylphosphate to aspartic semialdehyde and a second one after condensation with pyruvate. The reactions below, show the pathway leading to diaminopimelic acid and lysine, the sequence of events involved in the N-acylation and condensation being unspecified at present:

Aspartic acid
↕ *kinase*
Aspartyl phosphate
↕ *semialdehyde dehydrogenase*
Aspartic semialdehyde + pyruvate
↓ *condensation and succinylation*
N-succinyl-α-amino-ε-ketopimelate
transaminase
↓ + L-glutamic acid
N-succinyl-L-diaminopimelic acid
↓ *deacylase*
LL-diaminopimelic acid
↕ *racemase*
meso-Diaminopimelic acid
↓ *decarboxylase*
L-lysine

The steps involved in the conversion of LL-diaminopimelic acid to L-lysine have been established by WORK and her colleagues. The racemase responsible for the interconversion of the LL-and meso- isomers of diaminopimelic acid was investigated by ANITA, HOARE AND WORK (1957) and it was found to occur quite widely in bacteria. The widespread distribution of the racemase suggested that the LL-isomer might have been synthesized first and as shown in the above series of reactions, this indeed was later shown to be the case. The other important enzyme involved in L-lysine production is the diaminopimelic acid decarboxylase which was detected by DEWEY AND WORK (1952) at the time DAVIS (1952) had established biosynthetic interrelationships of lysine, diaminopimelic acid and threonine with *Escherichia coli* mutants.

Both the racemase and the decarboxylase attack the meso-isomer of DAP at the D 'end' of the molecule and since the DD-isomer is not a substrate for either enzyme, it appears that the L-configuration is necessary for the attachment of DAP to either the enzyme or coenzyme (pyridoxal phosphate). The decarboxylases of *Escherichia coli* and *Aerobacter aerogenes* were studied in detail by DEWEY, HOARE AND WORK (1954). The partially purified enzymes were highly specific for natural diaminopimelic acid.

That the function of diaminopimelic acid decarboxylase is to provide the bacterial cell with a source of lysine was first suggested from the discovery that certain lysine-requiring mutants of *Escherichia coli* lacked the decarboxylase (DEWEY AND WORK, 1952). Confirmation that lysine originated from diaminopimelic acid was obtained when MEADOW AND WORK (1959) investigated the incorporation of [^{14}C]-diaminopimelic acid into amino acids of growing *Escherichia coli*. The results indicated that the radioactive label in diaminopimelic acid only appeared in lysine and moreover, neither amino acid interchanged its carbon with other amino acids.

The proportion of lysine originating from diaminopimelate was investigated in two mutants of *Escherichia coli*, 173–25 having a relative requirement for lysine and an absolute need for diaminopimelic acid (DAVIS, 1952), and D which had been trained to dispense with lysine (MEADOW, HOARE AND WORK, 1957). The labelled DAP contributed radioactive carbon to the extent of 80% of the cell-wall lysine in mutant 173–25, but only 55% with mutant D. A surprising result with these incorporation studies was the observation that DAP was not labelled when ^{14}C-glucose was used whereas labelling of the lysine and other amino acids occurred (MEADOW AND WORK, 1959). These results led to the possibility that two pathways operated for the biosynthesis of lysine in *Escherichia coli*.

The suggestion by MEADOW AND WORK (1959) of a pathway for lysine formation other than that occurring via diaminopimelic acid was at variance with the results obtained by VOGEL (1959a, b), indicating that lysine biosynthesis in all of the bacteria tested involved DAP as an intermediate. This problem was considered further by RHULAND AND HAMILTON (1961) who used γ-methyl-diaminopimelic acid to supply the requirement of diaminopimelic acid by *Escherichia coli* 173–25. Nutritional studies showed that γ-methyl-diaminopimelic acid replaced diaminopimelic acid only when lysine was supplied in the medium. Although the γ-methyldiaminopimelic acid was incorporated into the cell walls it could not substitute for DAP as a precursor of lysine. RHULAND AND HAMILTON (1961) concluded that the only functional pathway of lysine biosynthesis in *Escherichia coli* is *via* diaminopimelic acid.

The racemases and decarboxylases for diaminopimelic acid are reactions of obvious importance in the biosynthesis of bacterial walls. In addition to these enzymes, L-amino acid oxidases attacking the meso- and LL-isomers have been investigated by WORK (1955) and transaminating systems have been studied in a variety of bacteria by MEADOW AND WORK (1958). While these latter enzymes are of general significance for the metabolism of this amino acid they probably have little influence on the steps involved in cell-wall formation. The transamination studies showed that the DD-isomer of DAP could participate in the enzymic reactions. This isomer had hitherto been found to be metabolically inert (MEADOW AND WORK, 1958).

At a more biological level, the importance of diaminopimelic acid as a building block for the rigid layer of the cell wall has been emphasized by the growth experiments performed by MEADOW, HOARE AND WORK (1957). Lysis of the DAP-requiring mutant of *Escherichia coli* occurred after exhaustion of the amino acid from the medium. During the period of lysis, phase-contrast microscopy revealed the presence of spherical bodies which later lysed. Electron micrographs prepared after lysis showed cell envelopes from which the protoplasmic contents had escaped. MCQUILLEN (1958) confirmed the observations that DAP-deprival leads to 'protoplast' or spheroplast formation in *Escherichia coli* 173–25 and their appearance in both the phase-contrast and electron microscopes were very similar indeed to those found in cultures grown in the presence of penicillin.

The consequences of limiting or depriving the cell of an amino acid which is an essential compound of the wall glycosaminopeptide structure have been clearly established. Thus deprival, limitation or withdrawal of the essential amino acid will lead to cell death, lysis or spheroplast formation depending on the particular environmental conditions. This suicidal property of inducing spheroplast formation by DAP-deprival was put to good effect by BAUMAN AND DAVIS (1957) for the isolation of auxotrophic mutants of *Escherichia coli.* The parental cells will grow and lyse when DAP is exhausted due to the formation of unstable spheroplasts; the mutant auxotrophs will survive in the minimal medium because of their inability to grow.

The phenomenon of cell death and lysis resulting from limitation or deprival of an amino acid essential for the formation of a fully functional wall is now well established. The results of limiting growth of an organism possessing specific amino acid requirements were first investigated in detail by TOENNIES AND GALLANT (1949), TOENNIES AND SHOCKMAN (1953). The dramatic lysis following lysine limitation was not fully appreciated until later investigations indicated that cell-wall synthesis was involved (SHOCKMAN, KOLB AND TOENNIES, 1958). Lysine was one of the principal amino acids of the cell wall of the strain of *Streptococcus faecalis* used by SALTON (1952) and the lysis observed by SHOCKMAN, KOLB AND TOENNIES (1958) was in accord with the results expected from limiting one of the wall 'building blocks'. There are therefore at least three instances where amino acid limitation results in cell lysis because of an inadequate supply of the particular amino acid for wall formation; diaminopimelic acid is required for wall of mutant strains of *Escherichia coli* as first shown by MEADOW, HOARE AND WORK (1957), lysine is required for streptococcal walls (SHOCKMAN, KOLB AND TOENNIES, 1958) and D-glutamic acid by *Bacillus subtilis* (MOMOSE, 1961a).

Peptide synthesis

The only direct peptide synthesis investigated in relation to the cell wall is the D-alanyl-D-alanine synthetase studied by NEUHAUS (1960, 1961, 1962). Following the discovery of the

D-alanine activating enzyme in bacteria (BADDILEY AND NEUHAUS, 1960), NEUHAUS (1960) demonstrated the presence in *Streptococcus faecalis* of an enzyme, independent of the D-alanine activating system, which catalysed the formation of D-alanyl-D-alanine. The synthetase was purified by NEUHAUS (1960, 1961) and was shown to be responsible for the following reaction:

$$2 \text{ D-alanine} + \text{ATP} \xrightarrow[\text{K}^+]{\text{Mg}^{2+}} \text{D-alanyl-D-alanine} + \text{ADP} + \text{P}_i$$

The amino acids which were effective as donors in the reaction were D-alanine and D-aminobutyric acid, whereas D-alanine, D-aminobutyric acid, D-serine, D-threonine and D-norvaline were effective as acceptors. The kinetics of this synthetase have been investigated in some detail by NEUHAUS (1962) and the formation of four dipeptides studied when varying ratios of D-alanine and D-α-amino-*n*-butyric acid were incubated with the enzyme.

It has been known for several years that inhibition of bacterial growth by D-cycloserine (D-4-amino-3-isoxazolidone) can be reversed by D-alanine (BONDI, KORNBLUM AND FORTE, 1957; SHOCKMAN, 1959). Moreover, the uridine nucleotides accumulating in *Staphylococcus aureus* grown in the presence of D-cycloserine are devoid of the D-alanyl-D-alanine peptide (CIAK AND HANN, 1959; STROMINGER, THRENN AND SCOTT, 1959). The antibiotic inhibited the D-alanyl-D-alanine synthetase and D-alanine racemase from *Staphylococcus aureus* (STROMINGER, ITO AND THRENN, 1960; STROMINGER, 1962). NEUHAUS AND LYNCH (1962) investigated the inhibitory action of D-cycloserine and related analogues on the purified *Streptococcus faecalis* synthetase. Their observations were consistent with the suggestion that both D-alanine sites bind D-cycloserine. It is quite interesting to note that the D-alanine activating enzyme which is probably concerned with the introduction of the alanine into the teichoic acid polymer (BADDILEY AND NEUHAUS, 1960) is not inhibited by D-cycloserine.

The inhibitory action of D-cycloserine on the alanyl-alanine synthetase offers an adequate explanation for the inhibition of the incorporation of [1-^{14}C]-DL-alanine into the cell walls of both Gram-positive and Gram-negative bacteria (STROMINGER, THRENN AND SCOTT, 1959; BARBIERI, DI MARCO, FUOCO AND RUSCONI, 1960) by this antibiotic.

Amino acid activating enzymes

An amino acid-activating enzyme for D-alanine was detected in *Lactobacillus arabinosus*, *Lactobacillus casei*, *Bacillus subtilis* and *Staphylococcus aureus* H, by BADDILEY AND NEUHAUS (1960). The enzyme from *Lactobacillus arabinosus* was purified 25-fold. The enzyme catalysed a [^{32}P] pyrophosphate-adenosine triphosphate (ATP) exchange in the presence of D-alanine and Mg^{2+} ions and D-alanine hydroxamate was formed in the presence of ATP, Mg^{2+} ions, hydroxylamine and D-alanine. The reactions shown below are consistent with the formation of an adenosine monophosphate-D-alanine anhydride in combination with the enzyme.

$$\text{ATP} + \text{D-alanine} + \text{hydroxylamine} \rightarrow \text{D-alanine hydroxamate} + \text{AMP} + \text{pyrophosphate}$$

$$\text{ATP} + [^{32}\text{P}] \text{ pyrophosphate} \rightleftharpoons [^{32}\text{P}] \text{ ATP} + (\text{D-alanine}) \text{ pyrophosphate.}$$

NEUHAUS (1960) has suggested that this activating enzyme may be concerned with the introduction of the D-alanine into the teichoic acid and may not be involved in the synthesis of wall peptide components or the addition of amino acids to the nucleotides.

The two algae, *Chlorella vulgaris* and *Prototheca zopfii* were found to be capable of activating α, ε-diaminopimelic acid by an ATP-pyrophosphate exchange (CIFERRI, GIROLAMO AND GIROLAMO BENDICENTI, 1961). Of the stereoisomers of diaminopimelic acid, the meso-form and the LL-isomer were activated at the same rate by the enzyme from *Prototheca zopfii.* Although CIFERRI, GIROLAMO AND GIROLAMO BENDICENTI (1961) studied the DAP-activating enzymes from algae in a note added in proof they indicated the presence of such enzymes in *Escherichia coli* K–12 and 26-26. The DAP-activating enzyme should therefore, be present in bacteria possessing DAP as a cell-wall constituent.

Amino sugars

The two amino sugars N-acetylglucosamine and N-acetylmuramic acid occur in the cell walls of most of the bacterial species so far examined. The biosynthesis of these two substances is, therefore, of prime importance to the bacterial cell. Investigations of the biochemistry of amino sugars that form part of the cell wall have been largely concerned with muramic acid. Although there is indeed a great deal known about amino sugar metabolism in bacteria (ROSEMAN, 1959) most of the studies have not been specifically directed towards finding out the relevance of the reactions to cell-wall synthesis.

The pathways for the formation of N-acetylglucosamine and N-acetylgalactosamine have been studied by ROSEMAN and his colleagues and other investigators and the principal enzymic reactions leading from glucose to the acetylated amino sugars have been found in extracts of several microorganisms (ROSEMAN, 1959). It is interesting to note that acetylation with acetylcoenzyme A occurs via the amino sugar phosphates, rather than with the free amino sugars (BROWN, 1955; DAVIDSON, BLUMENTHAL AND ROSEMAN, 1957). The pathway for the biosynthesis of D-glucosamine and its N-acetyl derivative has been investigated more fully in microorganisms than has the route of formation of the other amino sugars. The principal steps are given below:

Glucose–6–phosphate
↓
Fructose-6-phosphate
NH_3 ↕ glutamine
Glucosamine-6-phosphate
↕ Acetyl CoA
N-acetylglucosamine-6-phosphate
↕
N-acetylglucosamine-1-phosphate

Experiments on the *in vivo* biosynthesis of D-glucosamine, using labelled glucose have shown that in group A streptococci the glucose carbon chain was preserved during the conversion to hexosamine (ROSEMAN *et al.*, 1953, 1954; TOPPER AND LIPTON, 1953). One of the key reactions, the L-glutamine-D-fructose-6-phosphate transamidase yielding glucosamine-6-phosphate was investigated by GHOSH, BLUMENTHAL, DAVIDSON AND ROSEMAN (1960). Recent studies by BARKULIS, BOLTRALIK, HEYMANN AND ZELEZNICK (1962) have shown that extracts of group A streptococci will catalyse an ATP-dependent phosphorylation of N-acetylglucosamine, yielding N-acetyl-glucosamine-6-phosphate. Kinase activity for D-glucosamine was much less and no kinase activity was detected for muramic acid. The latter observation is in agreement with unpublished results performed by the author with extracts of *Micrococcus lysodeikticus.*

In addition to carrying out the steps in amino sugar biosynthesis, several other enzymes have been reported in bacteria that may have some bearing on the formation of amino sugars associated with wall or wall compounds. GLASER (1959) found an epimerase in *Bacillus subtilis* which catalysed the conversion of uridine diphosphate-N-acetylgalactosamine. This bacterial enzyme apparently differed from the rat liver, epimerase enzyme system, which not only produced free N-acetylgalactosamine (CARDINI AND LELOIR, 1957) but also formed N-acetylmannosamine from uridine diphosphate-N-acetylglucosamine (COMB AND ROSEMAN, 1958). Bacteria have also been sources of active amino sugar deaminases (COMB AND ROSEMAN, 1958) and deacetylases (ROSEMAN, 1957).

The *in vivo* labelling of the N-acetylglucosamine of cell walls has not been specifically determined, although SALTON AND GHUYSEN (1959, 1960) reported that the amino sugars (both glucosamine and muramic acid) of the wall of *Micrococcus lysodeikticus* were selectively labelled when exponentially growing cells were transferred to a medium containing [^{14}C]-fructose or [^{14}C]-glucose and yeast extract. Direct experiments on the preservation of the sequence of carbon atoms in glucose during its conversion into cell-wall amino sugar have yet to be performed but it seems likely to be so from some of the existing steps already demonstrated in *Escherichia coli* and group A streptococci.

The origin of N-acetylmuramic acid (N-acetylglucosamine-3-O-lactic acid ether) is of special interest because of its unique occurrence in the amino sugar-peptide heteropolymers of bacterial walls and those of closely related organisms. Although this amino sugar has been chemically synthesized by four different groups of investigators (see Chapter 4) very little has been done on its biosynthesis. STROMINGER (1958, 1960) was the first worker to suggest that pyruvate (through phosphoenolpyruvate) was the source of the side chain of muramic acid. He reported the reactions involving uridine diphosphate-N-acetylglucosamine and phosphoenolpyruvate in extracts of *Staphylococcus aureus* as shown below:

Uridine diphosphate-N-acetylglucosamine + phosphoenolpyruvate → Uridine diphosphate-N-acetylglucosamine-pyruvate + P_i

Uridine diphosphate-N-acetylglucosamine-pyruvate → Uridine diphosphate-N-acetylglucosamine-lactic acid

Although the activities in the staphylococcal extracts were somewhat low, STROMINGER

(1960) calculated that they were sufficient to synthesize the necessary amount of amino sugar for the cell wall in ten minutes, a period well inside the mean generation time of the organism.

RICHMOND AND PERKINS (1960, 1962) have investigated the formation of muramic acid *in vivo* by following the flow of radioactive carbon into the pyranose ring and side chain of muramate under cultural conditions favouring selective synthesis of the cell wall of *Staphylococcus aureus*. Glucose was found to be a precursor of all carbon atoms in muramic acid. In the absence of exogenous alanine, [^{14}C]-glucose heavily labelled the glucosamine, muramic acid and alanine of the newly synthesized glycosaminopeptide. The extent of labelling of each of these three components was not very different. RICHMOND AND PERKINS (1962) found that in a medium containing 200 μg of glucose and 400 μg of DL-alanine/ml, more than 80% of the carbon atoms for C–1 to C–6 positions of muramic acid came from exogenous glucose, whereas about 2% of the carbon atoms for C–7, C–8 and C–9 positions came from exogenous alanine and more than 80% from glucose. They concluded that under their experimental conditions the side chain of muramic acid is more closely related metabolically to the products of glucose breakdown than it is to alanine.

These studies by RICHMOND AND PERKINS (1962) established that radioactive DL-alanine, pyruvate, lactate or aspartate labelled the side chain of muramic acid (C–7, C–8 and C–9) in preference to the pyranose ring (C–1 to C–6) when the incubation medium contained non-radioactive glucose. With the C–3 compounds the identity of the carbon atoms was maintained. The experiments suggested that phosphoenolpyruvate is the most likely immediate precursor of the side chain of muramic acid but as the study was an *in vivo* one, the authors rightly pointed out that it was not possible to exclude any of the compounds of the glycolytic pathway between triosephosphate and phosphoenolpyruvate.

There seems little doubt that the pyranose ring of muramic acid originates directly from glucose or glucosamine. Using the organism *Lactobacillus bifidus* var. *pennsylvanicus* with its specific growth requirement for β-glucosaminides, ZILLIKEN (1959) and O'BRIEN, GLICK AND ZILLIKEN (1960) were able to show an impressive incorporation of [1-^{14}C] β-methyl-N-acetyl-D-glucosaminide carbon into cell-wall muramic acid. The specific activity in the muramic acid was 19,000 cpm/mole compared to 21,000 cpm/mole for the starting material and it was concluded that D-glucosamine was a direct precursor of muramic acid.

Although the metabolic origins of the carbon atoms of the sugar ring and side chain of muramic acid are now quite evident, the mechanism of biosynthesis of muramic acid has not been fully established. Whether or not the main pathway for muramic acid synthesis is at the nucleotide level as suggested by STROMINGER (1958, 1960) is still open to question. To test the possibility of a more direct route of synthesis in accord with the reactions already described in other microorganisms (ROSEMAN, 1959), SALTON (1962) examined the ability of extracts of *Micrococcus lysodeikticus* to utilize the amino sugar phosphate (glucosamine-6-phosphate) for muramic acid biosynthesis. The results suggested a condensation of phosphoenolpyruvate and glucosamine-6-phosphate to give the following reaction:

Glucosamine-6-phosphate + phosphoenolpyruvate → muramic acid-6-phosphate + P_i

The nature of the intermediate in this condensation has not been established, nor have the cofactor requirements. The condensation of phosphoenolpyruvate to form an ether linkage is not unique for this type of reaction has been investigated with extracts of *Escherichia coli* in the conversion of shikimic acid to precursors of the pathway for the biosynthesis of aromatic amino acids (LEVIN AND SPRINSON, 1960).

The compound behaving as muramic acid-6-phosphate in the extracts of *Micrococcus lysodeikticus* was identical to that prepared by incubating yeast hexokinase (Boehringer), ATP, Mg^{2+} and muramic acid. It is perhaps relevant to this problem that muramic acid-6-phosphate had already been isolated from bacterial products by ÅGREN AND DE VERDIER (1958). In addition to the bound form of muramic acid-6-phosphate, a uridine nucleotide, probably UDP-muramic acid-6-phosphate was also detected. Thus the problems of biosynthesis of muramic acid from glucosamine, its acetylation and condensation with uridine triphosphate to form the nucleotide, uridine diphosphate-N-acetyl-glucosamine-lactic acid, will have to be examined more fully in the future. Just which one of the suggested pathways for muramic acid biosynthesis operates *in vivo* has yet to be decided upon and it is likely that isolation of auxotrophic mutants may well give the final answer to this problem.

Sugars

Very little direct work has been done on the biosynthesis of sugars in the bacterial cell wall despite the fact that these structures may have characteristic monosaccharides such as pentoses, deoxyhexoses, dideoxyhexoses and aldoheptoses.

The origin of L-rhamnose in group A streptococcal cell wall was studied by SOUTHARD, HAYASHI AND BARKULIS (1958, 1959). Experiments were performed with [1–^{14}C] glucose and [6–^{14}C] glucose and the L-rhamnose was isolated from the walls of the streptococci grown in media containing the radioactive compounds. The specific activity of the isolated rhamnose was comparable to that of the added glucose indicating that the glucose acted as the principal carbon source for the rhamnose. With the [1–^{14}C] labelled glucose, the rhamnose contained 90% of the recovered radioactivity in C–1. In the [6–^{14}C] glucose experiment 80–90% of the recovered radioactivity was found in C–5 + C–6 of the rhamnose. Thus the rhamnose appears to be derived from the glucose without a scission of the sugar chain. The conversion of glucose to rhamnose via the thymidine diphosphate nucleotides (PAZUR AND SHUEY, 1960) would be compatible with the conservation of label in the experiments with cell-wall rhamnose.

It appears likely that other wall sugars may arise by interconversions of sugar nucleotides. Thus HEATH AND ELBEIN (1962) have achieved the enzymic synthesis of guanosine diphosphate-colitose by direct conversion from guanosine diphosphate mannose in cell-free extracts of a mutant strain of *Escherichia coli*. Guanosine diphosphate-fucose was not an intermediate in the conversion of the GDP-mannose to GDP-colitose.

There seems little doubt that many of the present gaps in our knowledge of the specific origins of cell-wall sugars will be filled in as the numbers of interconversions of monosaccharides at the nucleotide level steadily increase.

2. Isolation, properties and formation of nucleotide intermediates

The importance of nucleotides in polymer biosynthesis had been recognized for some time prior to the suggestion that nucleotide-peptides from bacteria may have a role in cell-wall formation. The possibility that these nucleotide anhydrides may be involved in wall biosynthesis was not recognized until 1957 when PARK AND STROMINGER (1957) proposed a mechanism for the mode of action of penicillin. Following the observations by PARK AND JOHNSON (1949) that uridine nucleotides accumulated in *Staphylococcus aureus* inhibited by penicillin, PARK (1952) isolated and chemically characterized several of the nucleotides, although the structure of the amino sugar component of these compounds remained unknown at that time. The detection of an unidentified amino sugar (later characterized and called muramic acid) in spore peptides (STRANGE AND POWELL, 1954) and in bacterial walls (CUMMINS AND HARRIS, 1956; SALTON, 1956) led to the recognition of the amino sugar moiety of the nucleotides as muramic acid (PARK AND STROMINGER, 1957). The sequence of the amino acids of the peptide portion of the nucleotide was established by STROMINGER (1959) and STROMINGER AND THRENN (1959) and the structure of the most complicated and principal nucleotide accumulating in penicillin-treated *Staphylococcus aureus* is given in Fig. 68.

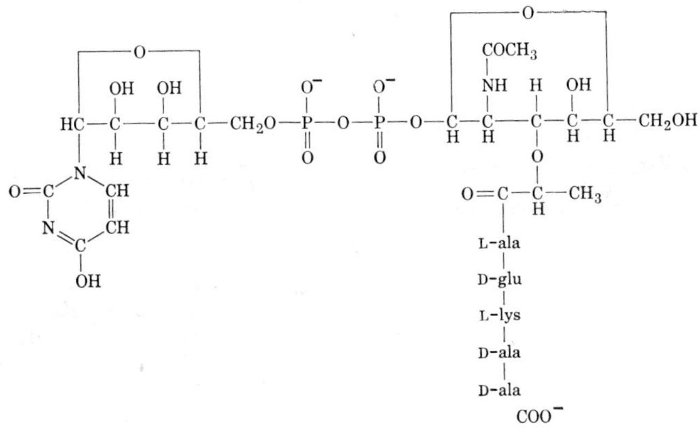

```
                                                          ┌──────────O──────────┐
       ┌─────O─────┐                                      │   COCH3              │
       │  OH   OH  │                O⁻       O⁻           │   |                  │
       │  |    |   |                |        |            │   NH    H    OH      │
      HC──C────C───C──CH2O──P──O──P──O──C──C────C────C──C──CH2OH
       |  |    |   |                ||       ||           H   H     |    H   H
       |  H    H   H                O        O                      O
      N                                                             |
  O=C    CH                                               O=C──C──CH3
    |    ||                                                 |  H
   N     CH                                               L-ala
     C                                                      |
     |                                                    D-glu
    OH                                                      |
                                                          L-lys
                                                            |
                                                          D-ala
                                                            |
                                                          D-ala
                                                              COO⁻
```

FIGURE 68. Structure of the uridine nucleotide isolated from penicillin – inhibited *Staphylococcus aureus*.

The detection of N-acetylmuramic acid, D-alanine and D-glutamic acid in these uridine nucleotides established a close biochemical relationship to the bacterial wall and offered strong circumstantial evidence that these compounds were biosynthetic wall intermediates. Indeed, the analysis of the 'complete' uridine nucleotide from antibiotic (novobiocin and penicillin) inhibited *Staphylococcus aureus* gave percentages of the L- and D-isomers of alanine very similar to those observed with cell-wall preparations (STROMINGER AND THRENN, 1959). The L-alanine of the walls and nucleotides accounted for 27.3–34.0% of the total alanine with corresponding values for the D-isomer varying from 65.4–68.0%. In retrospect it is perhaps surprising that the proportions of the alanine isomers in wall and nucleotide are so similar, considering the presence of the teichoic acid with only D-alanine.

Although the nature of the nucleotides first isolated by PARK (1952) and the wide variety characterized in the past ten years or so, strongly suggest a direct role in wall biosynthesis, the successful transfer to the wall of labelled amino sugar-peptide fragments of the nucleotides, has not been conclusively established. Experimental verification of the 'hypothesis' that these nucleotides are indeed intermediates has proved difficult and there are many technical reasons why attempts to demonstrate incorporation of wall components from nucleotides have not succeeded (STROMINGER, 1960; SALTON, 1960). However, the isolation of nucleotides containing characteristic components of the wall and associated polymers, such as the amino sugar-peptides with lysine or diaminopimelic acid, ribitol, glycerol, rhamnose and colitose, strengthens the probability that the nucleotides are directly involved in the biochemical reactions leading to cell-wall synthesis. For the present it will be assumed that the nucleotides are indeed 'wall precursors' and it seems likely that some evidence on this question will be forthcoming before this book appears in print. Indeed, ITO AND SAITO (1963) have stated in the discussion of a recent paper that radioactivity from a ^{14}C-labelled uridine-nucleotide-peptide was incorporated into mucopeptide of disrupted *Micrococcus lysodeikticus* and *Staphylococcus aureus*. The details of these experiments will be awaited with great interest.

Many of the nucleotides containing cell-wall constituents have been isolated from bacteria exposed to antibiotics such as penicillin, bacitracin, novobiocin, vancomycin, 5-fluorouracil, gentian violet and D-cycloserine. These antibacterial agents have interfered with some stage of wall biosynthesis and have resulted in an accumulation of these compounds in the cell's metabolic pool. Deprivation of amino acids required as specific building blocks for the wall has also led to nucleotide accumulation. Thus interference with wall biosynthesis either with certain antibiotics or by depriving the cell of compounds specifically required for the wall will result in a building up of the internal pool of nucleotides.

The time course of penicillin-induced accumulation of uridine nucleotides in *Staphylococcus aureus* was investigated by STROMINGER (1957) and maximum nucleotide values were found in two hours of exposure to the antibiotic. SAUKKONEN (1961) has also shown a very marked nucleotide accumulation one hour after the addition of penicillin to *Staphylococcus aureus* grown to approximately half maximal growth. Although the nucleotide patterns in normal and penicillin-treated bacteria were generally similar, there was a massive accumulation of both uridine and cytidine nucleotides in the presence of the penicillin. The amount of cytidine diphosphate ribitol was in many cases greater than the quantities of individual uridine compounds.

One observation made by SAUKKONEN (1961) is of considerable importance for future studies on 'wall nucleotides'; it was found that any delay in chilling the cultures prior to extraction in the cold, resulted in a considerable reduction in the amount of nucleotide. This finding may be relevant to the very low nucleotide levels found in untreated, normal bacteria. Such organisms would be synthesizing walls and should contain the necessary supply of nucleotide precursors and it may well be that much of the normal complement of nucleotides had been 'lost' during the handling procedures used in the past. Nevertheless, nucleotides with wall constituents have been isolated from cells which have not been exposed to antibiotics. The variety of nucleotide intermediates of obvious biochemical relationship to the cell wall, that have been isolated from normal, antibiotic-inhibited bacteria and 'wall' auxotrophs, is summarized in Table 63.

Initially only uridine and cytidine nucleotides were found in bacteria as probable wall intermediates. However, the variety of bases has increased as more and more compounds have been isolated and characterized and as shown in Table 63 the same wall constituents (muramyl peptides, sugars and amino sugars) may possess different bases in the nucleotide compounds. The detection of both uridine and adenylmuramyl peptides is of considerable interest but at present it is difficult to say what special biochemical significance should be attached to this discovery. Although the muramyl nucleotides may contain different bases it is almost certain that both types are concerned with wall synthesis as muramic acid seems to be exclusively in the wall (with one possible exception – see capsules – Chapter 1). On the other hand, N-acetylglucosamine and rhamnose nucleotides may not be concerned solely with the biosynthesis of wall compounds and at this stage it is not possible to say which of the various nucleotides are the wall intermediates.

Cytidine diphosphate glycerol was implicated in polymer biosynthesis by its isolation from *Lactobacillus arabinosus*, in which organism a polyglycerophosphate compound was also detected (see Chapter 5). This nucleotide is probably involved in the biosynthesis of the more widely distributed 'intracellular teichoic acids' and may occasionally have a dual function in those organisms having a glycerol teichoic acid in the cell wall. CDP-glycerol has been included under *Lactobacillus arabinosus* in Table 63 for historical completeness, although it is probably not a wall intermediate in this organism; the isolation of this nucleotide from organisms possessing glycerol teichoic acid in the wall (e.g. *Staphylococcus albus*) has not been reported so far.

Very little work has been published on the confirmation of the nucleotide structures by chemical synthesis and the investigations are confined almost exclusively to those of BADDILEY and his colleagues on the cytidine polyol nucleotides. The proposed structures for the bacterial cytidine diphosphate glycerol and cytidine diphosphate ribitol were confirmed by total chemical synthesis in the studies of BADDILEY, BUCHANAN AND SANDERSON (1958) and BADDILEY, BUCHANAN AND FAWCETT (1959) respectively. CDP-glycerol was synthesized from cytidine-5′phosphate and 2,3-O-isopropylideneglycerol-1-phosphate by treating with dicyclohexylcarbodi-imide in aqueous pyridine, removing the isopropylidene residue and separating on Dowex-1 resin (BADDILEY, BUCHANAN AND SANDERSON, 1958). Reaction of cytidine-5′ phosphate and ribose-5-phosphate with dicyclohexylcarbodi-imide gave cytidine diphosphate ribose and on reduction of this nucleotide with sodium borohydride at pH 8.5–9.2 the synthetic cytidine diphosphate ribitol was formed and could be isolated in good yield from the reaction products (BADDILEY, BUCHANAN AND FAWCETT, 1959).

With the more complicated muramyl-peptide nucleotides, the confirmation of the structures will have to await the development of suitable methods for the chemical synthesis of muramic acid-peptides and the conjugation of these compounds to the respective nucleoside-5′phosphate. However, despite the absence of chemical synthetic work on these 'wall' nucleotides, progress has been made at the biochemical level and some of the steps in the formation of the most complicated uridine nucleotides have been dissected (STROMINGER, 1960; 1962).

Enzymic reactions and nucleotide formation

The conversion of sugar-1-phosphates to uridine diphosphate nucleotides by reactions with uridine triphosphate in the presence of pyrophosphorylases has been known for some time (KALACKAR AND CUTOLO, 1952) and they conform to a general mechanism of nucleotide formation originally found by KORNBERG (1948, 1950). The formation of uridine diphosphate-N-acetylglucosamine, a compound which is undoubtedly concerned in the overall biosynthesis of wall, occurs by the following reaction:

$$\text{Uridine triphosphate} + \text{N-acetylglucosamine-1-phosphate} \rightarrow \text{Uridine diphosphate-N-acetylglucosamine} + \text{PP}$$

The pyrophosphorolysis shown in the above reaction was observed by SMITH AND MILLS (1954) and this enzyme from the organism *Staphylococcus aureus*, was partially purified by STROMINGER AND SMITH (1959). A number of reactions of this type have been reported for the formation of nucleotides containing other cell-wall constituents. SHAW (1957) suggested that that extracts of *Lactobacillus arabinosus* catalysed a reversible pyrophosphorolysis of cytidine diphosphate glycerol (CDP-glycerol) and cytidine diphosphate ribitol (CDP-ribitol). In subsequent studies SHAW (1962) purified an enzyme from *Lactobacillus arabinosus* and found a specific enzyme called cytidine diphosphate glycerol pyrophosphorylase responsible for the reaction in the following equation:

$$\text{Cytidine triphosphate} + \text{L-}\alpha\text{-glycerol phosphate} \rightarrow \text{cytidine diphosphate glycerol} + \text{PP}$$

This enzyme was not active towards CDP-ribitol although the crude extracts did possess some activity on the ribitol compound. An active CDP-ribitol pyrophosphorylase carrying out the reaction shown below was found in soluble extracts of *Staphylococcus aureus* H:

$$\text{Cytidine triphosphate} + \text{D-ribitol-5-phosphate} \rightarrow \text{cytidine diphosphate ribitol} + \text{PP}$$

Both pyrophosphorylases were detected in several Gram-positive bacteria and in the alga *Chlorella vulgaris*. In *Escherichia coli* only the CDP-glycerol pyrophosphorylase was found but negative results for both enzymes were given by *Micrococcus lysodeikticus*, *Bacillus cereus* and *Lactobacillus casei* (SHAW, 1962).

Of the more conspicuous cell-wall compounds, reactions involving rhamnose have been studied in extracts of *Streptococcus faecalis* by PAZUR AND SHUEY (1960). Two reactions have been described involving a pyrophosphorolysis to give thymidine diphosphate glucose which is then converted in the presence of reduced triphosphopyridine nucleotide to thymidine diphosphate rhamnose (TDP-rhamnose). It should be emphasized however, that the latter nucleotide has not been isolated from this organism so while it seems likely that the reactions described by PAZUR AND SHUEY (1960) may be relevant to the formation of a wall nucleotide intermediate no firm conclusion can be made at this stage. TDP-rhamnose has been detected in other bacteria and the enzymes converting TDP-glucose to TDP-rhamnose have also been found in several organisms (OKAZAKI, 1960; KORNFELD AND GLASER, 1960; BADDILEY, BLUMSON, DI GIROLAMO AND DI GIROLAMO, 1961). The TDP-rhamnose acts as a glycosyl donor in the biosynthesis of rhamnolipid in *Pseudomonas aeruginosa* (BURGER, GLASER AND BURTON, 1962).

Alternative series of nucleotides growing in the past few years have now been extended to amino sugars. KORNFELD AND GLASER (1962) have detected the enzymic synthesis of thymidine diphosphate-N-acetylglucosamine in extracts of *Pseudomonas aeruginosa*. The thymidine nucleotides in addition to the well known uridine series must now be considered as possible precursors of the amino sugars in bacterial products.

The first steps in the biosynthesis of the acetylmuramic acid nucleotide prior to the addition of the amino acid residues have not been conclusively established. The presence of a uridine diphosphate acetylglucosamine pyrophosphorylase in *Staphylococcus aureus* (STROMINGER AND SMITH, 1959) could provide the step prior to the formation of a pyruvate-enol ether of UDP-acetylglucosamine by a transfer reaction with phosphoenolpyruvate. STROMINGER (1958) suggested that a 3-O-substitution had occurred and enzymes from several bacteria were apparently capable of carrying out this reaction. However, as STROMINGER (1962) has pointed out the products of the reaction:

$$\text{UDP-acetylglucosamine-pyruvate} \rightarrow \text{UDP-acetylmuramic acid}$$

have not been definitely identified and this reaction should not be regarded as having been established.

Thus the pathway for UDP-acetylmuramic acid formation is still uncertain. SALTON (1962) has suggested that the transfer reaction involving the addition of the side chain from phosphoenolpyruvate may occur with the glucosamine-6-phosphate rather than at the nucleotide level as suggested by STROMINGER (1958). No pyrophosphorylase for the formation of UDP-acetylmuramic acid (UDP-GNAc-lactic) from N-acetylmuramic acid -1-phosphate and UTP has so far been reported and this could be a key reaction preceding the subsequent additions of amino acids that have now been established by STROMINGER and his colleagues.

Whatever the route of biosynthesis of uridine diphosphate acetylmuramic acid may be, the sequential addition to this nucleotide, of L-alanine (ala), D-glutamic acid (glu), L-lysine (lys) and D-alanyl-D-alanine has been demonstrated by ITO AND STROMINGER (1960), STROMINGER (1962) in the following series of reactions:

$$\text{UDP-GNAc-lactic} + \text{L-ala} \xrightarrow[\text{Mn}^{2+}]{\text{ATP}} \text{UDP-GNAc-lactyl-L-ala}$$

$$\text{UDP-GNAc-lactyl.L-ala} + \text{D-glu} \xrightarrow[\text{Mn}^{2+}]{\text{ATP}} \text{UDP-GNAc-lactyl.L-ala.D-glu}$$

$$\text{UDP-GNAc-lactyl.L-ala.D-glu} + \text{L-lys} \xrightarrow[\text{Mn}^{2+}]{\text{ATP}} \text{UDP-GNAc-lactyl.L-ala.D-glu.L-lys}$$

$$\text{UDP-GNAc-lactyl.L-ala.D-glu.L-lys} + \text{D-ala-D-ala} \xrightarrow[\text{Mn}^{2+}]{\text{ATP}} \text{UDP-GNAc-lactyl.L-ala.D-glu.L-lys.D-ala.D-ala}$$

The specificity of the first three enzymic reactions responsible for the building up of the nucleotide muramyl peptide was investigated by following the incorporation of ^{14}C-labelled amino acids into the nucleotides and the data recorded by STROMINGER (1962) are shown in Table 64. Another specific enzyme was concerned with the addition of the dipeptide, D-alanyl-D-alanine, to the nucleotide and the reaction was highly specific for the D-stereoisomer of the dipeptide (STROMINGER, 1962) a fact which can be readily seen from the incorporation of ^{14}C-labelled compounds into the nucleotide (Table 65).

The distribution of the enzymes catalysing the addition of amino acids to the nucleotides has been investigated by STROMINGER, ITO AND THRENN (1961) and the sharp dichotomy existing for the additon of L-lysine and α, ε-diaminopimelic acid is of considerable biochemical interest. Those organisms containing lysine in the wall possessed a lysine-adding enzyme which was inactive with DAP as a substrate. Other bacteria with DAP in the cell walls contained a DAP-adding enzyme which was inactive towards L-lysine. Thus the specificity of peptide sequence is determined by a single enzyme which can add lysine or diaminopimelic acid.

The enzymic addition of the D-alanyl-D-alanine peptide to UDP-GNAc-lactyl-ala-glu-DAP has been investigated with extracts from *Escherichia coli* by COMB (1962). The enzyme has been purified 400-fold and it catalyses the following reaction:

$$\text{UDP-GNAc-lactyl-ala-glu-DAP} + \text{D-ala-D-ala} + \text{ATP} \rightarrow$$
$$\text{UDP-GNAc-lactyl-ala-glu-DAP.D-ala.D-ala} + \text{ADP} + \text{P}_i$$

It is not known whether the α or ε-carboxyl group of DAP is involved in the peptide bond formation. COMB (1962) has postulated a mechanism for the reaction involving the formation of an 'activated enzyme' before peptide bond formation. STROMINGER (1962) also discussed the mechanisms of synthesis of the peptides and in three of the cases studied, adenosine diphosphate and inorganic phosphate were identified as products of the reaction. In attempting to assess whether or not ribonucleic acid may be involved in the additions of amino acids, STROMINGER (1962) has mentioned that preincubation of *Staphylococcus aureus* enzymes with ribonuclease had no effect on their activity.

The inhibitory action of D-cycloserine on cell-wall formation is now clearly understood, for this antibiotic inhibits the alanine racemase and the D-alanyl-D-alanine synthetase reactions (STROMINGER, ITO AND THRENN, 1960; NEUHAUS, 1962) but has no action on the enzymic addition of the dipeptide to the nucleotide (STROMINGER, 1962).

Although the staphylococcal glycosaminopeptide contains a glycine peptide, no reactions leading to the addition of this amino acid to the most complex nucleotide (UDP-GNAc-lactyl-ala-glu-lys-ala-ala) have been reported, nor is there any suggestion as to the mechanism of polyglycine formation.

3. *Cell-wall biosynthesis and its inhibition by antibiotics and antibacterial agents*

At present there are few subcellular or isolated enzyme systems available for studying wall biosynthesis at the stages beyond the formation of the uridine nucleotides discussed in the preceding section. However, quite a lot of work has been done on the formation of cell-wall material by intact bacteria and many of these studies have involved the use of antibiotics and an investigation of the incorporation of radioactive compounds into the wall. It is evident that where specific radioactive labelling is to be used with intact cells, it is necessary to differentiate between incorporation into wall and intracellular components.

Investigations of the biosynthesis of wall at the intact cell level have been greatly facilitated by the relative ease of isolation of the wall as a stable, insoluble component and analysis for characteristic substances not in the intracellular structures and by the ability of chloramphenicol to dissociate wall and protein synthesis (MANDELSTAM AND ROGER, 1958). Moreover, in some organism the 'nutritional requirements' for wall production may be sufficiently different to those needed for total growth, a property used by SHOCKMAN and his colleagues in devising selective conditions for wall synthesis in *Streptococcus faecalis*. Further information on wall biosynthesis with intact bacteria has been gained also from studies with auxotrophic mutants dependent upon a specific cell-wall 'building block' for growth.

Synthesis of glycosaminopeptides by intact cells

The synthesis of cell-wall material by washed suspensions of two different strains of *Staphylococcus aureus* in defined incubation mixtures was reported independently by MANDELSTAM AND ROGERS (1958) and HANCOCK AND PARK (1958). Both groups of investigators found that cell-wall synthesis continued at a substantial rate in the presence of chloramphenicol, whereas cellular protein synthesis was inhibited to the extent of 95%. These studies also clarified the nature of the chloramphenicol-resistant incorporation of amino acids into the trichloroacetic acid insoluble fractions of disrupted *Staphylococcus aureus* (GALE AND FOLKES, 1955). This uptake of amino acids was largely into cell-wall material.

MANDELSTAM AND ROGERS (1959) reported that the increase in isolated wall 'mucopeptide' fraction (probably also contained the teichoic acid as no special steps were taken to remove this component) varied from 20–150% in one hour at 37° when *Staphylococcus aureus* was incubated in buffers containing glucose and combinations of the 'wall' amino acids. The results obtained with washed suspensions of *Staphylococcus aureus* are presented in Table 66. Although the nucleic acid content increased as much as 30% MANDELSTAM AND ROGERS (1959) concluded that mucopeptide synthesis was mainly responsible for the increased opacity of the bacterial suspensions. Under these conditions amino acid incorporation indicated that negligible protein synthesis had occurred and in these incubation mixtures chloramphenicol had little effects on the labelling of the staphylococcal protein fraction.

HANCOCK AND PARK (1958) used a trichloroacetic acid extraction procedure for the separation of the 'mucopeptide' fraction from *Staphylococcus aureus* and in their experiments they were able to show a doubling of the amount of wall in the presence of chloramphenicol. The incorporation of ^{14}C-labelled amino acids such as alanine, glutamic acid, lysine and glycine was inhibited only to the extent of 4–8% by chloramphenicol when the cells were transferred to a synthetic growth medium containing the radioactive amino acids. On the other hand, the inhibition of the incorporation into protein of the non-wall amino acids, leucine, proline and phenylalanine was as much as 85–98%. The method for isolating the teichoic-acid-free 'mucopeptide' from *Staphylococcus aureus* and the details of the incorporation studies were described more fully in a later publication (PARK AND HANCOCK, 1960).

With the washed suspension system, MANDELSTAM AND ROGERS (1959) were able to show that penicillin did indeed inhibit the biosynthesis of staphylococcal cell-wall material. However, cessation of glycosaminopeptide synthesis was not immediate and it continued for 45 minutes at a reduced rate before stopping. Rather similar results were obtained with bacitracin.

M. P. HATTON (unpublished observations) investigated the conditions for wall synthesis in *Micrococcus lysodeikticus* and found that in this organism the increase in optical density on incubation in a mixture of wall amino acids and glucose was by no means entirely due to wall synthesis, even in the presence of chloramphenicol. The increases in optical densities and the amounts of wall (by direct isolation) occurred much more slowly in this organism than in *Staphylococcus aureus*. These results and those of ALLISON *et al.* (1962) cast some doubt on the suggestion that the effects of chloramphenicol can be so clear-cut. Indeed, with *Escherichia coli* B/r, ALLISON *et al.* (1962) found that there was an increase in biopolymers and cell numbers in chloramphenicol-exposed cultures thus giving rise to inhomogeneous bacterial populations.

Following the studies of TOENNIES AND SHOCKMAN (1953, 1959) it became apparent that the amount of wall synthesized by *Streptococcus faecalis* depended upon the nutritional status of the growth medium. Under conditions of limitation of certain amino acids, post-exponential 'growth' occurred as indicated by increases in optical densities. Such post-exponential growth was traced to the bacterial cell wall and on limitation of the threonine supply, the net gain in wall substance was as high as 210%. The additional cell-wall substance accounted for about two-thirds of the increase in dry weight and nitrogen (SHOCKMAN, 1959). The depletion of threonine gave about twice as much wall in the post-exponential phase as that found for valine limitation. These investigations indicated the 'nutritional' independence of cell-wall synthesis from cytoplasmic protein synthesis. At least amino acids not synthesized by the organism could be dispensed with for wall synthesis.

SHOCKMAN *et al.* (1961) have established the minimal nutritional requirements for cell-wall synthesis in *Streptococcus faecalis*. The 'wall' medium contained inorganic salts, glucose, acetate, ammonium salts, glutamate, aspartate, lysine, alanine and cystine. 'Log phase' cells transferred to this simple medium gave increases in the turbidity of up to 40%. The increased optical density in this medium was accompanied by a gain of nitrogen, in the cell-wall fraction as well as an increase in the total cellular amount of rhamnose. The addition of chloramphenicol failed to reduce the observed turbidity increment.

It is evident that with organisms having complex nutritional requirements this property can be used in dissociating total growth from cell-wall synthesis. Whether the additional wall formed on transfer of the cells to a deficient, wall-synthesizing medium differs markedly from that already present, cannot be said. The increase in the amount of cell-wall nitrogen rather suggests a higher peptide content. This could conceivably be achieved by adding muramyl-peptide residues as branches or cross-linked structures to a backbone of amino sugars. Such post-exponential changes in the wall may accuount for the increased thickness and structural rigidity and may possibly be related to the greater resistance to lytic enzymes. There is obviously a great need for more chemical investigations of the changes occurring under the conditions of these very interesting observations by Shockman and his colleagues.

Action of inhibitors on cell-wall synthesis

Penicillin and related antibiotics

The events leading to the hypothesis that penicillin interferes with the biosynthesis of the cell wall are now so well known that they hardly need reviewing here. The key observations suggesting such a theory to account for penicillin action upon bacteria, were the structural changes in the wall believed to accompany the morphological abnormalities of bacteria grown in the presence of the antibiotic (DUGUID, 1946; LEDERBERG, 1956) and the recognition of the biochemical significance of the nucleotides accumulating in penicillin-treated *Staphylococcus aureus* (PARK AND STROMINGER, 1957). The inhibition by penicillin of a step or series of steps in the biosynthesis of the cell-wall glycosaminopeptide could account for the principal structural and biochemical changes observed in cells.

Although penicillin has now been shown to inhibit glycosaminopeptide formation in 'resting', washed cells (MANDELSTAM AND ROGERS, 1959), the structural consequences of the action of the penicillins are dramatically seen in bacteria grown in the presence of these antibiotics. The anatomical results of exposing *Staphylococcus aureus* to penicillin for three hours have been established by examining thin sections of normal and treated cells and the electron micrographs taken by MURRAY, FRANCOMBE AND MAYALL (1959) are presented in Fig. 69. While it it is not possible to determine the precise site of penicillin uptake other than in the membrane fraction (COOPER, 1955) the grossly distorted appearance of newly formed cross-wall in the penicillin-treated cells leaves little doubt as to the general location of the 'lesion'.

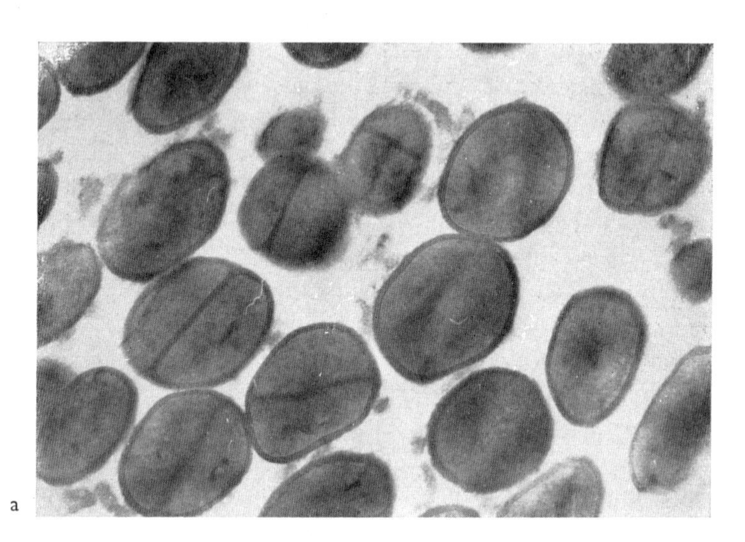

a

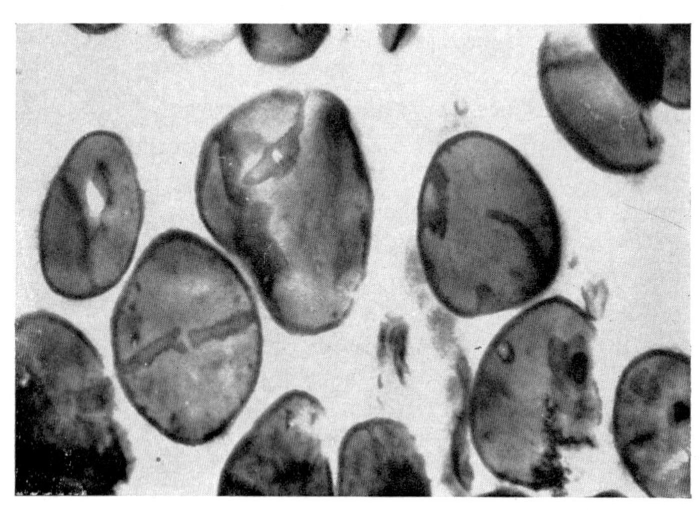

b

FIGURE 69. Effects of penicillin on the cell-wall structure of *Staphylococcus aureus*. (a) Thin sections of untreated cells, ×28,000. (b) sections of cells exposed to penicillin for three hours, ×31,500. From the study of MURRAY, FRANCOMBE AND MAYALL, 1958.

It seems likely that the structural changes occurring in penicillin-treated *Staphylococcus aureus* would be generally typical for penicillin-sensitive Gram-positive organisms. The cessation of wall formation in the presence of penicillin thus puts the Gram-positive cell in a 'straight-jacket' and death of the cell may ultimately occur from a mechanical breach in the cell wall. The loss of mechanical function of the wall may result from increased growth of intracellular components to the point where the wall can no longer contain the new cellular material, for protein and nucleic acid synthesis usually continue in the presence of the antibiotic (GALE, 1959). The final 'bursting at the seams' may be assisted by the cell's own wall-degrading enzymes. The level of activity of such enzymes may govern the time taken for lysis of the Gram-positive bacteria following the primary action of penicillin.

In Gram-positive bacteria, the morphological integrity of the cell and viability under normal cultural conditions (in distinction to those required for growth of the L-phase) are dependent upon the mechanical and biosynthetic continuity of the rigid structure of the wall. With the Gram-negative bacteria the presence of additional components in the wall or envelope makes them less vulnerable and in a sense they have a second line of defence. The effects of penicillin action on the surface structures of this group of bacteria have already been discussed in Chapter 1. Because of the more complex nature of the wall in the Gram-negative organisms and the possession of the protein-lipid-polysaccharide layer, they can tolerate the loss or biochemical block involving the rigid layer. Although there is some increase in osmotic fragility in the spheroplast formed by growth in the presence of penicillin, as a cellular system they can continue their activities for longer periods than the Gram-positive bacteria so treated and indeed, they possess a greater ability to survive and revert to the original cell type (MCQUILLEN, 1960).

At the biochemical level the action of penicillin in interfering with wall synthesis has been amply demonstrated and is in accord with the interpretations deduced from the more biological studies of the anatomical changes induced by the antibiotic. The biochemical evidence supporting the proposed mode of action of penicillin has come from three principal sources: (a) the nature of the nucleotide accumulation and the general continuation of other biosynthetic processes in the cell, (b) direct inhibition of glycosaminopeptide synthesis in washed cell suspensions placed in a 'wall-synthesizing' medium, and (c) the inhibition of incorporation of radioactive components known to be specific building blocks of the wall. The properties of the nucleotides accumulating in the presence of penicillin and other antibiotics have already been discussed and the other cellular syntheses occurring in bacteria exposed to penicillin have been admirably dealt with by GALE (1959).

One of the major criticisms of the hypothesis that penicillin's bactericidal activity is due to inhibition of wall biosynthesis, has been that concentrations used in obtaining the three types of evidence listed above have exceeded the minimal germicidal doses. These criticisms have been met in part, by careful quantitative work on the inhibition of wall biosynthesis in studies by ROGERS AND JELJASZEWICZ (1961). They have shown a close relationship between the relative bactericidal doses for a cell and the extent of inhibition of mucopeptide synthesis.

Several other investigators have disagreed with the proposed mode of action of penicillin. TRUCCO AND PARDEE (1958) based their conclusions on the lack of inhibitory effect of penicillin upon the incorporation of [^{14}C] glucose into the whole cell-wall structure of *Escherichia coli*. Such a result could have been anticipated, as the continued production of the protein, lipid and polysaccharide constituents with ^{14}C derived from glucose would mask the effects of penicillin on the rigid layer compound and no conclusion could be made about the latter, unless they were specifically examined. HUGO AND RUSSELL (1961) believed that the lethal action of penicillin on *Aerobacter cloacae* was not necessarily a consequence of osmotic fragility. The most marked discrepancy between viability and osmotic fragility was at high concentrations of penicillin, but at concentrations approaching the more usual antibiotic levels the differences began to disappear. The similarity of the biochemical lesion brought about by penicillin action in both Gram-positive and Gram-negative bacteria now seems to be established beyond doubt (ROGERS AND MANDELSTAM, 1962; STROMINGER, 1962).

There is abundant evidence that penicillin inhibits the formation of glycosaminopeptides and/or the incorporation of labelled compounds into this component of the cell walls of *Staphylococcus aureus* and *Escherichia coli* (HANCOCK AND PARK, 1958; MANDELSTAM AND ROGERS, 1959; NATHENSON AND STROMINGER, 1959; 1961). WYLIE AND JOHNSON (1962) also concluded that penicillin specifically inhibited the synthesis of the 'basal structure' of the wall of *Escherichia coli*. Incorporation of [^{14}C] 'invert sugar' into cell constituents and walls of *Bacillus subtilis* was investigated and penicillin inhibited uptake into whole cells to the extent of 20% and into walls by 88% (ROBERTS AND JOHNSON, 1962). Some idea of the marked difference in the inhibitory effects of penicillin on the incorporation of radioactive substances into wall and cellular protein and nucleic acid is given by the data presented in Table 67 from the studies of NATHENSON AND STROMINGER (1959).

ROGERS AND JELJASZEWICZ (1961) carried out an extensive investigation of the mode of action of benzylpenicillin, 6-(α-phenoxy-propionamido) penicillanic acid (Broxil), 6-(2,6-dimethoxybenzamido) penicillin (Methicillin, Celbenin) and 6-(D-α-amino-phenylacetamido) penicillanic acid (Ampicillin) on *Staphylococcus aureus* strains. The effects of these penicillins on incorporation of [^{14}C] glutamic acid into material insoluble in hot trichloroacetic acid were studied. Although inhibition of glycosaminopeptide synthesis never appeared to more than 70–80% complete, it was found that penicillin had little effect on the synthesizing system during the first 10–15 minutes; thereafter 0.6 μg benzylpenicillin/ml completely inhibited the system. ROGERS AND JELJASZEWICZ (1961) concluded that 'the concentration of benzylpenicillin required to inhibit cell-wall-mucopeptide formation by *Staphylococcus aureus* strain Oxford is of the same order as that required to prevent growth.'

The relative activities of the various penicillins, judged by lethal dosages and inhibitory action upon mucopeptide synthesis, were quite similar and although the data quoted were inadequate for a detailed comparison there was roughly a 10-fold difference between the antibiotic concentration inhibiting wall formation and that of the lethal dosage for each of the penicillins studied. These values conflict with the conclusion that the concentrations of benzylpenicillin required to prevent growth and wall synthesis were of the same order. However, the lethal doses quoted by ROGERS AND JELJASZEWICZ (1961) for the comparison with wall biosynthesis were taken from other investigations and their own growth inhibitory concentration of benzylpenicillin was at least five times higher than the independent values. While there is little doubt about the validity of the broad correspondence between lethal action and inhibition of glycosaminopeptide synthesis, the strict quantitative comparisons are not entirely satisfactory. Part of the problem of comparing the two manifestations of antibiotic activity arise from the differences between the two systems – e.g. cells dividing in a medium adequate for full growth compared to non-dividing cells in a simplified wall-synthesizing medium.

These studies on the mode of action of the penicillins were extended to *Escherichia coli* by ROGERS AND MANDELSTAM (1962) and from the incorporation of [^{14}C] glucose into bound α, ε-diaminopimelic acid it was evident that both benzylpenicillin and Ampicillin (6[D-α-aminophenylacetamido] penicillanic acid) are inhibitors of cell-wall glycosaminopeptide synthesis. The concentration of Ampicillin necessary for 50% inhibition of formation of wall material was one tenth the amount of benzylpenicillin required for this inhibitory level.

Thus investigations performed by several independent groups of investigators (HANCOCK AND PARK, 1958; MANDELSTAM AND ROGERS, 1959; NATHENSON AND STROMINGER, 1959; WYLIE AND JOHNSON, 1962; ROGERS AND MANDELSTAM, 1962) have firmly established the inhibition of glycosaminopeptide synthesis in certain Gram-positive and Gram-negative bacteria. The nature of the primary step or steps in the lethal action of penicillin still remains unknown and the quantitative aspects of the effects of the antibiotics on growth and wall synthesis have not been fully studied. One other interesting problem which seems to be completely unanswered is the difference in sensitivity between the Gram-positive and Gram-negative bacteria (excluding questions of penicillinase production and its substrate specificity, etc.). It is indeed surprising that the Gram-positive bacteria with much greater amounts of glucosaminopeptide in their walls are more vulnerable than Gram-negative organisms with as little as 5–20% mucopeptide component in the walls. These differences almost suggest that in the Gram-negative bacteria there is a greater multiplicity of synthesizing sites than in the Gram-positive organisms. A localized distribution of wall synthesizing or polymerizing centres in Gram-positive bacteria in contrast to a general distribution of sites over the whole area of the plasma membrane or protein-lipid-polysaccharide component could account for such a difference in sensitivity. Unfortunately there is little experimental evidence available to enable us to answer this fascinating problem. However, some clues may be forthcoming from attempts now being made to determine the sites of new wall formation.

The development of some of the new penicillins has helped to clarify the process of glycosaminopeptide biosynthesis in *Staphylococcus aureus* resistant to benzylpenicillin. With a penicillinase-producing strain of *Staphylococcus aureus*, ROGERS AND JELJASZEWICZ (1961) found that 80–90% inhibition of mucopeptide synthesis occurred at about the same concentrations of Methicillin required for the effect upon the sensitive Oxford strain. As may have been suspected from the similarity of cell-wall composition for both benzylpenicillin-resistant and sensitive strains of *Staphylococcus aureus*, the mechanism of biosynthesis of the walls must be similar and both are intrinsically sensitive to inhibition with penicillin. PARK AND GRIFFITH (1962) investigated the accumulation of nucleotides in parent-type and penicillin-resistant mutants of *Staphylococcus aureus* and they also concluded that the mutant organisms 'must make the same cell wall as the parent by the same pathway.'

The biosynthesis of the glycosaminopeptide component of the wall of *Escherichia coli* is much more sensitive to Ampicillin (6[D-α-aminophenylacetamido] penicillanic acid) than it is to benzylpenicillin, requiring 4 μg/ml compared to 40 μg/ml of the latter for a 50% inhibitory level (ROGERS AND MANDELSTAM, 1962). Before such effects as these and the general differences in levels of sensitivity between non-penicillinase producing Gram-positive and Gram-negative bacteria can be fully understood, much more will have to be known about the specific steps inhibited by penicillin and the nature and distribution of the wall synthesizing sites in the membrane.

COLLINS AND RICHMOND (1962) have pointed out the structural similarity between N-acetylmuramic acid and penicillin and have considered these facts in relation to the mode of action of the antibiotic at the molecular level. Solid models of benzylpenicillin and N-acetylmuramic acid (3-O-D-carboxyethyl-D-glucosamine or 3-O-D-lactyl ether of D-glucosamine) are illustrated in Fig. 70 and show the similar distribution of reactive groups. Thus

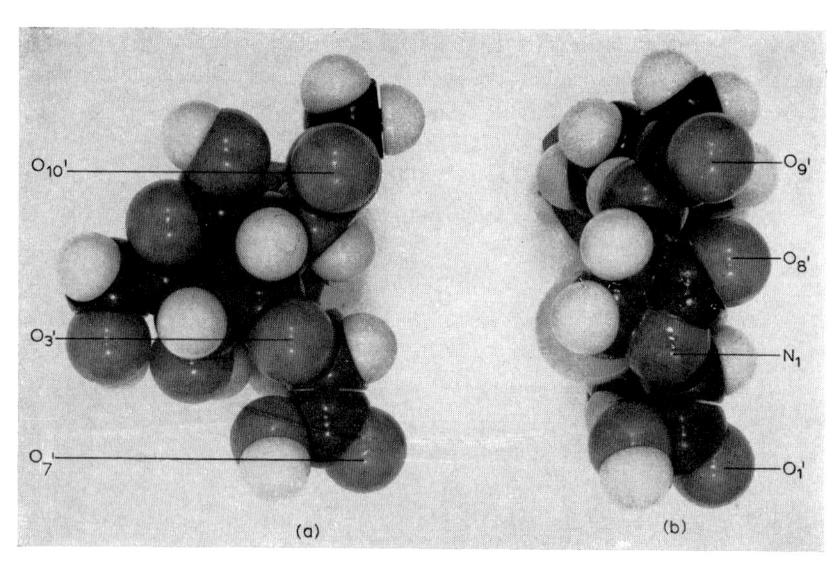

FIGURE 70. Solid models of the molecules of (a) N-acetylmuramic acid (3-O-carboxylethyl-D-glucosamine). and (b) benzyl penicillin, showing the similar distribution of reactive group. N-acetylmuramic acid: O_3', ether oxygen of the lactyl side-chain; O_7', carboxyl group; O_{10}', carbonyl oxygen of N-acetyl group. Penicillin: triply substituted nitrogen of β-lactam ring; O_1', carboxyl group; O_9', carbonyl of benzoyl side-chain; O_8', β-lactam carbonyl group (COLLINS AND RICHMOND, 1962).

identical positions could be occupied by O_1', N_1 and O_9' of penicillin and O_7',O_3' and O_{10}' of N-acetylmuramic acid. COLLINS AND RICHMOND (1962) suggested that penicillin could bind with part of an active centre normally binding a N-acetylmuramic acid residue.

This attractive hypothesis advanced by COLLINS AND RICHMOND (1962) was supported by the following structural considerations:

(1) all but one of the hydrogen bonding groups of penicillin are exposed on one face of the molecule which also resembles part of the face of N-acetylmuramic acid (see Fig. 70).
(2) the carboxyl groups of both have similar pK values and occupy similar positions in the configurations shown in the models.
(3) nitrogen atom N_1 of penicillin and O_3', in N-acetylmuramic acid occupy identical positions in the two models and the nitrogen atom can form a hydrogen bond in a similar direction to one of the two that can be formed by the oxygen atom.
(4) although the carbonyl groups C_9O_9', in penicillin and $C_{10}O_{10}'$, in N-acetylmuramic acid arise in slightly different directions, the two oxygen atoms occupy identical positions in the two molecules and can form hydrogen bonds in a common direction.
(5) there are no large protruding groups which would interfere with the binding of the penicillin molecule to the active centre.

As pointed out by COLLINS AND RICHMOND (1962) alterations in the structure of penicillin that do not affect the positions of the groups discussed above do not necessarily result in loss of antibiotic activity. On the other hand, loss of the reactive carbonyl group and changes in the positions of the binding groups lead to a total loss of activity.

It seems likely that the cephalosporins have a similar mode of action to penicillin and it is of great interest that they also possess the same binding groups in almost identical conformations (ABRAHAM AND NEWTON, 1958; HODGKIN AND MASLEN, 1961).

While the similarity of the conformation of 'binding groups' in penicillins and cephalosporins can be correlated with the structure of N-acetylmuramic acid, the similar biochemical and biological effects of the cyclic polypeptide antibiotic bacitracin cannot, at present, be readily fitted into this hypothesis. However, it seems quite feasible that binding sites immediately adjacent to the active centre for penicillin action could ultimately lead to the same overall effects. The accumulation of uridine nucleotides in the presence of penicillin, cephalosporin and bacitracin (ABRAHAM, 1957) suggests that the sites of action of these antibiotics are in close proximity to one another, though not necessarily identical. Whether the possession of an ε (β-aspartyl)-L-lysine residue in the bacitracin molecule has any real structural similarity to a corresponding part of the wall, uridine nucleotide or membrane component remains to be shown conclusively.

Thus the primary reaction involved in the lethal action of penicillin on microorganisms still eludes us, although many of the structural and biochemical consequences are quite well understood. The enzymic reactions resulting in the addition of the amino acids and the peptide, D-alanyl-D-alanine, to the nucleotides of muramic acid are not inhibited by penicillin, so it must be assumed that the block is in the transfer reactions involving existing wall as acceptor and probably a variety of nucleotide precursors. The quantitative aspects of nucleotide accumulation in *Staphylococcus aureus* recently studied by ITO AND SAITO (1963) are in agreement with penicillin action at this level.

Other antibiotics and antibacterial agents inhibiting wall synthesis

Of all the antibiotics having a primary or secondary effect upon cell-wall formation, the actions of penicillins and D-cycloserine (oxamycin) have been studied in the greatest detail. The antibiotic, D-cycloserine, has been shown to exert its action by inhibiting the D-alanyl-D-alanine synthetase and the alanine racemase.

The mode of action of vancomycin has been investigated by JORDAN (1961), JORDAN AND INNISS (1961) AND REYNOLDS (1961) and this antibiotic now joins the penicillins in its ability to inhibit the biosynthesis of the glycosaminopeptide of *Staphylococcus aureus*. Within 2–5 minutes after the addition of the vancomycin the production of glycosaminopeptide (judged by incorporation of [^{14}C] glycine or [^{14}C] glutamic acid into wall) was inhibited, whilst interference with RNA synthesis was not apparent until 20 minutes later. Protein synthesis was unimpaired. REYNOLDS (1961) also observed accumulation of nucleotides in the presence of vancomycin.

With the knowledge that uridine nucleotide precursors were probably involved as intermediates in wall synthesis, it seemed worthwhile investigating the action of the pyrimidine analogue, 5-fluorouracil upon bacteria. 5-Fluorouracil was originally developed as a 'carcinostatic' agent (DUSCHINSKY, PLEVEN AND HEIDELBERGER, 1957) but it also possessed germicidal activity for various microorganisms (HEIDELBERGER *et al.*, 1957). It was suggested that the loss of viability of *Escherichia coli* exposed to 5-fluorouracil was due to the cessation of DNA synthesis and subsequent 'thymine-less death' (COHEN *et al.*, 1958).

TOMASZ AND BOREK (1959) found that the lethal action of this pyrimidine analogue on *Escherichia coli* K–12 was accompanied by the induction of osmotic sensitivity and could not be abolished by thymine. In addition to the reduced osmotic stability of *Escherichia coli* exposed to 5-fluorouracil, TOMASZ AND BOREK (1960) also reported an inhibition of the incorporation of diaminopimelic acid into the cell walls and the accumulation of acid-soluble N-acetylhexosamine esters. It was concluded from these studies that 5-fluorouracil induced some defect in the cell wall of *Escherichia coli*. OTSUJI AND TAKAGAKI (1959) had similarly proposed that the accumulation of N-acetylhexosamine compounds in the presence of 6-azauracil was due to a selective inhibition of wall synthesis by this drug. SELLS (1959) found that 6-azauracil inhibited formation of wall in *Bacillus cereus*.

TOMASZ AND BOREK (1962) confirmed the action of 5-fluorouracil as an inhibitor of glycosaminopeptide formation in *Escherichia coli*. This analogue inhibited the incorporation of [^{3}H]-diaminopimelic acid into the wall to the extent of 64%. Of the 5-substituted uracil analogues studied, 5-fluorouracil induced the greatest accumulation of N-acetylhexosamine compounds. The presence of 'spheroplast-like' forms accompanying the biochemical effects of the drug are also symptomatic of an action on the biosynthesis of the rigid component of the wall.

ROGERS AND PERKINS (1960) investigated the action of 5-fluorouracil on the synthesis of glycosaminopeptide in *Staphylococcus aureus*, by determining the effect of the drug on incorporation of [^{14}C]-wall amino acids into the hot trichloroacetic acid insoluble fraction. The inhibition of incorporation varied somewhat from one amino acid to another and at a 5-fluorouracil level of 0.5 μ mole/ml about 20–35% inhibition occurred. This inhibitory effect was prevented by the simultaneous addition of uracil or uridine. As with penicillin, compounds containing bound amino sugars accumulated in the presence of the pyrimidine analogue and these substances were not readily utilized by the cell when the fluorouracil was removed. Fractionation of the trichloroacetic acid soluble material from the treated cells gave several compounds:

(a) containing fluorouracil, phosphorus, muramic acid, alanine, glutamic acid and lysine in the molecular proportions of 1:2:1:3:1:1;

(b) a fluorouridine-phosphorus-N-acetylglucosamine compound with molar ratios of 1:2:1; and

(c) a mixture of fluorouridine-muramic acid-alanine compounds.

The accumulation of compounds identical to those found in penicillin-treated cells (apart from the presence of the fluorouracil) was of considerable biochemical significance. ROGERS AND PERKINS (1960) also separated a compound containing muramic acid, alanine, glutamic acid and glycine and an unidentified base. This is the only reported instance of a nucleotide from *Staphylococcus aureus* containing glycine in addition to the usual compounds found in certain of the uridine nucleotides. Its detection is of considerable interest, for if it appears consistently and proves to be a regular nucleotide intermediate it could throw some light on the mechanism of formation and addition of the glycine residues to the staphylococcal wall. Despite the effects of 5-fluorouracil on wall synthesis, ROGERS AND PERKINS (1960) suggest that it is more likely that growth inhibition is due to the action on DNA synthesis.

Pathways for cell-wall biosynthesis

The conclusion that the major pathways for the biosynthesis of cell-wall glycosaminopeptides, teichoic acids and polysaccharides occur via transglycosylation reactions involving nucleotide precursors seems inescapable. Direct proof of this generally accepted hypothesis has recently been reported and it would now appear likely that the various nucleotides with 'wall constituents' found in bacteria are indeed intermediates in wall biosynthesis. If the widely adopted views of polymer biosynthesis apply to wall formation then it may occur, at least in part, according to the following general reaction:

$$\text{Acceptor (wall or polymer)} + \text{nucleotide-wall component} \rightarrow \text{Acceptor-wall component} + \text{nucleotide diphosphate}$$

Direct evidence that nucleotide-bound muramyl-peptide is a natural precursor of wall mucopeptide has now been obtained in a particulate cell-free system from disrupted *Staphylococcus aureus* by CHATTERJEE AND PARK (1964) and under special conditions related to growth the synthesis of the mucopeptide in this system is inhibited by penicillin. These studies have now provided preparations which will lead to a rapid advance in the problem of investigating wall biosynthesis and similar particulate enzyme systems possessing glycopeptide glycosyl-transferase activity have also been found by MEADOW, ANDERSON AND STROMINGER (personal communication).

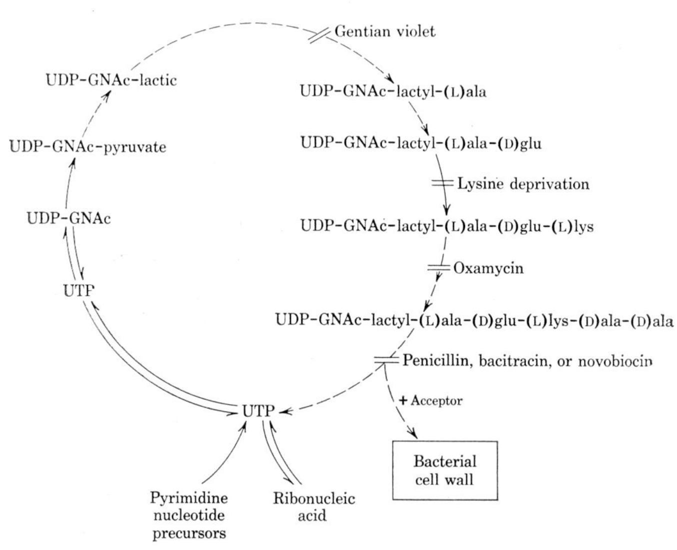

FIGURE 71. A scheme for the pathway of biosynthesis of part of the cell wall of *Staphylococcus aureus* suggested from studies by STROMINGER AND THRENN, 1959.

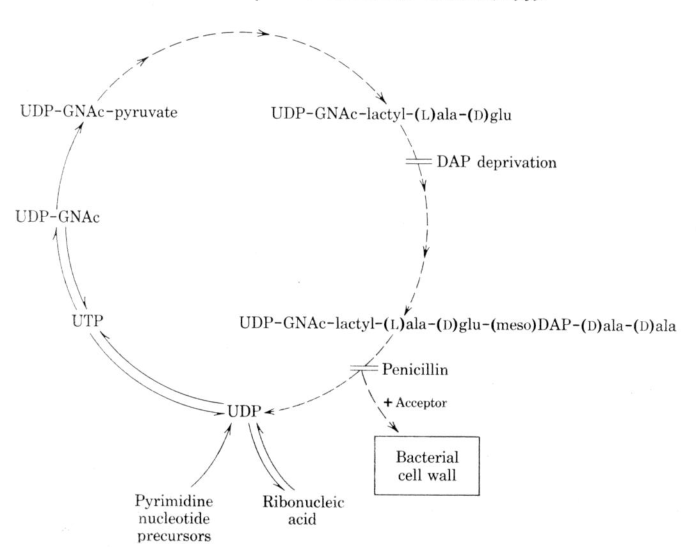

FIGURE 72. A scheme for the pathway of biosynthesis of part of the glycosaminopeptide layer of *Escherichia coli* cell wall as suggested by STROMINGER, 1960.

Many of the building blocks of the wall polymers have now been detected in nucleotide form and the direct biosynthesis of such nucleotides by pyrophosphorolysis reactions involving sugars, amino sugars and polyols has been established. The manner in which the wall peptide is formed by sequential addition of amino acids and dipeptide to the uridine diphosphate N-acetylmuramic acid provides a fascinating contrast to the biochemical events in the assembly of amino acids into protein. The action of antibiotics and the identification of the products accumulating in bacteria have led to the formulation of the early steps in the pathways for glycosaminopeptide biosynthesis. The schemes proposing the route of synthesis of these components in *Staphylococcus aureus* and *Escherichia coli* have been suggested by STROMINGER AND THRENN (1959) and STROMINGER (1960) and they are illustrated in Figs. 71 and 72.

The sequence of events leading to the assembly of the N-acetyl compounds of muramyl peptides, muramic acid and glucosamine into the structures recognized as 'wall material' is not known. It is clear from what we know of the chemistry of walls that the addition of muramyl peptides to the wall must be co-ordinated with the transfer of N-acetylglucosamine residues as well as N-acetylmuramic acid groups in some, if not in all cell walls. To bring the various nucleotide intermediates together SALTON (1960) proposed the co-ordinating hypothetical pathway shown in Fig. 73. ITO AND SAITO (1963) have recently obtained evidence from nucleotide accumulation indicating the feasibillity of the suggested pathways in Figs. 71–73.

Apart from the assembly of the peptide chain on UDP-N-acetylmuramic acid, much of the proposed biosynthetic pathway remains hypothetical and it will be of interest to see the speed with which the many gaps in our knowledge will be filled in. These tentative pathways are just the beginning of our knowledge, for the mechanisms involved in the formation of insoluble cell-wall polymer, complete with additional components such as poly-glycine peptide and other amino acids (e.g. D-aspartic acid) have not yet been probed.

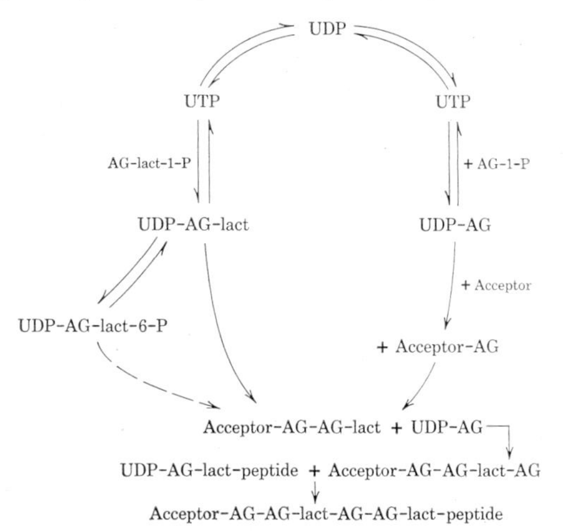

FIGURE 73. A hypothetic pathway for the biosynthesis of the glycosaminopeptide of a bacterial cell wall as proposed by SALTON, 1960.

Some of the technical problems involved in the demonstration that the wall components from the uridine nucleotides can be transferred to wall acceptor or suitable model acceptor compounds have been discussed by STROMINGER (1960) and SALTON (1960). Problems of penetration of externally added nucleotides, orientation of the membrane at the zone of biosynthesis and size and nature of wall acceptors are all relevant to the process of cell wall formation and the right experimental conditions would have to be found to meet their individual requirements for wall synthesis. No doubt some of the difficulties will be rapidly overcome and there are already suggestions from recent experiments in several laboratories that these have been partially resolved with disrupted cell preparations.

Even the walls of Gram-positive bacteria are now highly complicated structures with a variety of polymers probably located in a fairly specific manner. The degree of co-ordination of biosynthesis giving a wall containing a basal structure of glycosaminopeptide with attached polysaccharide and loosely bound teichoic acid seems a remarkable feat. However, when we consider the great complexity of the 'wall' of a Gram-negative bacterium with the specific polysaccharides oriented at the surface and protein-lipid structures arranged in an orderly fashion together with an underlying rigid component we are almost forced to conclude that the structure is self-duplicating with its biosynthetic mechanisms 'built in'. It is here that the biosynthetic mechanisms and the 'structural' problems of growth of the wall become interwoven and it seems most likely that the final stages of wall synthesis will involve a fuller understanding of the anatomical and biochemical relationships of wall and cell membranes.

For wall biosynthesis, two main possibilities exist, each of which would meet the general requirements of conserving the catalytic transglycosylation mechanisms as well as the requirements of biochemical control of growth of the wall. Biosynthesis at a localized area in the membrane would readily meet such needs. It is difficult, for example to visualize how a teichoic acid molecule could be assembled outside the rigid layer by transfer of building blocks from nucleotide precursors and at the same time conserve the CDP for further reactions. The other possibility for the complex structure of the Gram-negative bacteria could involve a 'built-in' complement of protein-lipid and polysaccharide synthesizing units. Again, unless growth occurs from a point it is difficult to visualize the nature of the mechanisms involved in building a highly complex layer on the outside of the rigid layer without some indigenous synthetic equipment. No doubt many of these interesting problems will be clarified as detailed studies of the site of wall formation progress.

Site of cell-wall formation

Until recently the manner in which the cell wall is laid down in bacteria has been largely a matter for speculation, although there have been observations suggesting a localized wall-formation process in some organisms. The presence of equatorial bands of thickening on isolated walls of dividing cells is in accord with centripetal growth of the cross-wall (DAWSON AND STERN, 1954). Thin sections of various Gram-positive bacteria have also supported the conclusion that cell-wall growth occurs in a limited zone of the dividing cell (CHAPMAN AND

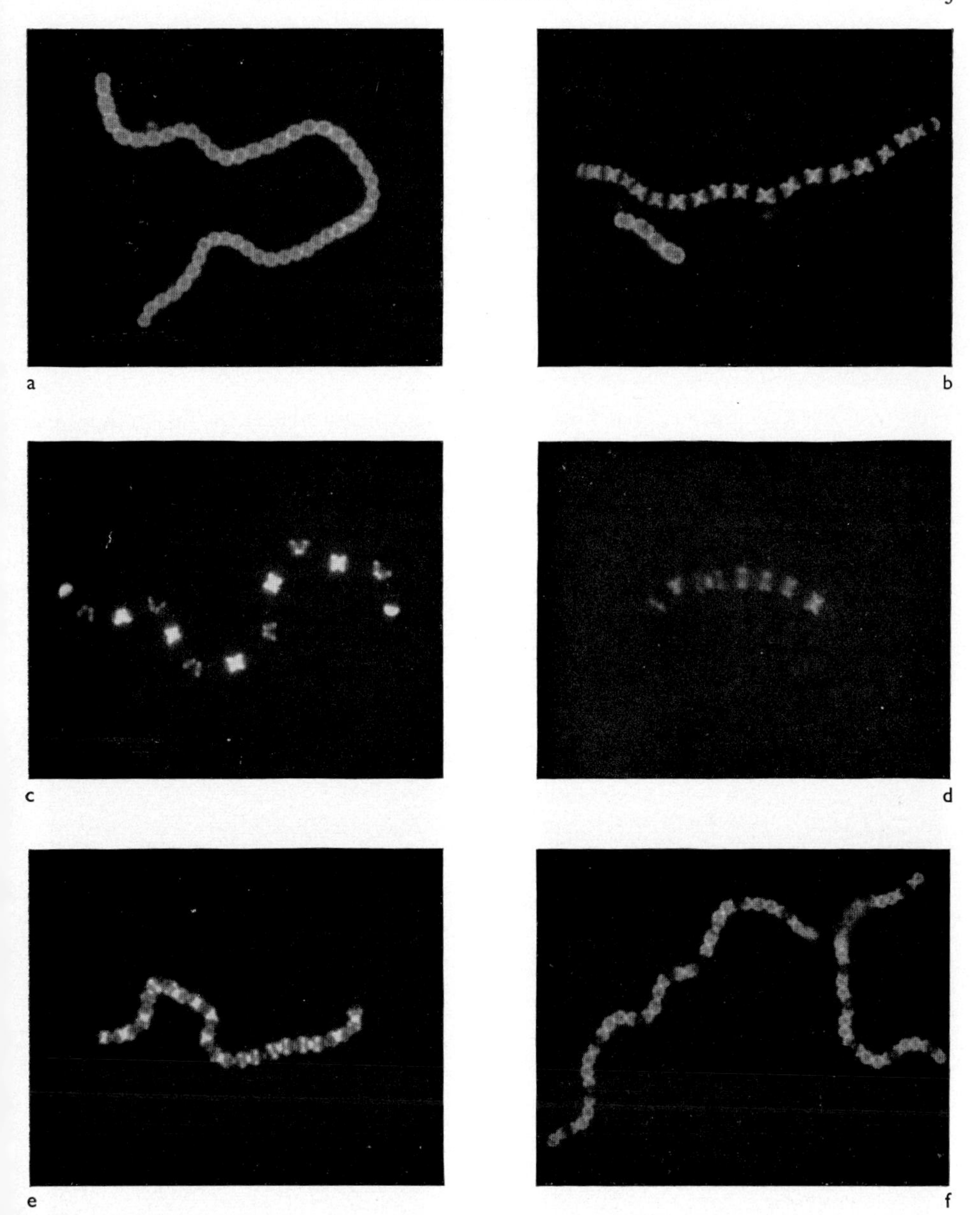

FIGURE 74. Ultraviolet photomicrographs showing sites of new cell-wall replication in Group A, type 19 streptococcus from the studies of COLE AND HAHN, 1962. Direct method of labelling by growth in homologous fluorescein-labelled globulin and examined at intervals after addition of unlabelled homologous globulin. (a) 0 minutes; (b) 15 minutes; (c) 60 minutes. Reverse method of staining by growing in unlabelled homologous globulin and removing at intervals after precipitation of antibody and reacting with homologous fluorescein-labelled globulin. (d) 15 minutes; (e) 30 minutes; (f) 60 minutes.

HILLIER, 1953; MURRAY, FRANCOMBE AND MAYALL, 1959; CHAPMAN, 1960). Thin sections of *Escherichia coli* obtained from dividing cells by CONTI AND GETTNER (1962) have indicated, however, that cell division occurred by centripetal growth of the cell wall. With flagellated Gram-negative bacteria it is possible that the cell wall grows by intercalation restricted to a specialized growing point (STOCKER, 1956).

The cell walls of most bacteria appear to be of fairly uniform thickness when viewed in thin sections of dividing cells. There is no critical information available which would enable us to conclude whether or not the wall undergoes 'post-exponential' thickening. Thus as COLE AND HAHN (1962) have emphasized 'the points of new wall and cross-wall origin relative to 'old' wall – and the directions of new wall growth – are also subjects about which little is known.'

We are now reaching the stage where our biochemical knowledge of wall biosynthesis and structural knowledge of the sites of wall formation will inevitably merge and give us a greater insight into the whole mechanism of new cell-wall synthesis. The recent application of such elegant techniques as fluorescent-antibody labelling (COLE AND HAHN, 1962; MAY,

FIGURE 75. Fluorescence photomicrograph of *Schizosaccharomyces pombe* showing conservation of cell-wall label during growth. The cells were stained by the indirect method with fluorescein-conjugated antibody 120 minutes after labelling with homologous antibody. $\times$ 2,500 (May, 1962).

1962) and radioautographic methods (VAN TUBERGEN AND SETLOW, 1961) is yielding valuable information on the distribution of new wall at the cell surface and in the cell envelope.

Two independent studies have shown how fluorescent-labelled antibody can be used in detecting newly synthesized wall in group A streptococci (COLE AND HAHN, 1962) and in the yeast, *Schizosaccharomyces pombe* and several bacteria (MAY, 1962; unpublished observations). Both studies used a 'direct' labelling method where the cells were either grown in the fluorescein-antibody conjugate (COLE AND HAHN, 1962) or treated with homologous antiserum and labelled immediately with fluorescein-conjugated anti-rabbit globulin. The cells were examined by fluorescence miroscopy after a period of growth to determine whether the newly synthesized wall carried any of the original fluorescent label. As shown in Figs. 74 and 75 unlabelled, localized areas indicated the sites of new wall formation in group A streptococci and in *Schizosaccharomyces pombe* respectively. Further treatment of the same yeast cells (Fig. 75) with homologous antiserum and fluorescent antiserum showed the extent of previously unstained wall.

In addition to detecting new wall by the presence of non-fluorescent gaps in the 'direct' abelling of group A streptococcal chains, COLE AND HAHN (1962) confirmed their observations with a 'reverse' technique. The latter method involved the adsorption of unlabelled antibody on to the surface of the streptococci, removal of unadsorbed antibody followed by washing and application of the fluorescein-conjugated globulin. The areas corresponding to the gaps in Fig. 74 (b, c) were stained with the fluorescent antibody as shown in Fig. 74 (d, e, f). COLE AND HAHN (1962) concluded that new cell wall in group A streptococci was not diffusely intercalated but its formation was initiated equatorially at the site of the next cross-wall formation. The results indicated that there were at least two sites of simultaneous activity, one involved in cross-wall formation to complete the previous division and the other site of peripheral wall and cross-wall development initiated for the 'current division'.

While this method of wall formation may occur with some organisms, it is already apparent that in other bacteria a more or less uniform extension of the wall takes place. Dispersion of the labelled wall surface has been established in *Salmonella typhimurium* as illustrated in MAY's unpublished data presented in Fig. 76. The rate of extension of wall in a chain of cells of *Escherichia coli* was also determined by time-lapse photography (MAY, unpublished observations) and the uniformity of growth and extension of the cell surface was confirmed with 'knobbed' cells as shown in Fig. 77. A similar conclusion has also been reached by VAN TUBERGEN AND SETLOW using radioautography of *Escherichia coli* cells exposed to [^{3}H]-diaminopimelic acid which would be selectively incorporated into the rigid component of the wall.

Although the fluorescent-labelling methods are of the greatest value in distinguishing the modes of wall formation as indicated by the presence of surface antigenic components, they cannot be expected to give information about the thickening of walls. The knowledge that there are at least two principal ways of distributing new wall material is of considerable importance and it will be of interest to see to what extent they can be correlated with sensitivity to penicillin and the capacity of cells to form the plasma membrane systems. The multiplicity of wall forming sites in Gram-negative bacteria would account for their greater resistance to penicillin, despite the smaller amount that the cells need to synthesize. It should, however, be emphasized that the fluorescent-labelling method is specific only for surface antigens and may not give valid results for unexposed layers. Whether the greater propensity for many Gram-positive bacteria to form mesosomes is a reflection of the different mode of wall formation cannot be said but it is certainly a fascinating property which merits further investigation.

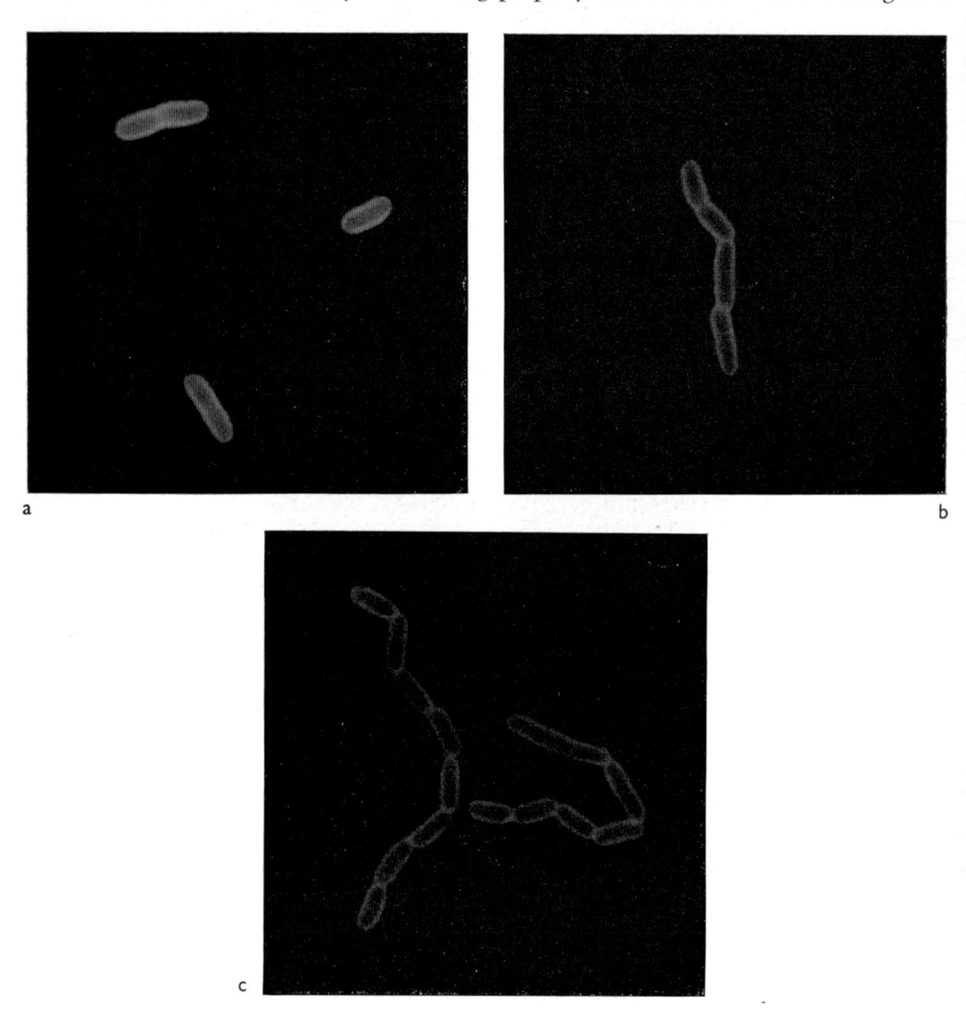

FIGURE 76. Fluorescence photomicrographs of *Salmonella typhimurium* LT2 showing dispersion of cell-wall label during growth and division. The cells were stained by the indirect method with fluorescein-conjugated antibody (a) immediately, (b) 37 minutes and (c) 60 minutes after labelling with homologous antibody. Each interval corresponds to a doubling in the turbidity of the culture. $\times$ 2,500 (May, unpublished observations).

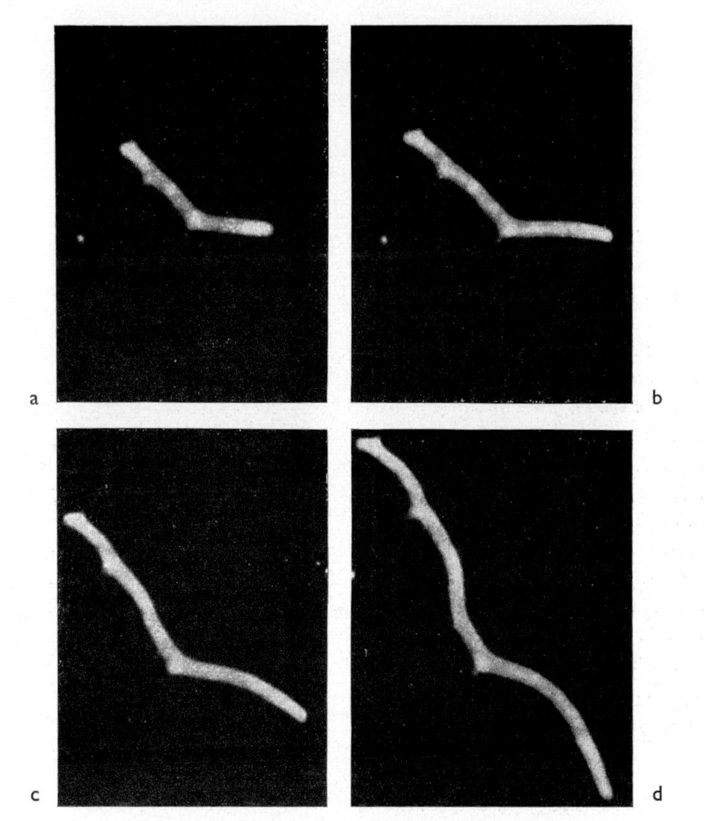

FIGURE 77. Time-lapse phase-contrast photomicrographs of a 'knobbed' cell of *Escherichia coli* 173–25 showing increase in length between the knobs as well as in the terminal segments. The interval between successive photographs was 10 minutes. (May, unpublished observations).

It is assumed that wall synthesis involves an intimate contact between membrane and the wall or its 'unfinished' acceptor end and the types of studies discussed above will enable us to clarify the relationships of these two structures to one another during the vital process of wall biosynthesis. There seems little doubt that the next ten years of accelerated endeavour in this field will be even more rewarding than the past preliminary decade which has developed the outlines of the problems of the nature and biosynthesis of the bacterial cell wall.

REFERENCES

ABELSON, P. H., E. BOLTON, R. BRITTEN, D. B. COWIE and R. B. ROBERTS, *Proc. Nat. Acad. Sci., U.S.*, 39 (1953) 1020.

ABRAHAM, E. P., *Biochemistry of Some Peptide and Steriod Antibiotics*, John Wiley, New York, 1957.

ABRAHAM, E. P. and G. F. NEWTON, *Amino acids and Peptides with Antibiotic Activity*, CIBA Found. Symp., Churchill, London, 1958.

ÅGREN, G. and C. H. DE VERDIER, *Acta Chem. Scand.*, 12 (1958) 1927.
ALLISON, J. L., R. E. HARTMAN, R. S. HARTMAN, A. D. WOLFE, J. CIAK and F. E. HAHN, *J. Bacteriol.*, 83 (1962) 609.
ANITA, M., D. S. HOARE and E. WORK, *J. Biochem.*, 65 (1957) 448.
ARCHIBALD, A. R., J. J. ARMSTRONG, J. BADDILEY and J. B. HAY, *Nature*, 191 (1961) 570.
BADDILEY, J., N. L. BLUMSON, A. DI GIROLAMO and M. DI GIROLAMO, *Biochim. Biophys. Acta*, 50 (1961) 391.
BADDILEY, J., J. G. BUCHANAN and C. P. FAWCETT, *J. Chem. Soc.*, (1959) 2192.
BADDILEY, J., J. G. BUCHANAN and A. R. SANDERSON, *J. Chem. Soc.*, (1958) 3107
BADDILEY, J. and F. C. NEUHAUS, *Biochem. J.*, 75 (1960) 579.
BARBIERI, P., A. DI MARCO, L. FUOCO and A. RUSCONI, *Biochem. Pharmacol.*, 3 (1960) 101.
BARKULIS, S. S., J. J. BOLTRALIK, H. HEYMANN and L. D. ZELEZNICK, *Abstracts Sth. Intern. Congr. Microbiol.* (1962) A1.9
BAUMAN, N. and B. D. DAVIS, *Science*, 126 (1957) 170.
BONDI, A., J. KORNBLUM and C. FORTE, *Proc. Soc. Exptl. Biol. Med.*, 96 (1957) 270.
BROWN, D. H., *Biochim. Biophys. Acta*, 16 (1955) 429.
BURGER, M., L. GLASER and R. M. BURTON, *Biochim. Biophys. Acta*, 56 (1962) 172.
BUTLER, J. A. V., A. R. CRATHORN and G. D. HUNTER, *Biochem. J.*, 69 (1958) 544.
CARDINI, C. E. and L. F. LELOIR, *J. Biol. Chem.*, 225 (1957) 317.
CHAPMAN, G. B., *J. Bacteriol.*, 79 (1960) 132.
CHAPMAN, G. B. and J. HILLIER, *J. Bacteriol.*, 66 (1953) 362.
CHATTERJEE, A. N. and J. T. PARK, *Proc. Natl. Acad. Sci. (USA)*, (1964), in the press.
CIAK, J. and F. E. HAHN, *Antibiotics and Chemotherapy*, 9 (1959) 47.
CIFERRI, O., M. DI GIROLAMO and A. DI GIROLAMO BENDICENTI, *Biochim. Biophys. Acta*, 50 (1961) 405.
CLARKE, P. H., P. GLOVER and A. P. MATHIAS, *J. Gen. Microbiol.*, 20 (1959) 156.
COHEN, S. S., J. G. FLAKS, H. D. BARNER, M. R. LOEB and J. LICHTENSTEIN, *Proc. Nat. Acad. Sci., U.S.*, 44 (1958) 1004.
COLE, R. M. and J. J. HAHN, *Science*, 135 (1962) 722.
COLLINS, J. F. and M. RICHMOND, *Nature*, 195 (1962) 142.
COMB, D. G., *J. Biol. Chem.*, 237 (1962) 1601.
COMB, D. G., W. CHIN and S. ROSEMAN, *Biochim. Biophys. Acta*, 46 (1961) 394.
COMB, D. G. and S. ROSEMAN, *Biochim. Biophys. Acta*, 29 (1958) 653.
CONTI, S. F. and M. E. GETTNER, *J. Bacteriol.*, 83 (1962) 544.
COOKSEY, K. E., R. A. ANWAR, C. ROY and R. W. WATSON, *Arch. Biochem. Biophys.*, 94 (1961) 541.
COOPER, P. D., *Bacteriol. Revs.*, 20 (1956) 28.
CUMMINS, C. S. and H. HARRIS, *J. Gen. Microbiol.*, 14 (1956) 583.
DAVIDSON, E. A., H. J. BLUMENTHAL and S. ROSEMAN, *J. Biol. Chem.*, 226 (1957) 125.
DAVIS, B. D., *Nature*, 169 (1952) 534.
DAWSON, I. M. and H. STERN, *Biochim. Biophys. Acta*, 13 (1954) 31.
DEWEY, D. L., D. S. HOARE, E. WORK, *Biochem. J.*, 58 (1954) 523.
DEWEY, D. L. and E. WORK, *Nature*, 169 (1952) 533.
DORFMAN, A. and J. A. CIFONELLI, in *Chemistry and Biology of Mucopolysaccharides*, CIBA Foundation Symp. Churchill, London, 1958, p. 64.
DUGUID, J. P., *Edinburgh Med. J.*, 53 (1946) 401.
DUSCHINSKY, R., E. PLEVEN and C. HEIDELBERGER, *J. Amer. Chem. Soc.*, 79 (1957) 4559.
FITZ-JAMES, P. C., *J. Biophys. Biochem. Cytol.*, 4 (1958) 257.
FREIMER, E. H., R. M. KRAUSE and M. MCCARTY, *J. Exptl. Med.*, 110 (1959) 853.
GALE, E. F., *Synthesis and Organisation in the Bacterial Cell*, John Wiley, New York, 1959.
GHOSH, S., H. J. BLUMENTHAL, E. DAVIDSON and S. ROSEMAN, *J. Biol. Chem.*, 235 (1960) 1265.
GILVARG C., *Fed. Proc.*, 15 (1956) 261.

GILVARG, C., *Biochim. Biophys. Acta*, 24 (1957) 216.
GILVARG, C., *J. Biol. Chem.*, 233 (1958) 1501.
GILVARG, C., *J. Biol. Chem.*, 236 (1961) 1429.
GILVARG, C., *J. Biol. Chem.*, 237 (1962) 482.
GILVARG, C., J. EDELMAN and S. H. KINDLER, *Proc. 4th Intern. Congr. Biochem.*, (1958).
GINSBURG, V., P. J. O'BRIEN and C. W. HALL, *J. Biol. Chem.*, 237 (1962) 497.
GLASER, L., *Biochim. Biophys. Acta*, 31 (1959) 575.
GLASER, L., *J. Biol. Chem.*, 235 (1960) 2095.
GOODER, H. and W. R. MAXTED, *Soc. Gen. Microbiol. Symposium* No. 11 (1961) 151.
HANCOCK, R and J. T. PARK, *Nature*, 181 (1958) 1050.
HEATH, E. C. and A. D. ELBEIN, *Proc. Nat. Acad. Sci., U.S.*, 48 (1962) 1209.
HEIDELBERGER, C., N. K. CHAUDARI, P. DANNENBERG, D. MOOREN, L. GRIESBACH, R. DUSCHINSKY, R. J. SCHNITZER, E. PLEVEN and J. SCHNEINER, *Nature*, 179 (1957) 663.
HIJMANS, W. J., *J. Gen. Microbiol.*, 28 (1962) 177.
HILL, P. B., *Biochim. Biophys. Acta*, 57 (1962) 386.
HODGKIN, D. C. and E. N. MASLEN, *Biochem. J.*, 79 (1961) 393.
HUGO, W. B. and A. D. RUSSELL, *J. Bacteriol.*, 82 (1961) 411.
IKAWA, M. and E. E. SNELL, *Biochim. Biophys. Acta*, 60 (1962) 186.
IKAWA, M., E. E. SNELL and E. LEDERER, *Nature*, 188 (1960) 658.
ITO, E., N. ISHIMOTO and M. SAITO, *Arch. Biochem. Biophys.*, 80 (1959) 431.
ITO, E. and M. SAITO, *Biochim. Biophys. Acta*, In the press (1963).
ITO, E. and J. L. STROMINGER, *J. Biol. Chem.*, 235 (1960) PC5, PC7.
JONSEN, J. and S. LALAND, *Adv. in Carbohydrate Chem.*, 15 (1960) 201.
JORDAN, D. C., *Biochem. Biophys. Res. Comm.*, 6 (1961) 167.
JORDAN, D. C. and W. E. INNISS, in *Antimicrobial Agents and Chemotherapy*. 1961, p, 218.
KALCKAR, H. M. and E. CUTOLO, *Résumes des Communications*, 2nd Intern. Congr. Biochimie, 1952, p. 260.
KORNBERG, A., *J. Biol. Chem.*, 176 (1948) 1475.
KORNBERG, A., *J. Biol. Chem.*, 182 (1950) 779.
KORNFERD, S. and L. GLASER, *Biochim. Biophys. Acta*, (1962)
LANDMAN, O. E. and S. HALLE, *Abstracts. 8th Intern. Congr. Microbiol.*, (1962) 28.
LARK, C. and K. G. LARK, *Bacteriol. Proc.*, (1958) 108.
LEDERBERG, J., *Proc. Nat. Acad. Sci., U.S.*, 42 (1956) 574.
LEDERBERG, J. and ST. J. CLAIR, *J. Bacteriol.*, 75 (1958) 143.
LEVIN, J. G. and D. B. SPRINSON, *Biochem. Biophys. Res. Comm.*, 3 (1960) 157.
LILLY, M. D., *J. Gen. Microbiol.*, 28 (1962) ii.
MANDELSTAM, J. and H. J. ROGERS, *Nature*, 181 (1958) 956.
MANDELSTAM, J. and H. J. ROGERS, *Biochem. J.*, 72 (1959) 654.
MARKOVITZ, A. and A. DORFMAN, *J. Biol. Chem.*, 237 (1962) 273.
MAY, J. W., *Exptl. Cell Res.*, 27 (1962) 170.
MCQUILLEN, K., *Biochim. Biophys. Acta*, 18 (1955) 531.
MCQUILLEN, K., *J. Gen. Microbiol.*, 18 (1958) 498.
MCQUILLEN, K., in *The Bacteria*, Ed. I. C. Gunsalus and R. Y. Stanier, Academic Press, New York, 1960, Vol. 1, p. 249.
MEADOW, P., D. S. HOARE, and E. WORK, *Biochem.* J., 66 (1957) 270.
MEADOW, P. and E. WORK, *Biochim. Biophys. Acta*, 28 (1958) 596.
MEADOW, P. and E. WORK, *Biochem. J.*, 72 (1959) 400.
MITCHELL, P. and J. MOYLE, *Soc. Gen. Microbiol. Symposium* No. 6 (1956)
MOMOSE, H., *J. Gen. Appl. Microbiol.* Suppl. 1, 7 (1961a) 359.

MOMOSE, H., *J. Gen. Appl. Microbiol.* Suppl. 1,7 (1961b) 365.
MOMOSE, H. and Y. IKEDA, *J. Gen. Appl. Microbiol., Suppl.* 1, 7 (1961) 370.
MORRISON, T. H. and C. WEIBULL, *Acta. Pathol. Microbiol. Scand.*, 55 (1962) 475.
MURRAY, R. G. E., W. H. FRANCOMBE and B. H. MAYALL, *Canad. J. Microbiol.*, 5 (1959) 641.
NARROD, S. A. and W. A. WOOD, *Arch. Biochem. Biophys.*, 35 (1952) 462.
NATHENSON, S. G. and J. L. STROMINGER, *Fed. Proc.*, 18 (1959) 426.
NATHENSON, S. G. and J. L. STROMINGER, *J. Pharmacol. Exptl. Therap.*, 131 (1961) 1.
NEUHAUS, F. C., *Biochem. Biophys. Res. Comm.*, 3 (1960) 401.
NEUHAUS, F. C., *Fed. Proc.*, 20 (1961) 1.
NEUHAUS, F. C., *J. Biol. Chem.*, 237 (1962) 3128.
NEUHAUS, F. C. and J. L. LYNCH, *Biochem. Biophys. Res. Comm.*, 8 (1962) 377.
NIKAIDO, H. and K. JOKURA, *Biochem. Biophys. Res. Comm.*, 6 (1961) 304.
O'BRIEN, P. J., M. C. GLICK and F. ZILLIKEN, *Biochim. Biophys. Acta*, 37 (1960) 357.
OKABAYASHI, T., *J. Bacteriol.*, 84 (1962) 1.
OKAZAKI, R., *Biochim. Biophys. Acta*, 44 (1960) 478.
OSBORN, M. J., S. M. ROSEN, L. ROTHFIELD and B. L. HORECKER, *Proc. Nat. Acad. Sci., U.S.*, 48 (1962) 1831.
OTSUJI, N. and Y. TAKAGAKI, *J. Biochem. Japan*, 46 (1959) 791.
PANOS, C. and S. S. BARKULIS, *J. Bacteriol.*, 78 (1959) 247.
PARK, J. T., *J. Biol. Chem.*, 194 (1952) 877, 885, 897.
PARK, J. T. and M. E. GRIFFITH, *Bacteriol. Proc.*, (1962) G110.
PARK, J. T. and R. HANCOCK, *J. Gen. Microbiol.*, 22 (1960) 249.
PARK, J. T. and M. J. JOHNSON, *J. Biol. Chem.*, 179 (1949) 585.
PARK, J. T. and J. L. STROMINGER, *Science*, 125 (1957) 99.
PAZUR, J. H. and E. W. SHUEY, *J. Amer. Chem. Soc.*, 82 (1960) 5009.
PETERKOFSKY, B. and C. GILVARG, *Fed. Proc.*, 18 (1959) 301.
REYNOLDS, P. E., *Biochim. Biophys. Acta*, 52 (1961) 403.
RHULAND, L. E., *J. Bacteriol.*, 73 (1957) 778.
RHULAND, L. E., *Nature*, 185 (1960) 224.
RHULAND, L. E. and B. BANNISTER, *J. Amer. Chem. Soc.*, 78 (1956) 3548.
RHULAND, L. E. and R. D. HAMILTON, *Biochim. Biophys. Acta*, 51 (1961) 525.
RICHMOND, M. H. and H. R. PERKINS, *Biochem. J.*, 76 (1960) 1P.
RICHMOND, M. H. and H. R. PERKINS, *Biochem. J.*, 85 (1962) 580.
ROBERTS, J. and M. J. JOHNSON, *Biochim. Biophys. Acta*, 59 (1962) 458.
ROGERS, H. R. and J. JELJASZEWICZ, *Biochem. J.*, 81 (1961) 576.
ROGERS, H. J. and J. MANDELSTAM, Biochem. *J.*, 84 (1962) 299.
ROGERS, H. J. and H. R. PERKINS *Biochem. J.*, 77 (1960) 448.
Rogers, H. J. and H. R. PERKINS, *Biochem. J.*, 82 (1962) 35P.
ROSEMAN, S., *J. Biol. Chem.*, 226 (1957) 115.
ROSEMAN, S., *Ann. Rev. Biochem.*, 28 (1959) 545.
ROSEMAN, S., J. LUDOWIEG, F. E. MOSES and A. DORFMAN, *J.Biol. Chem.*, 206 (1954) 665.
ROSEMAN, S., F. E. MOSES, J. LUDOWIEG and A. DORFMAN, *J. Biol. Chem.*, 203 (1953) 213.
SALTON, M. R. J., *Biochim. Biophys. Acta.*, 8 (1952) 510.
SALTON, M. R. J., *Soc. Gen. Microbiol. Symp.* No. 6, 1956.
SALTON, M. R. J. *Microbial Cell Walls*, John Wiley and Sons, New York, 1960.
SALTON, M. R. J., *J. Gen. Microbiol.*, 29 (1962) 15.
SALTON, M. R. J. and J. M. GHUYSEN, *Biochim. Biophys. Acta.* 36 (1959) 552.
SALTON, M. R. J. and J. M. GHUYSEN, *Biochim. Biophys. Acta*, 45 (1960) 355.
SAUKKONEN, J. J., *Nature*, 192 (1961) 816.

SHARP, J. T., W. HIJMANS, and L. DIENES, *J. Exptl. Med.*, 105 (1957) 153.
SHAW, D. R. D., *Biochem. J.*, 66 (1957) 56P.
SHAW, D. R. D., *Biochem. J.*, 82 (1962) 297.
SHOCKMAN, G. D., *Proc. Soc. Exptl. Biol. Med.*, 101 (1959a) 693.
SHOCKMAN, G. D., *J. Biol. Chem.*, 234 (1959b) 2340.
SHOCKMAN, G. D., M. J. CONOVER, J. J. KOLB, L. S. RILEY and G. TOENNIES, *J. Bacteriol.*, 81 (1961) 44.
SHOCKMAN, G. D., J. J. KOLB and G. TOENNIES, *J. Biol. Chem.*, 230 (1958) 961.
SMITH, E. E. B., B. GALLOWAY and G. T. MILLS, *Biochim. Biophys. Acta*, 33 (1959) 276.
SMITH, E. E. B. and G. T. MILLS, *Biochim. Biophys. Acta*, 13 (1954) 386.
SOUTHARD, W. H., J. A. HAYASHI and S. S. BARKULIS, *Bacteriol. Proc.*, (1958) P90.
SOUTHARD, W. H., J. A. HAYASHI and S. S. BARKULIS, *J. Bacteriol.*, 78 (1959) 79.
STOCKER, B. A. D., *Soc. Gen. Microbiol. Symp.* No. 6, 1956, p. 19.
STRANGE, R. E. and J. F. POWELL, *Biochem. J.*, 58 (1954) 80.
STROMINGER, J. L., *J. Biol. Chem.*, 224 (1957) 509.
STROMINGER, J. L., *Biochim. Biophys. Acta.* 30 (1958) 645.
STROMINGER, J. L., *Compt. Rend. Trav. Lab. Carlsberg*, 31 (1959) 181.
STROMINGER, J. L., *Physiol. Revs.*, 40 (1960) 55.
STROMINGER, J. L., *Fed. Proc.*, 21 (1962) 134.
STROMINGER, J. L., E. ITO, and R. H. THRENN, *J. Amer. Chem. Soc.*, 82 (1960) 998.
STROMINGER, J. L., E. ITO, and R. H. THRENN, *Fed. Proc.*, 20 (1961) 228.
STROMINGER, J. L. and M. S. SMITH, *J. Biol. Chem.*, 234 (1959) 1822.
STROMINGER, J. L. and R. H. THRENN, *Biochim. Biophys. Acta*, 33 (1959) 280.
STROMINGER, J. L., R. H. THRENN and S. S. SCOTT, *J. Amer. Chem. Soc.*, 81 (1959) 3803.
TANAKA, M., Y. KATO and S. KINOSHITA, *Biochem. Biophys. Res. Comm.*, 4 (1961) 114.
THORNE, C. B., *Soc. Gen. Microbiol. Symposium No. 6*, 1956, p. 68.
THORNE, C. B., C. G. GOMEZ and R. D. HOUSEWRIGHT, *J. Bacteriol.*, 69 (1955) 357.
THORNE, C. B. and D. M. MOLNAR, *J. Bacteriol.*, 70 (1955) 420.
TOENNIES, G. and D. L. GALLANT, *Growth*, 13 (1949) 21.
TOENNIES, G. and G. D. SHOCKMAN, *Arch. Biochem. Biophys.*, 45 (1953) 447.
TOENNIES, G. and G. D. SHOCKMAN, *4th Intern. Congr. Biochem.* Vol. XIII, (1959) 365.
TOMASZ, A. and E. BOREK, *Proc. Nat. Acad. Sci., U.S.*, 45 (1959) 929.
TOMASZ, A. and E. BOREK, *Proc. Nat. Acad. Sci., U.S.*, 46 (1960) 324.
TOMASZ, A. and E. BOREK, *Biochemistry*, 1 (1962) 543.
TOPPER, Y. J. and M. M. LIPTON, *J. Biol. Chem.*, 203 (1953) 135.
TRUCCO, R. E. and A. B. PARDEE, *J. Biol. Chem.*, 230 (1958) 435.
TUTTLE, A. L. and H. GEST, *J. Bacteriol.*, 79 (1960) 213.
VAN TUBERGEN, R. P. and R. B. SETLOW, *Biophys. J.*, 1 (1961) 589.
VOGEL, H. J., *Fed. Proc.*, 18 (1959a) 345.
VOGEL, H. J., *Biochim. Biophys. Acta*, 34 (1959b) 282.
WEIBULL, C., *J. Bacteriol.*, 66 (1953) 688.
WEIBULL, C., *Ann. Rev. Microbiol.*, 12 (1958) 1.
WICKEN, A. J., S. D. ELLIOTT and J. BADDILEY, *J. Gen Microbiol.*, (1963), in the Press.
WOOD, W. A. and I. C. GUNSALUS, *J. Biol. Chem.*, 190 (1951) 403.
WORK, E., *Biochim. Biophys. Acta*, 17 (1955) 410.
WORK, E., *Colloques Intern. C.N.R.S.*, 92 (1959) 143.
WORK, E., *J. Gen. Microbiol.*, 25 (1961) 167.
WYLIE, E. B. and M. J. JOHNSON, *Biochim. Biophys. Acta*, 59 (1962) 450.
ZILLIKEN, F., *Fed. Proc.*, 18 (1959) 966.

On looking back

The study of the nature of the microbial cell has progressed rapidly in the last few decades and has kept pace with the exciting advances being made in the more general field of cellular biology. The knowledge gained from the investigations of microorganisms has given us a new insight into biochemical, genetic and structural processes in living cells and as a group, the 'microbes' merit the special plea by KLUYVER AND VAN NIEL in *The Microbes Contribution to Biology* to remember that these small creatures have contributed much in broadening our understanding of many fundamental biochemical processes which in concert together constitute what we call 'life''. Bacteria have been conspicuous in elucidating mechanisms of RNA, DNA and protein biosynthesis and have presented us with new fascinating details of genetic transformations and control of cellular activities. Many of these properties mapped in detail with bacteria are probably common to all types of living organisms. Some of the more unusual properties of microorganisms have enabled us to discover the basis of drug and antibiotic action at the molecular and enzymic level and many of these studies have been intimately concerned with the fate of surface structures of microbial cells.

The 'solid facts' of the past 10–15 years work on bacterial cell walls, briefly summarized in the foregoing chapters, represent one of the many facets of our rapidly expanding knowledge of cell structure and function at the chemical and biochemical level. Perhaps the most interesting single feature to emerge from these studies has been the discovery of several chemical entities 'unique' to the bacterial cell wall. The isolation and identification of compounds such as muramic acid, diaminopimelic acid and the teichoic acids can now be used in defining our concepts of bacterial cells and certain closely related groups of organisms such as the blue-green algae and rickettsias.

To those interested in the fascinating problems of the origins of life and cellular organisation and the subsequent events now recognizable in phylogenetic relationships, it has always been a disappointment that the bacteria have set themselves apart and we have no fossil remains on which to base our speculations about the evolution of these organisms. Even the most robust parts of microbial cells, the rigid cell walls, were likely to have survived for relatively short geological periods and then only in 'mummified' form. However, some information about the existence of bacteria in recent geological periods would be infinitely better than none and with the delicate techniques of fluorescent labelling of antibodies for surface components it would be of great interest to determine whether microorganisms could be detected in hitherto undisturbed sediments and ice deposits. The microflora of the gut contents of the ancient mammoths for long buried in the ice of Siberia would have presented a unique opportunity for some microbiological 'prospecting' and if man has the good fortune

The tables are printed together at the end of the book.

of discovering such material again, no effort should be spared in attempting to identify ancient microbial remains. It is true that the most one could expect from these studies would be the determination of the similarity or dissimilarity of antigenic determinant groups on the surface structures of modern and ancient microorganisms. It may well be that such material is still too close in geological time to the species in existence at present, to show any marked difference.

In the bacterial world, apart from nutritional considerations, the only feature resembling anything like an evolutionary or phylogenetic thread is the possession of the cell-wall glycosaminopeptides with muramic acid as the most conspicuous substance. As already pointed out cell-wall heteropolymers can be traced through Gram-positive bacteria to the majority of Gram-negative organisms, into the blue-green algae and in the rickettsias. The extensive work of CUMMINS on the composition of cell walls of Gram-positive bacteria has established the usefulness of chemical criteria in the taxonomy of these organisms. Although the taxonomic affinities of related organisms can be reinforced and indeed clarified by determining the chemical constitution of their cell walls, there are no clear lines of 'evolution' from one bacterial group to another, when this property is considered. Abrupt changes in the nature of the 'characteristic' wall components can occur within genera and in moving from one group to another group of organisms related on morphological and biochemical criteria (e.g. corynebacteria, mycobacteria, propionibacteria, *Streptomyces* and *Micromonospora* spp.).

The chemical make-up of the cell wall is a 'fixed' property of the organism, at least qualitatively speaking, and there is little doubt that the polymers associated with the cell wall are chemically the most individualistic components of the cell. From the discussions of the chemistry of the bacterial cell wall it will be apparent that it is a much easier task to determine the individuality of a bacterial species by chemical analysis of the wall than by studying any of the other major macromolecular substances of the cell e.g. proteins, RNA and DNA. More sophisticated methods are required for the differentiation of the latter cellular substances.

The investigations with bacterial protoplasts pioneered by WEIBULL have shown that the cell loses a number of its most characteristic properties upon loss of the cell-wall structure. Along with the loss of the most conspicuous chemical compounds of the cell, the bacterium deprived of its wall also loses its sensitivity to bacteriophages, may lose its ability to biosynthesize wall, and it displays changed patterns of sensitivity to antibiotics and the 'residual cell' may enter a new pathway of survival as an L-phase organism. The loss of wall on protoplast formation and conversion of bacterial cells into the L-phase cultures is one of the most interesting biological features stemming as an offshoot from the advances made in the cell wall – bacterial anatomy field. It could well be that the L-phase cells and the pleuropneumonia-like organisms represent the evolutionary 'dustbin' or 'dead-end' for bacteria and other microorganisms. One avenue of genetic change has been sealed off with the loss of cell-wall receptors for transducing phage in the case of Gram-positive bacteria undergoing this conversion. The frequency of reversion of the L-phase organism to the original cell type is not well authenticated and little can be said as to whether the L-phase organisms with one less surface layer than a bacterial cell can be more readily transformed. Perhaps one of the interesting lines of investigation which these intrigueing organisms will stimulate, will be to discover in more detail the biochemical changes accompanying the conversion to the L-phase and the extent to which the genetic information, presumably still in the L-phase cell, can express itself.

The general loss of ability to form a wall on conversion to protoplasts and the apparent complete loss in the case of several Gram-positive bacteria fully stabilized in the L-phase, poses many interesting problems for the future. The wall apparently confers some autonomy upon the cell but it is difficult at present to know what else is lost during the transformation to protoplasts and L-phase. Attention has already been drawn to the loss of the glycerol teichoic acid (the group antigen) of streptococci belonging to Lancefield group D and it may well be that this compound has some special function in wall synthesis and could be the 'gap substance' assumed to be sandwiched between wall and plasma membrane. Many of these problems will undoubtedly be explored in the future.

It has been something of a biological surprise to learn that bacteria are sufficiently flexible not to require a wall for survival. Enzymes capable of degrading the glycosaminopeptides of cell walls (muramidases) are widely found in Nature and as the author has pointed out, conversion of bacteria to protoplasts undoubtedly occurs under natural conditions. In specialized environments these protoplasts would probably become 'stabilized' as L-phase organisms and the latter may be more widely found in Nature than we suspect. It may well be that we are all carrying around our own particular brand of L-phase organism following the curing of acute infections with antibiotics such as penicillin. The extent to which these organisms are associated with chronic complaints does not seem to have been assessed, but it is of interest to note that the L-phase group A streptococci continue to make the M-proteins *in vitro*. It would be unfortunate for man if these forms also synthesize endo- and exo-toxins responsible for the pathogenicity of particular organisms.

If for a mental exercise we were to sit down and design a progenitor of the present bacterial groups we are familiar with, we would be forced to select an organism similar to any Gram-positive species and preferably a coccus for structural simplicity. On nutritional grounds a Gram-positive organism would be a good candidate, for it possesses the ability to accumulate amino acids from its external environment, a property which would precede complete synthetic ability for these compounds, if one assumes that the ancestral bacteria evolved in a 'primordial soup' rich in the building blocks for the major macromolecules of the cell. Apart from the more limited synthetic ability and the potentiality for taking up pre-formed compounds from without, the rest of such a progenitor-type cell would have to be very complex and biochemically sophisticated if it were to succeed in growing and surviving. However, again on chemical grounds a wall of the type present in Gram-positive bacteria could conceivably precede the more complex structures of the Gram-negative bacteria. Unfortunately there is no information about the possible occurrence of D-amino acid isomers in the amino acid pool formed chemically prior to the evolution of living cells. By loss of glycosaminopeptide, bacteria could move in an evolutionary sense, from a Gram-positive type of organism to Gram-negative species, ultimately terminating in organisms such as the PPLO and halophilic species devoid of this heteropolymer.

It must be admitted that the most compelling factor in suggesting this sequence of evolution in bacteria is based on the nutritional and synthetic abilities in going from heterotrophic to chemolithotrophic organisms but it is paralleled so far as our present knowledge permits, by these changes from a simple wall heteropolymer to a very complex structure. On purely structural grounds on the other hand there would be excellent reasons for believing that as the supply of nutrients in the 'primordial soup' diminished somewhat and organisms began to develop proteolytic activities, selection pressure would favour the evolution of Gram-positive organisms with their walls containing the D-isomers of amino acids and the possible protection against enzymic breakdown this property may confer on the organisms. This would displace the Gram-positive organism a little further away from the initial ancestral type of bacterial cell. Moreover, it is tempting to suggest that something like a primitive protoplast with a membrane enclosing the genetic and replicating information and mechanisms would precede the development of an organism with a highly specialized wall structure such as that seen in Gram-positive bacteria. There is obviously much scope for speculation on the possible origins, evolution and interrelationships of the bacteria and it is our great misfortune that in all probability we shall never be able to put our speculations to the scientific test.

Whether other microorganisms such as fungi, yeasts, algae and protozoa evolved from bacterial predecessors cannot be said. The phylogenetic 'thread' of glycosaminopeptides does not appear to overlap into the yeasts, fungi and protozoa so far examined. Several algae produce diaminopimelic acid but it is not known if walls other than those of bacteria, blue-green algae, the rickettsias and close relatives possess the 'key' substances of the amino sugar – peptide polymers. The search for the 'microbial missing links' in these groups could throw some light on the possible relationships within the Protista.

At the structural level it has been pointed out in Chapter 1 that bacteria and blue-green algae are less differentiated than other microbial groups. Their chromatinic structures are naked and are not surrounded by a nuclear membrane. Moreover, their electron transport – respiratory mechanisms are located in simple membrane systems and are not organized in the form of mitochondrial structures. A similar parallel also exists for the photosynthetic apparatus. The most conspicuous biochemical difference is the widespread inability of bacteria and blue-green algae to produce sterols. In these three features then, the bacteria and closely related microorganisms could be regarded as the more primitive.

Studies of the bacterial cell wall have given us new knowledge of the occurrence of substances largely confined to these microorganisms. Although much remains to be done on the structure of wall polymers some of the chemical and biochemical features of bacteria are now more readily understood. The next decade of research in this field will undoubtedly lead to a solution of the major problems of wall biosynthesis and the nature of the repeating units of the glycosaminopeptides. The wall is but one of the principal structures of the bacterial cell and in order to fully understand the processes concerned with its biosynthesis it is inevitable that our attention will be drawn towards the more biochemically functional structure of the cell, the membrane system. Much less is now known about the nature of the bacterial membrane than the cell wall and we have now reached the stage where the progress of further work on wall formation will require a greater knowledge of the chemistry and enzymic properties of the cell membranes. From the future investigations of the nature of membranes and our present knowledge of the bacterial walls we will be in a strong position to trace the similarities and dissimilarities of the major structures of a wide variety of cells and it would be intrigueing to discover whether any unusual components of the bacterial membrane can also be traced into structures of the cells of other microbial groups and higher organisms.

APPENDIX

TABLES 1 to 67

TABLE 1

A COMPARISON OF STRUCTURAL AND BIOCHEMICAL FEATURES OF CERTAIN MICROORGANISMS WITH THOSE OF PLANT AND ANIMAL CELLS

Cells from:	Nuclei	Reticulum
Bacteria	nucleoplasm, no enveloping membrane	endoplasmic reticulum no detectable
Blue-green Algae	nucleoplasm, no enveloping membrane	endoplasmic reticulum no tinguishable from photosy tic lamellae
Yeasts and Fungi	nuclei with well-defined membranes*	reticulum present at least in species
Plants	nuclei with well-defined membranes	endoplasmic reticulum pres
Animals	nuclei with well-defined membranes	endoplasmic reticulum pres

* Membranes around nuclei detected in thin sections of fungi but not in stained preparation of some species (ROB 1957).

:hondria	Walls and Membranes	Chemical features
ized mitochondria ab- membranes and invagi- membranous 'mesoso- chromatophore mem- s in photosynthetic or- ms	rigid walls and plasma mem-branes or multilayered envelope structures	cell-wall glycosamino-peptides (mucopeptides) present;** sterols not produced
ized mitochondria ab- membranes, photosyn- lamellae	rigid walls or multilayered envelope structures	cell-wall glycosamino-peptides (mucopeptides) present;** sterols not produced
hondria present; orga- mitochondria absent in)bic yeasts – membra- :eticulum found	rigid walls and plasma mem-branes	glycosaminopeptides absen cell-wall polysaccharides; sterols produced
hondria; chloroplasts	rigid walls and plasma mem-branes	glycosaminopeptides ab-sent; cell-wall polysac-charides; sterols produced
hondria	limiting surface 'unit mem-brane'	lipo-protein surface mem-branes, sterols produced

ls possessing glycosaminopeptides show some degree of sensitivity to the penicillins; other cells are unaffected.

TABLE 2

A COMPARISON OF CHEMICAL CONSTITUENTS OF CAPSULAR SUBSTANCES AND CELL WALLS OF BACTERIA

	Characteristic components identified
CAPSULES	
Bacillus anthracis	Polypeptide – D-glutamic acid
Bacillus megaterium	Polysaccharide – amino sugars, glucosamine, galactosamine and muramic acid?
Klebsiella-Aerobacter	Polysaccharides – uronic acid, fucose and hexose
Pneumococcus (various types) type VI	Polysaccharides – hexoses, uronic acids, deoxyhexoses and amino sugars Polysaccharide – rhamnose, galactose, ribitol-phosphate
Staphylococcus aureus	Polysaccharide? – amino sugar, 'wall' amino acids, phosphorus
Streptococcus sp.	Polysaccharide – rhamnose, galactose, uronic acid
WALLS	
Bacillus spp.	Amino sugar-peptide – Amino acids including D-glutamie acid, glucosamine, muramic acid Polyol-phosphate – ribitol phosphate, sugar and D-alanine (in some spp.)
Staphylococcus aureus	Amino sugar-peptide – amino acids including D-glutamic acid, glucosamine and muramic acid Polyol-phosphate – ribitol phosphate, amino sugar, D-alanine
Streptococci	Amino sugar-peptide – amino acids including D-glutamic acid, glucosamine, muramic acid Polysaccharide – rhamnose and other sugars or amino sugars

TABLE 3

A COMPARISON OF THE COMPOSITION OF WALLS FROM NORMAL CELLS AND PENICILLIN-INDUCED SPHEROPLASTS OF *Vibrio metchnikovi* and *Salmonella gallinarum*

	% dry weight	
	Cell walls	Spheroplast walls
Vibrio metchnikovi		
Polysaccharide	12.3	11.4
Lipid	11.2	10.0
Amino sugar	1.8	1.5
Diaminopimelic acid	0 6	0.3
Salmonella gallinarum		
Polysaccharide	28	28
Lipid	22	19.5
Amino sugar	3.5	2.3
Diaminopimelic acid	1.4	0.9

TABLE 4

PROPERTIES OF THE SURFACE ENVELOPE OF VARIOUS BACTERIA

	Organism	Structures present
GROUP 1		
	Mycoplasma spp.	Single multilayered membrane
	cocci in	Wall
	Mycoplasma sp.	Multilayered membrane
	Halobacterium spp.	Single multilayered membrane
	Protoplasts of *B. megaterium* *M. lysodeikticus*	Single multilayered plasma membrane
GROUP 2		
	Gram-positive bacteria (various species)	Wall
		Plasma membrane
	Nocardia calcarea	Microcapsule
		Wall
		Plasma membrane
GROUP 3		
	Gram-negative bacteria	Double 'membrane' envelope
	e.g. *Escherichia coli*	Multilayered 'wall' or compound memb
		Multilayered plasma membrane
	Spirillum spp.	Multilayered 'wall' or compound meml
		Multilayered plasma membrane
	Spheroplasts and L-forms of Gram-negative bacteria	Multilayered 'wall' or compound meml
		Multilayered plasma membrane
GROUP 4		
	Lampropedia hyalina	Outer envelope with mesh and punctate
		Intercalating layer
		Wall
		Membrane
	Tetrad-forming coccus and *Micrococcus radiodurans*	Complex envelope, cell wall and plasma brane

TABLE 4 (continued)

Thickness	Special features
75 Å	Glycosaminopeptide constituents absent
150 Å	
75 Å	
c. 75 Å	Glycosaminopeptide constituents absent
	Spherical macromolecules (c. 100 Å) present
~75 Å	Glycosaminopeptide constituents absent
150–800 Å	Glycosaminopeptide component in all walls; wall usually amorphous, may show some differences in electron density
75 Å	
50 Å	Microcapsular layer of very uniform thickness
150 Å	
~75 Å	
60–80 Å	Glycosaminopeptide as rigid component
~75 Å	
~100 Å	Glycosaminopeptide as rigid component;
	Hexagonally packed spherical macromolecules, 80–120 Å*
~75 Å	
~75 Å	Glycosaminopeptide constituents reduced, outer component no longer rigid
~75 Å	
	Repeat distance in mesh-like layer 145 Å
	No chemical evidence about the nature of the various components
	Hexagonally packed sub-units

:tron micrographs of MURRAY (1963b) show even greater complexity of this layer.

TABLE 5

ENZYMIC ACTIVITIES LOCALIZED IN ENVELOPES (CELL-WALL MEMBRANES) AND MEMBRANES OF GRAM-POSITIVE BACTERIA

ENVELOPES FROM:	ENZYMIC ACTIVITIES	REFERENCE
Lactobacillus arabinosus	ATP ase? Hexokinase	HUGHES, 1962
MEMBRANES FROM:		
Bacillus megaterium	$NADH_2$ oxidase, succinic, malic and lactic dehydrogenases	STORCK AND WACHSMAN, 1957 WEIBULL, BECKMAN AND BERGSTROM, 1959
	Incorporation of ^{32}P into lipids	HILL, 1962
Micrococcus lysodeikticus	Succinic dehydrogenase	LUKOIANOVA, GELMAN AND BIRIUSOVA, 1961
	$NADH_2$ oxidase, ATP ase	SALTON (unpublished)
Staphylococcus aureus	Succinic, malic, lactic, formic and α-glycerophosphoric dehydrogenases, acid phosphatase	MITCHELL AND MOYLE, 1956
Streptococcus (group A)	Synthesis of hyaluronate	MARKOVITZ AND DORFMAN, 1962
Streptococcus faecalis	ATP ase, polynucleotide phosphorylase	ABRAMS, MCNAMARA AND JOHNSON, 1960 ABRAMS AND MCNAMARA 1962

TABLE 6

ENZYMIC ACTIVITIES LOCALIZED IN ENVELOPE PREPARATIONS (WALL – MEMBRANES) FROM GRAM-NEGATIVE BACTERIA

ENVELOPES FROM:	ENZYMIC ACTIVITIES
Alcaligenes faecalis	Oxidative phosphorylation, ATP ase and dehydrogenases
Azotobacter agilis	Hydrogenase, lactic, malic and succinic dehydrogenases, $NADH_2$ and $NADPH_2$ oxidases
Escherichia coli	Hydrogenase, dehydrogenases, ATP ase, $NADH_2$ oxidase
*Pseudomonas aeruginosa**	Glucose oxidase, succinic dehydrogenase, $NADH_2$ oxidase, ATP ase
Pseudomonas fluorescens	Nicotinic hydroxylase, succinic dehydrogenase, $NADH_2$ oxidase, ATP ase and malic oxidase

Data from summaries by HUGHES, 1962 and MARR, 1960b.
* Data from CAMPBELL, HOGG AND STRASDINE, 1962.

TABLE 7

A COMPARISON OF THE COMPOSITION OF CELL WALL AND 'MEMBRANE' FRACTIONS OF *Micrococcus lysodeikticus* AND *Bacillus megaterium*

Constituent	*Micrococcus lysodeikticus* % dry weight		*Bacillus megaterium* % dry weight	
	Wall	Membrane* fraction	Wall	Membrane fraction
Nitrogen	7.6	8.4	7.4–7.8	10.3–10.9
Phosphorus	0.22	1.16	3.4–3.5	
Lipid	0	28.0	4.2–5.9	15.9–20.9
Mannose	0	18.9		
Glucose	3.5–5.8	0	0.3–0.9**	1.8–9.8**
Amino sugar	16–22	2.7	7–9	⟨ 0.1

Data from GILBY, FEW AND MCQUILLEN, 1958; WEIBULL AND BERGSTRÖM, 1958.
* Contains plasma and mesosome membrane structures as shown by SALTON AND CHAPMAN, 1962.
** 'Hexose' values

TABLE 8

MAJOR CLASSES OF CELL-WALL CONSTITUENTS AND THE GRAM REACTION OF CERTAIN MICROORGANISMS

Gram-positive organisms	Major chemical components of cell walls
YEASTS	
Saccharomyces cerevisiae	Polysaccharide, protein, lipid
Candida spp.	Polysaccharide, protein
BACTERIA	
Micrococcus lysodeikticus	Glycosaminopeptide, polysaccharide
Staphylococcus aureus	Glycosaminopeptide, teichoic acid
Streptococci	Glycosaminopeptide, teichoic acid, polysaccharide
Corynebacterium spp.	Glycosaminopeptide, polysaccharide
Mycobacterium spp.	Glycosaminopeptide, glycolipid, peptidoglycolipid
GRAM-NEGATIVE ORGANISMS	
Escherichia coli *Aerobacter aerogenes* *Pseudomonas* spp. *Spirillum serpens*	Protein, polysaccharide, lipid, glycosaminopeptide

TABLE 9

THE RELATIONSHIP BETWEEN THE GRAM REACTION AND THE LEAKAGE OF ^{32}P COMPOUNDS FROM LABELLED BACTERIA EXPOSED TO ABOUT 100% ENTHANOL

Gram reaction	Organisms	^{32}P released (%) relative to maximum*
—	*Pseudomonas* sp.	96
—	*Proteus vulgaris*	90
—	*Escherichia coli*	84
—	*Alcaligenes faecalis*	78
—	*Vibrio metchnikovi*	75
—	*Salmonella gallinarum*	75
—	*Aerobacter aerogenes*	75
—	*Chromobacterium prodigiosum*	69
—	*Neisseria catarrhalis*	65
—	*Pseudomonas fluorescens*	60
—	*Spirillum serpens*	56
—	*Escherichia dispar*	54
—	*Bacillus brevis*	54
+	*Leuconostoc mesenteroides*	43
+	*Clostridium perfringens*	36
+	*Micrococcus lysodeikticus*	35
+	*Candida pulcherrima*	34
+	*Streptococcus faecalis*	33
+	*Clostridium sporogenes*	30
+	*Lactobacillus arabinosus*	30
+	*Staphylococcus aureus*	26
+	*Bacillus megaterium*	22
+	*Corynebacterium hoffmanni*	17
+	*Corynebacterium xerosis*	13
+	*Bacillus cereus*	11
+	*Saccharomyces cerevisiae*	10

* Maximum ^{32}P released from organism usually occurred in 50–75% (v/v) ethanol in water solution.

TABLE 10

CELL-WALL CONTRIBUTION TO CELL DRY WEIGHT

Organisms	Cell walls as % dry weight bacteria
Bacillus megaterium	20–25
Escherichia coli	15
Micrococcus lysodeikticus	20–35
Myxococcus xanthus	7–8
Staphylococcus aureus	20
Streptococcus faecalis	27 (exponential phase) 38 (stationary phase)

TABLE 11

DETERMINATIONS OF THE THICKNESS OF BACTERIAL CELL WALLS

Organisms	Wall thickness (Å)
Escherichia coli	~75–100
Halobacterium halobium	~75
Lactobacillus acidophilus	800
Marine pseudomonad (NCMB 845)	~75
Micrococcus lysodeikticus	400–500
Mycobacterium tuberculosis	350
Myxococcus xanthus	250
Staphylococcus aureus	150–200
Streptococcus faecalis	200

TABLE 12

FINE STRUCTURE OF MICROBIAL CELL WALLS AND ENVELOPES REVEALED BY ELECTRON MICROSCOPY

Organisms	Type of fine structure
ALGA	
Chlorella pyrenoidosa	Double-layered, micro-fibrillar polysaccharide (fibres at 90° to one another) + amorphous matrix
YEAST	
Saccharomyces cerevisiae	Multi-layered wall, micro-fibrallar layer (fibres 90° to one another); fibres oriented around bud scars
BACTERIA	
Escherichia coli	Multilayered (2 electron dense; 1 electron transparent layers); continuous rigid layer present
Halobacterium halobium *Spirillum* sp. *Spirillum serpens* *Rhodospirillum rubrum*	Multilayered structures with 'spherical' macromolecules (80–145 Å spacings) visible; hexagonal packing; layers of considerable complexity revealed in PTA preparations of *S. serpens*
Lampropedia hyalina	Outer envelope with a punctate layer and another structured layer having a crystalline lattice appearance
Bacillus megaterium	Fibrous ?
Staphylococcus aureus *Streptococcus faecalis*	Amorphous structure – thickened bands at zone of wall formation

TABLE 13

DISSOLUTION OF THE WALLS OF GRAM-NEGATIVE BACTERIA BY SODIUM DODECYL SULPHATE (0.2% W/V) AT PH 7 AND 8

Organisms	% Turbidity decrease (in 30 min at 37°)	
	pH 7	pH 8
Escherichia coli	70.6	79.7
Pasteurella pseudotuberculosis	83.4	92.0
Proteus vulgaris	84.4	92.4
Pseudomonas aeruginosa	70.6	84.4
Salmonella gallinarum	80.6	92.6
Vibrio metchnikovi	38.5	93.7

TABLE 14

SOLUBILITY PROPERTIES OF ISOLATED WALLS AND CELL-WALL POLYMERS IN VARIOUS REAGENTS

GRAM-POSITIVE ORGANISMS	
Cell walls and glycosaminopeptide	Solubilized by: 0.5 *N* NaOH in N_2 atmosphere; 10% sodium hypochlorite in the cold. Insoluble in hot formamide, 90% phenol, hot saturated picric acid, sodium dodecyl sulphate solutions
Teichoic acids	Soluble in trichloroacetic acid (5–10%) in the cold
Oligo and polysaccharides	Soluble in hot formamide, hot saturated picric acid, hot trichloroacetic acid
GRAM-NEGATIVE ORGANISMS	
Cell walls	Disaggregated by sodium dodecyl sulphate, 90% phenol (incomplete)
Protein-lipid-polysaccharide complexes	90% phenol, hot 45% phenol, sodium dodecyl sulphate, trichloroacetic acid, diethylene glycol (partial solubility depending on the particular component)
Glycosaminopeptide	Insoluble in phenol and dodecyl sulphate

TABLE 15

COMPOSITION OF THE CELL WALLS OF SEVERAL GRAM-POSITIVE BACTERIA

	% N	% P	% Reducing* substances
Bacillus megaterium	5.3	0.42	48
Bacillus subtilis	5.1	5.35	34
Micrococcus lysodeikticus	8.7	0.09	45
Sarcina lutea	7.6	0.22	46
Streptococcus faecalis	5.6	1.88	61

* Estimated after hydrolysis with 2 *N* HCl for 2 h, expressed as glucose equivalent.

TABLE 16

COMPOSITION OF THE CELL WALLS OF SEVERAL GRAM-NEGATIVE BACTERIA

	% N	% P	% Reducing substances*
Escherichia coli	10.1	1.52	16
*Pseudomonas aeruginosa***	8.4	1.7	8.0
*Pseudomonas aeruginosa***	10.1	1.6	12.1
Rhodospirillum rubrum	8.3	—	23
*Salmonella bethesda***	8.0	1.7	18.5
Salmonella pullorum	6.4	0.88	46

* Estimated after hydrolysis with 2 *N* HCl for 2 h, expressed as glucose equivalent
** Data from Collins, 1963.

TABLE 17

CHEMICAL COMPOSITION OF BACTERIAL ENDOSPORE WALLS ('COATS')

Organism	% N	% P	% Lipid	Reference
Bacillus cereus	13.2	1.2	0.9	Strange and Dark, 1956
Bacillus cereus	10.7	0.7	7.3	Yoshida *et al.*, 1956
Bacillus megaterium	12.8	0.3	—	Strange and Dark, 1956
Bacillus subtilis	13.1	1.6	1.1	Strange and Dark, 1956
Bacillus subtilis	12.9	1.4	3.0	Salton and Marshall, 1959

TABLE 18

A COMPARISON OF THE CHEMICAL COMPOSITION OF SPORE WALLS AND WALLS OF VEGETATIVE CELLS OF *Bacillus subtilis*

	Dry weight (%)	
	Spore walls	Cell walls
Total N	12.9	4.6
Total P	1.4	4.2
Total amino sugars (as glucosamine)	2.2	10.7
Glucosamine	1.5	7.9
Muramic acid	0.6 (1.1)*	2.3 (4.1)*
Diaminopimelic acid	c. 1	5.6
Lipid		
a. ether extractable	1.4	0.2
b. total**	3.0	0.7
O-ester groups (as O-acetyl)	0.027	0.255

* Values in brackets represent muramic acid contents corrected by the factor of 1.78 (STRANGE AND DARK, 1956).
** 'Total lipid' estimated by ether extraction after hydrolysis according to SALTON, 1953.

TABLE 19

CALCIUM AND MAGNESIUM CONTENTS OF WHOLE CELLS AND ISOLATED CELL WALLS OF *Rhizobium trifolii* (FROM HUMPHREY AND VINCENT, 1962)

	mM/g dry weight	
	Calcium	Magnesium
Whole cells (Normal)	0.062	0.064
Cell walls (Normal)	0.081	0.016
Cell walls (EDTA-treated)	0.010	0.005
Whole cells (Ca^{2+} deprived*)	0.007	0.154
Cell walls (Ca^{2+} deprived*)	0.047	0.024

* Cells grown in the absence of added Ca^{2+}

TABLE 20

A COMPARISON OF CELL-WALL 'POLYSACCHARIDE' CONTENTS DETERMINED BY THREE DIFFERENT METHODS

Walls from	'Polysaccharide' contents as % dry wt cell walls			
	Batch No.	Reducing value* Hagedorn & Jensen	Anthrone*	Glucose oxidase
B. megaterium	1	45.6	19.4	13.7
	2	40.4	12.1	9.4
M. lysodeikticus	1	42.6	7.5	3.5
	2	43.1	7.5	4.3
	3	44.7	—	5.8
S. lutea	1	42.5	17.1	9.5
	2	42.2	15.0	8.5

* Results expressed in terms of glucose standards.

TABLE 21

PRINCIPAL CLASSES OF CHEMICAL CONSTITUENTS FOUND IN CELL WALLS OF BACTERIA, BLUE-GREEN ALGAE AND RICKETTSIAS

BACTERIA	
Gram-positive	Glycosaminopeptides Oligosaccharides Polysaccharides Teichoic acids Teichuronic acids Glycolipids and Mycosides
Gram-negative	Glycosaminopeptides Teichoic acid (*Escherichia coli*) Proteins Lipids Polysaccharides Lipo-polysaccharides Lipo-protein ?
BLUE-GREEN ALGAE	Glycosaminopeptides Protein in some Carotenoids
RICKETTSIAS AND PSITTACOSIS VIRUSES	Glycosaminopeptides and other components

TABLE 22

AMINO ACID COMPOSITION OF CELL WALLS OF VIRULENT AND AVIRULENT STRAINS OF GROUP A STREPTOCOCCI BEFORE AND AFTER DIGESTION WITH TRYPSIN

Amino acid	Virulent Q43 cell wall		Avirulent Q496 cell wall	
	Untreated μ mole/mg N	Trypsin-residue μ mole*	Untreated μ mole/mg N	Trypsin-residue μ mole*
Glutamic acid	6.14	2.80	5.79	2.88
Lysine	4.56	2.76	4.19	2.68
Alanine	13.30	10.60	12.70	9.50
Glucosamine	4.86	4.83	4.74	4.86
Muramic acid	0.88	0.90	0.80	0.78
Aspartic acid	2.98	0.53	3.37	0.57
Threonine	1.38	0.31	1.77	0.32
Serine	1.31	0.32	1.67	0.32
Glycine	1.77	0.38	1.97	0.35
Valine	0.92	—	1.50	—
Isoleucine	0.66	0.17	1.22	0.17
Leucine	1.75	0.25	1.87	0.24
Tyrosine	0.62	0.15	0.80	0.18
Phenylalanine	0.58	0.15	0.62	0.21
Arginine	1.02	—	1.25	—
Methionine	0.30	—	0.52	—
Proline	1.04	—	1.30	—
Ammonia	6.93	4.80	10.20	5.59

* Values adjusted to represent residues of walls analyzed as 'untreated' walls.
Data from TEPPER, HAYASHI AND BARKULIS, 1960.

TABLE 23

PRINCIPAL COMBINATIONS OF MAJOR AMINO ACID CONSTITUENTS FOUND IN WALLS OF GRAM-POSITIVE BACTERIA

GROUPS	AMINO ACIDS
Actinomyces spp.	Alanine, glutamic acid and lysine
Aerococci	
Arthrobacter spp.	
Bacillus sphaericus	
Corynebacterium insidiosum	
Corynebacterium pyogenes	
Lactobacilli	
Nocardia salivae	
Streptococci	(aspartic acid in some)
Corynebacterium tritici	Alanine, glutamic acid and glycine
Corynebacterium spp.	Alanine, glutamic acid, lysine and glycine
Lactobacillus bifidus	
Micrococci	
Staphylococci	(serine in some)
Bacilli	Alanine, glutamic acid, diaminopimelic acid
Corynebacteria	
Leptotrichia spp.	
Listeria monocytogenes	
Mycobacteria	
Nocardia	
Propionibacterium shermanii	
Arthrobacter spp.	Alanine, glutamic acid, diaminopimelic acid and glycine
Clostridia	
Corynebacterium simplex	
Micrococcus varians	
Micromonospora	
Nocardia pelletieri	
Propionibacteria	
Streptomyces	

TABLE 24

THE AMINO ACID COMPOSITION OF THE ISOLATED CELL WALLS OF SEVERAL GRAM-POSITIVE BACTERIA

Amino acid	*Lactobacillus casei*★ (g/100 g)	*Streptococcus faecelis*★ (g/100 g)	*Streptococcus* group A★★ (μ mole)
Alanine	8.4	12.0	10.60
Arginine	—	—	—
Aspartic acid	3.1	0.8	0.53
Diaminopimelic acid	6.7	—	—
Glutamic acid	7.2	5.4	2.80
Glycine	1.0	0.2	0.38
Histidine	—	—	—
Isoleucine	1.4	0.4	0.17
Leucine	1.4	0.4	0.25
Lysine	2.3	4.5	2.76
Methionine	—	—	—
Phenylalanine	—	—	0.15
Proline	—	—	—
Serine	0.6	0.2	0.32
Threonine	1.1	0.2	0.31
Tyrosine	—	—	0.15
Valine	1.4	0.24	—

★ Data from IKAWA AND SNELL, 1956.
★★ Data from TEPPER, HAYASHI AND BARKULIS, 1960 for strain Q43.

TABLE 25

AMINO ACID COMPOSITION OF CELL WALLS OF SEVERAL GRAM-NEGATIVE BACTERIA

Amino acid	g/100 g wall			μ moles†
	Escherichia★ *coli*	*Pseudomonas*★★ *aeruginosa*	*Salmonella*★★ *bethesda*	*Aerobacter cloacae*
Alanine	5.6	5.1	4.3	0.75
Arginine	3.8	1.3	0.4	0.25
Aspartic acid	7.1	9.3	14.5	0.75
Diaminopimelic acid	+	3.2	2.1	0.20
Glutamic acid	6.9	5.8	4.9	0.62
Glycine	3.1	7.1	5.5	0.44
Histidine	0.9	—	—	0.06
Isoleucine	3.7	} 2.8	} 1.6	0.18
Leucine	5.3			0.37
Lysine	4.0	1.7	0.5	0.29
Methionine	0.7	—	—	—
Phenylalanine	3.0	7.3	4.1	0.18
Proline	1.5	—	—	0.17
Serine	3.7	5.4	2.6	0.32
Threonine	3.8	—	—	0.27
Tyrosine	3.3	4.2	2.5	0.17
Valine	3.4	3.9	2.0	0.36

'+' present; '—' not estimated.
★ SALTON, unpublished data.
★★ From COLLINS, 1963.
† Data from SCHOCHER *et al.*, 1962.

TABLE 26

AMINO ACID COMPOSITION OF SPORE COATS OF *Bacillus coagulans* AND *Bacillus subtilis*

Amino acids	μM/ml hydrolysate*	*B. coagulans**	*B. subtilis*
		molar ratios**	
Lysine	5.577	6	6
Histidine	2.941	3	1
Aspartic acid	2.851	3	1
Glycine	2.832	3	5
Alanine	2.637	3	4
Tyrosine	2.579	3	1.5
Glutamic acid	2.380	3	3
Arginine	1.688	1.5	tr.
Phenylalanine	1.640	1.5	tr.
Proline	1.594	1.5	tr.
Leucine	1.444	1.5	2
Valine	1.104	1	2
Serine	1.018	1	4
Cystine	1.011	1	2
Methionine	1.012	1	tr.
Isoleucine	0.929	1	2
Threonine	0.822	1	1
Diaminopimelic acid	no. std. deter.	1	1

* Determined by Spinco amino acid analyzer.
– From HUNNELL AND ORDALL, 1960.
** Determined by paper chromatography SALTON AND MARSHALL, 1959.

TABLE 27

MOLAR RATIOS OF AMINO ACIDS IN CELL WALLS OF VARIOUS STRAINS OF *Staphylococcus aureus*

Strain	Amino acid				Reference
	Alanine	Glutamic acid	Lysine	Glycine	
Duncan	2.8	1.0	1.9	6.3	HANCOCK AND PARK, 1958
Duncan	2.8	1.0	0.9	4.4	SALTON AND PAVLIK, 1960
H	3.2–3.7	1.0	0.9–1.2	–*	PARK AND STROMINGER, 1957
Oxford	1.9	1.0	1.0	4.9	ROGERS AND PERKINS, 1959
Oxford	1.4, 1.7	1.0	0.87, 0.71	2.50, 3.10	ROGERS AND JELJASZEWICZ, 1961
Oxford penicillin-resistant	1.94	1.0	1.04	3.43	ROGERS AND JELJASZEWICZ, 1961
11	1.4	1.0	0.5	0.05	ROGERS AND PERKINS, 1959
524/SC	1.8	1.0	0.6	3.2	MANDELSTAM AND ROGERS, 1959

* Not determined.

TABLE 28

MOLAR RATIOS OF THE PRINCIPAL AMINO ACIDS IN CELL WALLS

Walls from	Lysine	Glutamic acid	Glycine	Serine	Alanine
Bacillus sp*	1	1.7	0.5	0.3	2.3
Corynebacterium sp.	1	1	0**	0.7	3.9
Micrococcus citreus	1	3	0.8	0	2.1
— *lysodeikticus*	1	1	1	0	2.6
— *roseus*	1	1.1	0	0	5.1
— *tetragenus*	1	1.2	1.2	0	2.3
— *urea*	1	1.3	1	0	2.3
Sarcina flava	1	1.4	1	0	2.2
— *lutea*	1	1.6	1	0	2.0
Sporosarcina ureae	1	1.7	1	0	2.0
Staphylococcus albus	1	1.1	4.8	0.4	2.9
— *aureus*	1	1.1	4.8	0.45	3.0
— *citreus*	1	1	4	0.5	3.1
— *saprophyticus*	1	1	4.6	0.6	3.3
Streptococcus faecalis	1	0.9	0	0	4

* A high proportion of threonine (0.7) was also present.
** '0' used to designate absence or only faint traces of amino acids.
Data from SALTON AND PAVLIK, 196c serine values for the 4 staphylococci incorrectly shown have been corrected above.

TABLE 29

MOLAR RATIOS OF THE PRINCIPAL AMINO ACIDS IN CELL WALLS

Walls from	DAP*	Glutamic acid	Glycine	Alanine
Bacillus cereus	1	1.3	0	2.6
— *megaterium*	1	1.8	0	2.8
— *pumilis*	1	1.6	0	4.6
— *stearothermophilis*	1	2.0	0	3.8
— *subtilis*	1	2.4	0	4.3
— *thuringiensis*	1	1.4	0	2.8
Micrococcus varians	1	4.3	1.8	2.6
Lactobacillus arabinosus	1	1.1	0	2.9

*α, ε-diaminopimelic acid.
Data from SALTON AND PAVLIK, 1960.

TABLE 30

MOLAR RATIOS OF AMINO ACIDS IN ALKALI-INSOLUBLE WALL FRACTIONS FROM GRAM-POSITIVE BACTERIA STUDIED BY KANDLER AND HUND, 1959

Organism	Alanine	Glutamic acid	Lysine	DAP	Glycine
L. acidophilus	1.00	0.49	1.04	—*	0.25
L. plantarum	1.00	0.64	0.07	0.50	0.12
Strep. lactis	1.00	0.73	1.34	—	0.17
Strep. cremoris	1.00	0.58	1.16	—	0.61
Leuc. mesenteroides	1.00	0.46	1.08	—	0.15
Ped. cerevisiae	1.00	0.60	1.06	—	0.14
Prop. shermanii	1.00	0.53	0.15	0.56	0.23
Prop. zeae	1.00	0.75	0.24	1.00	0.65
Bacillus cereus	1.00	0.67	0.10	0.64	0.14
Bacillus licheniformis	1.00	0.84	0.44	0.54	0.26
Bacillus sphaericus	1.00	0.57	0.86	—	0.25

* Not present.

TABLE 31

MOLAR RATIOS OF AMINO ACIDS IN GLYCOSAMINOPEPTIDE FRACTIONS ISOLATED FROM WALLS OF GRAM-NEGATIVE BACTERIA

Organism	Amino acid				Reference
	Alanine	Glutamic acid	DAP	Lysine	
Escherichia coli ATCC 9637	1.50	1.0	0.71	0.19	MANDELSTAM, 1962
— *coli* B	1.47	1.0	0.74	0.23	MANDELSTAM, 1962
— *coli* K12	1.50	1.0	0.68	0.20	MANDELSTAM, 1962
Cit. freundii	1.50	1.0	0.70	0.21	MANDELSTAM, 1962
Pr. vulgaris	1.60	1.0	0.73	0.20	MANDELSTAM, 1962
Pseudomonas fluorescens KB1	1.50	1.0	0.76	0.18	MANDELSTAM, 1962
Ser. marcescens	1.60	1.0	0.78	0.12	MANDELSTAM, 1962
K. pneumoniae	1.46	1.0	0.76	0.15	MANDELSTAM, 1962
Aerobacter cloacae	1.5	1.0	1.0	—*	SCHOCHER, BAYLEY AND WATSON, 1962
Escherichia coli B	1.6	1.0	0.9	0.23	MARTIN AND FRANK, 1962
Spirillum sp.	2.2	1.0	1.0	—	MARTIN AND FRANK, 1962

* '–' Not reported.

TABLE 32

MOLAR RATIOS OF PRINCIPAL AMINO ACIDS IN CELL WALLS OF BLUE-GREEN ALGAE

Organism	Amino acid				
	Alanine	Glutamic acid	DAP	Glycine	Serine
Microcoleus vaginatus[1]	1.84	1.0	0.86	0.15	0.08
Phormidium uncinatum[2]	2.1	1.0	1.06	0.12	0.08

[1] Data from SALTON, unpublished results.
[2] Data from FRANK, LEFORT AND MARTIN, 1962.

TABLE 33

AMINO ACID COMPOSITION OF LACTIC ACID BACTERIA

	mg/100 mg cell wall		
	Streptococcus faecalis	*Lactobacillus plantarum*	*Leuconostoc citrovorum*
Glutamic acid *L*	0.6	0	0.9
D	4.6	7.6	10.4
Alanine (Total)	4.4	11.6	9.8
D	1.7	3.7	4.6
Aspartic acid (Total)	2.4	0.6	8.1
L	0.7		1.8
Lysine (Total)	2.5	0.4	5.6
L	2.4	0.5	6.2
DAP	0	5.2	0
α-Aminosuccinoyl-lysine	0	0	4.4
Ammonia	1.1	2.4	3.3

Data from IKAWA AND SNELL, 1960.

TABLE 34

PERCENTAGE OF GLUTAMIC ACID, ASPARTIC ACID, AND ALANINE IN THE D-CONFIGURATION IN CELL WALLS

	% of total in D-form		
	Glutamic acid	Aspartic acid	Alanine
S. faecalis	85	71	39
L. casei	100	50	61
L. plantarum	100		32
L. mesenteroides	73	67	54
L. pentosus	94		66
L. citrovorum	89	78	47
L. bulgaricus	87	72	40
L. lactis	94	78	61
L. acidophilus	91	67	48

Data from IKAWA AND SNELL, 1960.

TABLE 35

DIAMINOPIMELIC ACID ISOMERS IN BACTERIAL CELL WALLS

Organism	Isomer
Gram-negative bacteria (various species)	DL- α, ε-diaminopimelic acid
Bacillus spp.	
Corynebacteria	
Lactobacilli	
Leptotrichia spp.	
Micrococcus varians	
Mycobacteria	
Nocardia	
Propionibacterium shermanii	
Bacillus megaterium	DL- and LL-α, ε-diaminopimelic acid
Micromonospora spp.	
Nocardia pelletieri	
Actinomyces hominis	LL-α, ε-diamonopimelic acid
Arthrobacter spp.	
Clostridium perfringens	
Corynebacterium simplex	
Propionibacterium spp.	
Streptomyces spp.	

TABLE 36

FREE AMINO GROUPS OF CELL WALLS DETERMINED BY REACTION WITH FLUORODINITROBENZENE

Walls from	μ moles/g wall (corrected for losses)							
	Ala	Asp	Glu	Gly	Lys	DAP	ε-NH_2 groups	
							Lys	DAP
B. cereus	4	0	0	0	0	1	0	35
B. licheniformis	50	0	0	0	0	1	0	130
B. megaterium	11	0	0	0	0	1	0	230
L. mesenteroides	45	12	0	0	0	0	24	0
M. citreus	2	0	43	0	0	0	4	0
M. lysodeikticus	15	0	tr.	0	5	0	400	0
ML-NDF*	45	0	0	0	0	0	420	0
M. *varians*	1	0	19	0	0	0	0	0
S. lutea	10	0	—	—	2	0	360	0
S. albus	20	0	0	5	0	0	160	0
S. aureus	180	0	0	32	0	0	2	0
S. citreus	7	0	tr.	35	0	0	5	0
E. coli	2	0	tr.	tr.	0	0	—	—
K. aerogenes	1	—	—	tr.	—	—	55	35
Pseudomonas sp.	7	—	—	—	—	2	55	140
S. gallinarum	3	—	tr.	—	—	—	—	—

* non-dialysable fraction from lysozyme-digested walls; SALTON, 1961.

TABLE 37

FREE AMINO GROUPS IN CELL WALLS AND TCA-INSOLUBLE RESIDUES DETERMINED BY REACTION WITH FLUORODINITROBENZENE

	Free Amino groups (μ moles/g wall)					
	Staphylococcus aureus			*Lactobacillus arabinosus*		
	Ala	Gly	ε-Lys	Ala	Glu	mono-NH_2-DAP
Wall	180	32	2	126	16	122
Cold-TCA-insoluble residue*	161	36		78	15	—
Hot TCA-insoluble residue**	98	116	15	24	30	230

* Wall extracted with 5% (w/v) TCA, 24 h, 4°
** Wall extracted with 5% (w/v) TCA, 10 min, 100°
All values corrected for losses.
Data from SALTON, 1961.

TABLE 38

C-TERMINAL RESIDUES OF WALLS LIBERATED BY HYDRAZINOLYSIS AT 100° FOR 8 H.

Walls from	μ moles/g wall (Corrected for losses)					
	Ala	Glu	Gly	DAP	Lys	Unknown compounds
B. cereus	34	8	0	5	0	None
B. licheniformis	8	tr.	0	4	0	1 or 2
B. megaterium	68	20	0	125	0	3
*L. mesenteroides**	16	0	0	0	70	None
M. citreus	25	92	0	0	60	2
M. lysodeikticus	76	0	500	0	0	None
ML-NDF	57	0	520	0	0	None
M. varians	18	15	14	14	0	2
S. lutea	16	120	370	0	tr.	tr.
S. albus	94	0	tr.	0	26	None
S. aureus	18	0	8	0	0	None
S. citreus	58	0	7	0	tr.	None
E. coli	14	13	5	10	+	1
K. aerogenes	18	tr.	5	7	+	2 or 3
Pseudomonas sp.	17	11	6	10	+	1 or 2
S. gallinarum	20	8	0	15	tr.	2

* Yielded in addition 6 μ moles Asp/g wall on hydrazinolysis.
0, not detected; +, present.
SALTON, 1961.

TABLE 39

C-TERMINAL GROUPS IN CELL WALLS AND TCA-INSOLUBLE RESIDUES DETERMINED BY HYDRAZINOLYSIS

	C-terminal groups (μ moles/g wall)			
	Staphylococcus aureus		*Lactobacillus arabinosus*	
	Ala	Gly	Ala	Glu
Wall	18	8	7	tr.
Cold-TCA-insoluble residue*	24	12	12	—
Hot-TCA-insoluble residue**	44	94	19	5

* Wall extracted with 5% (w/v) TCA, 24 h, 4°
** Wall extracted with 5% (w/v) TCA, 10 min, 100°
All values corrected for losses.
Data from SALTON, 1961.

TABLE 40

AMINO SUGAR CONTENTS OF CELL WALLS OF GRAM-POSITIVE AND GRAM-NEGATIVE BACTERIA DETERMINED BY THE ELSON AND MORGAN REACTION

Organism	% Amino sugar*
GRAM-POSITIVE	
*Bacillus brevis***	14.5
Bacillus cereus	31
Bacillus cereus var. *mycoides*	32
Bacillus megaterium	18
Bacillus pumilis	9.2
Bacillus sp.	6.7
Bacillus stearothermophilus	12.7
Bacillus subtilis	6.8
Bacillus thuringiensis	9.8
Candida pulcherrima	1.2
Corynebacterium sp.	14.3
Lactobacillus arabinosus	11.5
Micrococcus citreus	21.7
Micrococcus lysodeikticus	16–22
Micrococcus roseus	16
Micrococcus tetragenus	13
Micrococcus urea	19
Micrococcus varians	10
Saccharomyces cerevisiae	1–2
Sarcina lutea	12–16
Sporosarcina ureae	14
Staphyolcoccus albus	16.8
Staphylococcus aureus	17
Staphylococcus citreus	10
Staphylococcus saprophyticus	14
Streptococcus faecalis	22
GRAM-NEGATIVE	
Aerobacter aerogenes	2
Alcaligenes faecalis	7.6
Chromobacterium prodigiosum	2
Escherichia coli	3
Escherichia dispar	5.0
Neisseria catarrhalis	8.1
Proteus morganii	4.1
Proteus vulgaris	5.1
Salmonella gallinarum	3.5
Spirillum serpens	6.8
Vibrio metchnikovi	1.8

* Determined after hydrolysis of walls with 2*N* HCl, 2 h at 100°, expressed as glucosamine equivalents.

** Cells frequently stained Gram-negative at time of harvesting for wall preparation.

TABLE 41

EFFECT OF HYDROLYSIS CONDITIONS ON THE AMINO SUGAR CONTENT OF THE WALL OF *Micrococcus lysodeikticus*

Normality of hydrochloric acid	Amino sugar content as % dry wt. walls Hours of hydrolysis at 100°				
	1	2	5.25	12	20
1 *N*	17.5	18.8	19.0	17.0	16.4
2 *N*	19.2	19.0	19.0	17.8	18.0
4 *N*	19.8	20.8	—	17.8	17.5
6 *N*	22.0	24.0	22 2	19.0	17.7

TABLE 42

OCCURRENCE OF AMINO SUGARS IN CELL WALLS AND BACTERIAL POLYSACCHARIDES

WALLS	COMPONENT	AMINO SUGARS
Gram-positive and Gram-negative bacteria	Glycosaminopeptides	N-acetyl-D-glucosamine N-acetyl-3-O-carboxy-ethyl-D-glucosamine
Group A streptococci	Polysaccharides	N-acetylglucosamine
Group C streptococci	Polysaccharide	N-acetylgalactosamine
Staph. aureus	Teichoic acid	N-acetylglucosamine
Staph. albus	Teichoic acid	N-acetylgalactosamine
Strep. faecalis	Teichoic acid	N-acetylgalactosamine
B. subtilis	Teichuronic acid	N-acetylgalactosamine
M. lysodeikticus	Polysaccharide	N-acetylmannuronic acid
Mycobacteria	Mycosides (Peptido-glycolipids)	Glucosamine Galactosamine
CAPSULAR AND SOMATIC SUBSTANCES		
Pneumococcus type 10	Polysaccharide	Glucosamine Galactosamine
type 14	Polysaccharide	N-acetyl-D-glucosamine
strain 39458	Polysaccharide	Mannosamine
B. subtilis	Polysaccharide	Diaminohexose
Gram-negative bacteria	Somatic	Glucosamine
(various species)	O-lipopolysaccharides	Galactosamine
Chr. violaceum	lipopolysaccharide	Fucosamine

TABLE 43

OPTICAL ROTATION AND CHROMATOGRAPHIC BEHAVIOUR OF NATURAL AND SYNTHETIC MURAMIC ACID AND THE STEREO-ISOMER

	Average values derived from several experiments		
	Optical rotation $[\alpha]_D^{20}$	R_F*	$R_{glucosamine\ value}$** on Zeo-Karb 225 column eluted with 0.33N HCl
Natural muramic acid	+109	0.53	1.10
Synthetic muramic acid	+109	0.53	1.10
Stereoisomer of muramic acid	+ 52	0.44	0.87

* Values obtained with Whatman no. 1 paper and phenol-water as solvent.

** Values in this column have been reported by CRUMPTON, 1958. The $R_{glucosamine\ value}$ relates the elution characteristics.

Data from STRANGE AND KENT, 1959.

TABLE 44

PRINCIPAL COMBINATIONS OF MONOSACCHARIDE CONSTITUENTS FOUND IN WALLS OF GRAM-POSITIVE BACTERIA

Groups	Monosaccharides (Occurring singly or in various combinations)
Micromonospora Sporosarcina Staphylococci Streptomyces	None
Actinomyces Aerococci Arthrobacteria Bacilli Micrococci Staphylococci Streptomyces	<u>Glucose, galactose</u>, mannose
Clostridia Lactobacilli Propionibacteria Streptococci	<u>Rhamnose</u>, glucose, galactose, mannose
Corynebacteria Mycobacteria Nocardia	<u>Arabinose</u>, glucose, galactose, mannose
Actinomyces bovis *Mycobacterium avium*★	6-Deoxytalose and other sugars

N.B. The above monosaccharide constituents are 'characteristic' for the majority of species of the above groups but non-conforming species will appear in other groups of sugar combinations. The predominant and conspicuous monosaccharides of the groups have been underlined.

★ Found in the isolated glycolipid.

TABLE 45

A COMPARISON OF THE SUGAR AND AMINO SUGAR COMPOSITION OF THE WALLS, FORMAMIDE-SOLUBLE CARBOHYDRATE AND FORMAMIDE-INSOLUBLE RESIDUES OF GROUP A AND A VARIANT WALLS OF STREPTOCOCCI

		Strains of streptococci			
		Group A		Group A-variant	
		D58	T12	T27A	K43 var.
		%	%	%	%
Cell walls	Rhamnose	21.0	34.0	35.0	36.0
	Glucosamine	10.4	18.8	5.9	7.2
	Muramic acid	3.6	6.4	5.9	5.0
Formamide-extracted carbohydrate	Rhamnose	60.0	58.8	85.0	80.5
	Glucosamine	30.0	25.0	3.0	1.4
	Muramic acid	—	—	—	—
Formamide residue	Rhamnose	2.4	4.9	4.0	1.2
	Glucosamine	10.0	11.0	6.9	6.1
	Muramic acid	9.3	9.4	6.8	10.7

From Krause and McCarty, 1961.

TABLE 46

MONOSACCHARIDE CONSTITUENTS OF ISOLATED WALL OR ENVELOPE FRACTIONS FROM GRAM-NEGATIVE BACTERIA

WALLS FROM	*MONOSACCHARIDES DETECTED*
Alcaligenes faecalis	Glucose, arabinose, fucose, rhamnose, unknown
Bacterium cadaveris	Galactose, glucose, rhamnose
Chlorobium thiosulphatophilum	Galactose, glucose, mannose, rhamnose
Chromobacterium kilense	Galactose, glucose, heptose, mannose
Chromobacterium prodigiosum	Glucose, heptose, mannose, traces pentose (?) and rhamnose
Escherichia alkalescens	Galactose, glucose, rhamnose
Escherichia dispar	Galactose, glucose, rhamnose
Halobacterium halobium	Galactose
Halobacterium salinarium	Glucose
Marine pseudomonad (NCMB 845)	Glucose, heptose
Organism LC I	Galactose, glucose, rhamnose
Photobacterium albensis	Glucose, heptose
Photobacterium fischeri	Galactose, glucose
Proteus vulgaris	Glucose, heptose
Pseudomonas sp.	Glucose, fucose, rhamnose
Rhizobium trifolii	Glucose, rhamnose
Rhodospirillum rubrum	Glucose, fucose, rhamnose, unknown
S. enteritidis (wild type)	Galactose, glucose, mannose, rhamnose, tyvelose
S. enteritidis (M mutant)	Glucose
Salmonella gallinarum	Galactose, glucose, mannose, rhamnose, tyvelose
Spirillum serpens	Glucose, heptose, rhamnose

Amino sugars detectable in all wall preparations.

TABLE 47

THE DISTRIBUTION OF DIDEOXYHEXOSES IN VARIOUS BACTERIAL SPECIES AND THEIR SEROLOGICAL GROUPING

3,6-dideoxyhexose	Organisms	Serological grouping and serotype	
Abequose	*Salmonella*	B	$C_{2/3}$
	Citrobacter	4,5	
	P. pseudotuberculosis	II	
Colitose	*Salmonella*	0	Z
	Escherichia coli	0111	0 55
	Arizona	20	9
Paratose	*Salmonella*	A	
	P. pseudotuberculosis		I and III
Tyvelose	*Salmonella*	D	
	P. pseudotuberculosis	IV	
Ascarylose	*P. pseudotuberculosis*	V	

From LÜDERITZ. 1960.

TABLE 48

THE DISTRIBUTION OF ALDOHEPTOSES IN POLYSACCHARIDE PRODUCTS FROM VARIOUS BACTERIAL SPECIES

Azotobacter indicum	glycero-gluco-	acidic polysacharide
B. pertussis	heptose*	lipopolysaccharide
Chr. violaceum (2 strains)	D-glycero-D-galacto-	somatic antigen and polysaccharide haptene
Chr. violaceum (7917)	D-glycero-D-manno-	polysaccharide haptene
Esch. coli B	L-glycero-D-manno	cell walls
P. pestis	glycero-manno	polysaccharide haptene
S. abortus-equi	heptose	lipopolysaccharide
Sh. flexneri (type 3)	L-glycero-D-manno	polysaccharide haptene

From data summarized by DAVIES, 1960.
* Where identity not indicated behaviour of heptose is compatible with a *glycero-manno* configuration.

TABLE 49

MONOSACCHARIDE UNITS OF POLYSACCHARIDES ISOLATED FROM GRAM-NEGATIVE BACTERIA

Organism	Component sugars										
	Galactose	Glucose	Mannose	Fucose	Rhamnose	3,6-Dideoxysugar	Aldoheptose	Amino sugar	Pentose	Uronic acid	Others
Azotobacter chroococcum	+	+	—	—	—	+	—	—	—	+	—
Bordetella pertussis	+	—	+	—	—	—	—	—	—	—	+
Chromobacterium violaceum											
'Birch'	+	—	—	—	+	—	+	+	—	—	—
7917	+	+	—	—	—	—	+	+	—	—	—
Escherichia coli (R)	+	+	—	—	—	—	+	—	—	—	—
,, ,, 0111:B4	+	+	—	—	—	+	—	—	—	—	—
,, ,, 018	+	+	—	—	+	—	—	—	—	—	—
Pasteurella pestis 'TS'	—	+	—	—	—	—	+	+	—	—	—
Proteus vulgaris 'X' (S)	+	+	+	—	—	—	—	+	+	+	—
,, ,, 'X' (R)	+	+	—	—	—	—	—		+	+	—
Proteus morgani	+	+	—	—	—	—	+		—	—	—
Pseudomonas aeruginosa	—	+	—	—	+	—	—	+	—	—	—
Pseudomonas fluorescens		+	—	+	+		—	—	—	—	—
Salmonella paratyphi A	+	+	+	—	+	+	+	+			—
Salmonella gallinarum	+	+	+	—	+	+	+	+			—
Salmonella adelaide	+	+	—	—	—	+	+	+			—
Shigella dysenteriae (S)	+	—	—	—	+	—	—	+	—	—	—
,, ,, (R)	+	+	—	—	—	—	+	+	—	—	—
Shigella flexneri 3 (Z)	—	+	—	—	+	—	+	+			
Shigella dispar	+	+	—	—	+	—	—	+			

Data from Davies, 1960 and Westphal and Lüderitz, 1960.

TABLE 50

LIPID CONTENTS OF CELL WALLS OF GRAM-POSITIVE AND OF GRAM-NEGATIVE BACTERIA

Walls from gram-positive bacteria	Total lipid (%)
B. cereus	0
B. megaterium	0
B. subtilis (vegetative)	0.7
B. subtilis (spores)	3.0
Corynebacterium diphtheriae★	30.5
Micrococcus lysodeikticus	0
Mycobacterium tuberculosis★★	64.4
Streptococcus faecalis	2.3
Walls from gram-negative bacteria	
Aerobacter aerogenes	14.6
Chromobacterium prodigiosum	12.8
Escherichia coli	20
Escherichia dispar	12.6
Neisseria catarrhalis	12.2
Proteus vulgaris	17.6
Pseudomonas aeruginosa★★★	15
Rhodospirillum rubrum	22
Salmonella bethesda★★★	12
Salmonella pullorum	19
Salmonella gallinarum	22
Vibrio costicolus	11.8
Vibrio metchnikovi	11.2

Data from SALTON, 1960, 1963; ★ MORI *et al*, 1960; ★★ KOTANI *et al.*, 1959; ★★★ COLLINS, 1963.

TABLE 51

THE AMINO ACID AND AMINO SUGAR COMPOSITION OF PURIFIED GLYCOSAMINOPEPTIDES ISOLATED FROM *Micrococcus lysodeikticus* CELL WALLS DIGESTED WITH LYSOZYME AND *Streptomyces* (F_I) N-ACETYLHEXOSAMINIDASE

Complexes			Approximate molar ratios*					
			Alanine	Glutamic acid	Lysine	Glycine	Glucos-amine	Muramic acid
Streptomyces F_I digest	F_I	I	3.1	1.0	1.3	1.1	1.2	0.9
	F_I	II B	3.5	1.0	1.1	1.0	1.2	9.5
	F_I	II C	1.6	1.0	1.0	0.6	0.9	0.8
Lysozyme	L	I	1.9	1.0	1.0	1.2	0.8	0.8
	L	II	3.0	1.0	1.2	0.9	0.9	1.1
Cell walls			3.0	1.0	1.2	0.9	0.9	1.1

* Estimated from ninhydrin colours; values relative to glutamic acid.
Data from GHUYSEN AND SALTON, 1960.

TABLE 52

CHEMICAL AND STRUCTURAL ANALYSIS OF *Micrococcus lysodeikticus* WALLS AND SOLUBLE PRODUCTS FROM LYSOZYME DIGESTS IN RELATION TO SUBUNIT SIZE

	Walls	Lysozyme NDF*	Glycos-amino-peptide**
MOLAR RATIOS			
Alanine	2.6	2.6	2
Glutamic acid	1	1	1
Lysine	1	1	1
Glycine	1	1	1
Muramic acid	1	<1	1
Glucosamine	1	<1	1
FREE AMINO GROUP (μ moles/g)			
Alanine	15	45	0
ϵ-NH_2-lysine	400	420	present
C-TERMINAL (μ moles/g)			
Glycine	500	520	—
MOLECULAR WEIGHTS	—	10,000 20,000	c. 1,200;*** c. 2,400
SUB-UNIT SIZE			
ϵ-NH_2-lysine	2,500	c. 2,500	
C-terminal glycine	2,000	c. 2,000	

* Non-dialysable fraction.
** Isolated by GHUYSEN, 1960.
*** Calculated values from the structures proposed, assuming low molecular weight behaviour on paper chromatograms.

TABLE 53

THE COMPOSITION OF SEVERAL COMPOUNDS ISOLATED FROM THE PRODUCTS OF PARTIAL ACID HYDROLYSIS OF WALLS OF *Micrococcus lysodeikticus*

Constituents	Composition (μ moles)*		
	Compound A	Compound B	Compound C
Glucosamine	1.0	1.0	Trace –0.29
Muramic acid	2.1, 5 2.	1.2, 1.1	4.0
Glycine	0.17, 0.20	0.93, 0.75	2.2–2.5
Alanine	0.08, 0.08	0.15, 0.16	2.9–3.9
Glutamic acid	Trace, 0.12	0.11, 0.14	1.6–1.7
Lysine	—	0.10, 0.20	1.1–1.6

* Composition determined after hydrolysis for 4 h with 4*N* HCl at 100.°
Data from PERKINS AND ROGERS, 1959

TABLE 54

ACTION OF THE MURAMIDASE, EGG-WHITE LYSOZYME, ON VARIOUS SUBSTRATES

'SUBSTRATES' DEGRADED BY LYSOZYME

Cell walls of bacteria
Insoluble glycosaminopeptide fractions from walls
Soluble glycosaminopeptide fractions isolated chemically
Tetrasaccharide from cell walls
[O-β-*N*-acetylglucosaminyl-(1→6)
-*O*-β-*N*-acetylmuraminyl-
(1→4)-*O*-β-*N*-acetylglucosaminyl
-(1→6)-β-*N*-acetylmuramic acid]
Chitin (colloidal)
Glycol chitin
Oligosaccharides of chitin
[β (1→4)-N-acetylglucosaminides]

COMPOUNDS NOT ATTACKED BY LYSOZYME

O-acetylated cell walls
β-phenyl-*N*-acetylglucosaminide
α-phenyl-*N*-acetylglucosaminide
de-acetylated oligosaccharides of chitin

TABLE 55

PRODUCTS OF ACID HYDROLYSIS OF TEICHOIC ACIDS FROM DIFFERENT BACTERIA

	Lactobacillus arabinosus	*Bacillus subtilis*	*Staphylococcus aureus*
Alanine	+	+	+
Glucose	+	+	—
Glucosamine	—	—	+
Inorganic phosphate	+	+	+
Anhydroribitol*	+	+	+
Anhydroribitol phosphate	+	+	+
Ribitol	+	+	+
Ribitol glucosaminide	—	—	+

* In addition a chloro-derivative of anhydroribitol is detectable when 6*N* HCl is used for hydrolysis.
Data from ARMSTRONG *et al.*, 1958.

TABLE 56

DISTRIBUTION OF TEICHOIC ACIDS IN BACTERIAL CELL WALLS

	Type of polymer	
	Glycerol	Ribitol
Lactobacillus arabinosus 17–5	—	+
L. casei (A.T.C. 7469)	+	—
L. delbrückii (N.C.I.B. 8608)	+	—
L. bulgaricus (N.C.I.B. 76)	+	—
Staphylococcus aureus H	tr.	+
Staph. aureus (Duncan)	tr.	+
Staph. aureus (Oxford)	+	+
Staph. citreus	+	—
Staph. albus (N.C.T.C. 7944)	+	—
Bacillus subtilis (vegetative form)	—	+
*Escherichia coli** (Type B)	tr.	—
Corynebacterium xerosis	+	—
Streptococcus faecalis (A.T.C. 9790)	+	+

* A ribitol-phosphate polymer has been detected in walls of this organism by LILLY, 1962.

TABLE 57

DISTRIBUTION OF TEICHOIC ACIDS IN WALLS AND CELLS OF VARIOUS BACTERIA

	Type of teichoic acid	
	Cell walls	Cellular
Lactobacillus arabinosus 17/5	Ribitol	Glycerol
L. plantarum NCIB 7220	Ribitol	Glycerol
L. brevis	Glycerol	Glycerol
L. bulgaricus NCIB 2889	Glycerol	Glycerol
L. bulgaricus NCIB 76	—	Glycerol
L. casei ATCC 7469	—	Glycerol
L. delbrückii NCIB 8608	—	Glycerol
L. delbrückii NCIB 8130	Glycerol	Glycerol
Bacillus cereus NCIB 2600	—	Glycerol
B. megaterium NCIB 7581	Ribitol (?)	Glycerol
B. megaterium KM	Ribitol (?)	—
B. subtilis	Ribitol	Glycerol
Staphylococcus aureus H	Ribitol	Glycerol
Staph. albus NCTC 7944	Glycerol	Glycerol
Streptococcus faecalis	Ribitol	Glycerol

Data from BADDILEY, 1961; BADDILEY AND DAVISON, 1961.

TABLE 58

CELL-WALL AGGLUTINATION TESTS SHOWING CROSS-REACTIONS BETWEEN STREPTOCOCCI OF DIFFERENT LANCEFIELD GROUPS

Antisera against group	Titre with cell-wall suspension from group								
	A	A variant (strain K 43)	B	C	E	F	K	L	O
A	640	320	—	40	20	20	—	20	—
A variant (K 43)	640	2560	—	320	—	—	20	40	—
B	—	—	80	—	—	—	—	—	—
C	40	80	—	160	—	—	—	—	20
E	40	—	—	—	320	—	—	—	—
F	—	—	—	—	—	320	—	—	—
K	40	20	—	20	20	80	5120	—	—
L	—	—	—	—	—	—	—	160	—
O	—	—	—	—	—	—	—	—	80

'—' Means no reaction in serum diluted 1/20.
Data from CUMMINS, 1962a.

TABLE 59

AGGLUTINATION OF CELL WALLS OF *Staphylococcus aureus* (COPENHAGEN) BY RABBIT ANTISERUM AND THE INFLUENCE OF CERTAIN TEICHOIC ACID HAPTENES ON THE REACTION

	Final Concentration (M)	Degree of agglutination			
		5 min	1 h	2 h	24 h
None		4+	4+	4+	4+
Acetylglucosamine	0.01	O	O	1+	3+
α-phenyl-acetylglucosaminide	0.0025	O	O	1+	3+
β-phenyl-acetylglucosaminide	0.0025	4+	4+	4+	4+
Teichoic acid (I)	0.0013	O	O	1+	3+
Alanine-free teichoic acid (II)	0.0008	O	O	1+	3+
II after treatment with β-acetylglucosaminidase	0.0003	O	O	O	O
Acetylglucosaminyl-ribitol phosphates	0.003	O	O	2+	4+
Acetylglucosaminyl-ribitol	0.003	O	O	2+	4+
D-ribitol-5-phosphate	0.01	4+	4+	4+	4+
Normal rabbit serum substituted for antiserum		O	O	O	O

Tubes were graded as follows: O = homogeneous milky suspension; 1+ = granular milky suspensions; 2+ = marked granularity; 3+ = cell walls clumped and partially settled; 4+ = completely clumped and settled, supernatant clear.
Data from SANDERSON, JUERGENS AND STROMINGER, 1961.

TABLE 60

CROSS-AGGLUTINATION REACTIONS BETWEEN *Salmonella gallinarum* ANTIGENS AND CORRESPONDING ANTISERA

Reciprocal agglutination titres

ANTISERA	ANTIGEN: O-suspension (whole cells)	Isolated cell walls	Isolated spheroplast walls	Lysozyme-residues of insoluble cell walls
O-suspension	400	360	200	80
Isolated cell walls	360	360	180	80
Isolated spheroplast walls	360	320	360	80
Lysozyme-insoluble residues of cell walls	360	320	360	360

Data from SHAFA AND SALTON (unpublished observations.)

TABLE 61

MONOSACCHARIDES OF THE O-ANTIGENS OF *Salmonella* SPECIES

es	Group A	O-Antigen	Galactosamine	Glucosamine	Heptose(s)	Galactose	Glucose	Mannnse	Fucose	Rhamnose	Abequose	Colitose	Paratose	Tyvelose
ratyphi A	A	1, 2, 12		+	+	+	+	+		+			+	
ratyphi A		2, 12		+	+	+	+	+		+			+	
. durazzo														
l		1, 2, 12		+	+	+	+	+		+			+	
ortus equi	B	4, 12		+	+	+	+	+		+	+			
ratyphi B		4, 5, 12		+	+	+	+	+		+	+			
hi murium		4, 5, 12		+	+	+	+	+		+	+			
ratyphi C	C_1	6, 7		+	+	+	+	+						
ntevideo		6, 7		+	+	+	+	+						
mpson		6, 7		+	+	+	+	+						
enchen	C_2	6, 8		+	+	+	±	+		+	+			
wport		6, 8		+	+	+	+	+		+	+			
ginia	C_3	(8)		+	+	+	±	+		+	+			
ntucky		(8), 20		+	+	+	+	+		+	+			
dai	D_1	1, 9, 12		+	+	+	+	+		+				+
ami		1, 9, 12		+	+	+	+	+		+				+
hi		9, 12		+	+	+	+	+		+				+
lo		9, 12		+	+	+	+	+		+				+
eritidis		1, 9, 12		+	+	+	+	+		+				+
linarum		1, 9, 12		+	+	+	+	+		+				+
asbourg	D_2	(9), 46		+	+	+	±	+		+				+
arlem		(9), 46		+	+	+	+	+		+				+

TABLE 61 (continued)

Species	Group A	O-Antigen	Galactosamine	Glucosamine	Heptose(s)	Galactose	Glucose	Mannose	Fucose	Rhamnose	Abequose	Colitose	Paratose
S. anatum	E	3, 10		+	+	+	+	+		+			
S. uganda		3, 10		+	+	+	±	+		+			
S. illinois		(3) ,(15), 34		+	+	+	+	+		+			
S. chittagong		(1), 3, 10, (19)		+	+	+	+	+		+			
S. aberdeen	F	11		+	+	+	+	+		+			
S. poona	G	13, 22	+	+	+	+	+		+				
S. mississippi		1, 13, 23	+	+	+	+	+		+				
S. carrau	H	6, 14, 24		+	+	+	+	+					
S. onderstepoort		(1), 6, 14, 25		+	+	+	+	+					
S. boecker		6, 14		+	+	+	+	+					
S. brazil	I	16	+	+	+	+	+	+	+				
S. kirkee	J	17		+	+	+	±						
S. berlin		17		+	+	+	+						
S. usumbura	K	18	+	+	+	+	+	+					
S. siegburg		6, 14, 18	+	+	+	+	+	+					
S. ghana	L	21	+	+	+	+	+						
S. minnesota		21	+	+	+	+	+						
S. halle	M	28_1, 28_2	+	+	+	+	+			(X)			
S. tel-aviv		28_1, 28_2	+	+	+	+	+			(X)			
S. urbana	N	30	+	+	+	+	+		+				
S. godesberg		30	+	+	+	+	+		+				

BLE 61 (continued)

ies	Group A	O-Antigen	Galactosamine	Glucosamine	Heptose(s)	Galactose	Glucose	Mannose	Fucose	Rhamnose	Abequose	Colitose	Paratose	Tyvelose
elaide	O	35		+	+	+	+					+		
verness	P	38	+	+	+	+	+							
andsworth	Q	39	+	+	+	+	+	+	+					
ogrande	R	40	+	+	+	+	+	+						
kavu		1,40	+	+	+	+	+	+						
aycross	S	41		+	+	+	+	+						
mpala	T	1,42		+	+	+	+			+				
eslaco		42		+	+	+	+			+				
ilwaukee	U	43	+	+	+	+	+		+					
arembe	V	44		+	+	+	+							
versoir	W	45		+	+	+	+		+					
rgen	X	47		+	+	+	+							
hlem	Y	48		+	+	+	+			(Colominic acid)				
eenside	Z	50	+	+	+	+	+					+		
forest		51	+	+	+	+	+							
recht		52		+	+	+	+							

ed data from KAUFFMANN, LÜDERITZ, STIERLIN AND WESTPHAL, 1960

TABLE 62

CHEMOTYPES OF THE *Salmonella* O-ANTIGENS BASED ON THE SUGARS FOUND IN THE LIPOPOLYSACCHARIDES

Chemo type	Number of sugars present	Amino-sugars		Hep-tose (s)	Hexoses			6-Deoxy Hexo[illegible]	
		Galacto-samine	Gluco samine		Galac-tose	Glucose	Mannose	Fucose	Rha no[illegible]
I	4		o	o	o	o			
II	5	o	o	o	o	o			
III	5		o	o	o	o	o		
IV	6	o	o	o	o	o	o		
V	5		o	o	o	o		o	
VI	6	o	o	o	o	o		o	
VII	5		o	o	o	o			o
VIII	6	o	o	o	o	o			o
IX	6	o	o	o	o	o			
X	5		o	o	o	o			
XI	6	o	o	o	o	o			
XII	7	o	o	o	o	o	o	o	
XIII	6		o	o	o	o	o		o
XIV	7		o	o	o	o	o		o
XV	7		o	o	o	o	o		o
XVI	7		o	o	o	o	o		o

X)	3, 6-Dideoxy Hexoses				Antigenic groupings
	Colitose	Abequose	Paratose	Tyvelose	
					J(17); V(44); X(47); Y(48), 52
					L(21); P(38); 51
					C_1(6, 7,); H(6, 14; 6, 14, 24; 1, 6, 14, 25); S(41)
					K(18; 6, 14, 18); R(40; 1, 40)
					W(45)
					G(1, 13, 23; 13 ,22); N(30); U(43)
					T(42; 1, 42)
					M(28_1, 28_3)
o					M(28_1, 28_2)
	o				O(35)
	o				Z(50)
					I(16), Q(39)
					E(3, 10; 3, 15; 1, 3, 19; 1, 3, 10, 19; 3, 15, 34); F(11) -
		o			B(4, 12; 4, 5, 12; 1, 4, 5, 12; 4, 12, 27); C_2 (6, 8); C_3 (8; 8, 20)
			o		A(2, 12; 1, 2, 12)
				o	D_1 (9, 12; 1, 9, 12); D_2 (9, 46)

Kauffmann, Lüderitz Stierlin and Westphal, 1960

TABLE 63

NUCLEOTIDES CONTAINING CHARACTERISTIC CONSTITUENTS OF BACTERIAL CELL-WALL POLYMERS

Organism	Inhibitor	Nucleotide
Staphylococcus aureus	None and Penicillin	UDP-GNAc UDP-GNAc-lactic UDP-GNAc-lactyl-ala UDP-GNAc-lactyl-ala-glu-lys-ala-ala UDP-GNAc-lactyl-ala-glu-lys-gly-asp*
	Oxamycin	UDP-GNAc-lactyl-ala-glu-lys
	Lysine-deprivation	UDP-GNAc-lactyl-ala-glu UDP-GNAc-lactyl-ala
	None Penicillin Gentian violet	CDP-ribitol
	5-fluorouracil	FUDP-GNAc-lactyl-ala-glu-lys-ala-ala FUDP-GNAc UDP- nucleotides also accumulate
Escherichia coli (DAP-dependent mutant)	None	UDP-GNAc-lactyl-ala-glu UDP-GNAc-lactyl-ala-glu-DAP-ala-ala
K 235	None	UDP-GNAc-lactic UDP-GNAc-lactyl-ala-glu UDP-GNAc-lactyl-ala-glu-DAP UDP-GNAc-lactyl-ala-glu-DAP-ala
26–26	None	CDP-ribitol
Escherichia coli	None	GDP-colitose

* Compound reported by Ito, Ishimoto and Saito, 1959.

TABLE 63 (continned)

Organism	Inhibitor	Nucleotide
Streptococcus (Group A)	None	UDP-GNAc UDP-GNAc-lactic
Streptococcus faecalis	Penicillin	UDP-GNAc-lactyl-ala-glu-lys-(ala-NH_3)-ala-ala
L. arabinosus	None	CDP-ribitol CDP-glycerol**
B. cereus	None	CDP-ribitol
Salmonella enteritidis	None	CDP-tyvelose***
Ps. aeruginosa	None	TDP-rhamnose
*Brevibacterium liquefaciens*****	None	UDP-GNAc; UDP-GNAc-lactic; UDP-GNAc-lactyl-ala; UDP-GNAc-lactyl-ala-glu-lys-ala-ala
Aerobacter cloacae	None	Glucosamine – muramic acid – adenylic-amino acid complex

** Although glycerol teichoic acid is not a component of the wall of this organism it is in other species and this nucleotide may be relevant to the biosynthesis of the polymer.
*** Isolated by NIKAIDO AND JOKURA, 1961.
**** Nucleotides found in the culture medium (OKABAYASHI, 1962).

N.B. It should not be assumed that all of the amino acid sequences of the proposed nucleotide structures have been confirmed.

ABBREVIATIONS:

GNAc = N-acetylglucosamine
GNAc-lactic = N-acetylmuramic acid
UDP– = uridine diphospho-, CDP-= cytidine diphospho-,
GDP– = guanosine diphospho-, FUDP– = fluorouridine diphospho-,
TDP– = thymidine diphospho-

TABLE 64

URIDINE NUCLEOTIDE REQUIREMENT IN SYNTHESIS OF PEPTIDE BONDS

Uridine Nucleotide Added	Substrate		
	C^{14}-L-Alanine	C^{14}-DL-Glutamic Acid*	C^{14}-L-Lysine
None	0	0	0
UDP-GNAc-lactic	1364	0	0
UDP-GNAc-lactyl.L-alanine		3720	0
UDP-GNAc-lactyl.L-Ala.D-Glu	8	0	3530
UDP-GNAc-lactyl.L-Ala.D-Glu.L-Lys	2200†	0	20
UDP-GNAc-lactyl.L-Ala.D-Glu.L-Lys.D-Ala.D-Ala	100	0	20

Reaction mixtures contained ATP as well as C^{14}-amino acids and uridine nucleotides indicated above. Data are recorded as c.p.m. incorporated into a charcoal adsorbable form. *AlthoughC^{14}-D L-glutamic acid was the substrate, it could be shown that only C^{14}-D-glutamic acid was enzymatically active. †This result is known to be due to the occurrence of alanine racemase and the enzymes catalyzing reactions of D-ala-D-alanine + its addition to nucleotides.
From STROMINGER, 1962.

TABLE 65

ENZYMIC ADDITION OF D-ALA.D-ALA TO UDP-GNAC-LACTYL.L-ALA.D-GLU.L-LYS

Additions	c.p.m. in UDP-GNAc-lactyl.L-Ala.D-Glu.L-Lys.D-Ala.D-Ala
None	105
D-Ala.D-Ala (synthetic)	4,790
D-Ala.D-Ala (isolated)	6,600
L-Ala.L-Ala	20
D-Ala.L-Ala	56
D-Alanine	81
L-Alanine	37

The substrates were UDP-GNAc-lactyl.L-ala.D-glu.C^{14}-L-lys (enzymatically synthesized), ATP, and dipeptides indicated. Radioactivity in the uridine nucleotide product was measured after separation of the products by paper chromatography.
Data from STROMINGER, 1962.

TABLE 66

CONDITIONS FOR THE SYNTHESIS OF MUCOPEPTIDE

Additions	Increase in mucopeptide %
None	0–10
DL-Lysine	20
DL-Glutamic acid	25
DL-Alanine	20
Glycine	55–80
Glycine + DL-Glutamic acid	60
Glycine + DL-Lysine	80
Glycine + DL-Lysine + DL-Glutamic acid + DL-Alanine	100–150

Washed staphylococci incubated 1 h in buffer containing 1% glucose and one or more amino acids at a final concentration of 400 μg/ml. Bacteria disintegrated and mucopeptide isolated.
Data from MANDELSTAM AND ROGERS, 1959.

TABLE 67

EFFECTS OF PENICILLIN ON INCORPORATION OF ISOTOPES INTO CELL WALL OR INTO CELL PROTEIN AND NUCLEIC ACID IN *Staphylococcus aureus* AND IN *Escherichia coli*

	Staphylococcus aureus				*Escherichia coli*	
	C^{14}-lysine		P^{33}-inorganic phosphate		H^3-DAP	C^{14}-glucose
	Cell wall	Protein	Cell wall	Nucleic acid	Cell wall	Cell wall
Control	34,800	5,100	155,000	11,600	1,040,000	389,000
+ Penicillin	3,290	4,960	48,900	11,600	297,000	334,000
% Inhibition	91%	2%	68%	0	72%	14%

Data are expressed as specific activities (cpm/mg).
From NATHENSON AND STROMINGER, 1959.

INDEX

PRINTED IN THE NETHERLANDS